MW01096962

CMS/CAIMS Books in Mathematics

Volume 9

CMS/CAIMS Books in Mathematics is a collection of monographs and graduate-level textbooks published in cooperation jointly with the Canadian Mathematical Society- Societé mathématique du Canada and the Canadian Applied and Industrial Mathematics Society-Societé Canadienne de Mathématiques Appliquées et Industrielles. This series offers authors the joint advantage of publishing with two major mathematical societies and with a leading academic publishing company. The series is edited by Karl Dilcher, Frithjof Lutscher, Nilima Nigam, and Keith Taylor. The series publishes high-impact works across the breadth of mathematics and its applications. Books in this series will appeal to all mathematicians, students and established researchers. The series replaces the CMS Books in Mathematics series that successfully published over 45 volumes in 20 years.

Stanislaus Maier-Paape • Pedro Júdice •
Andreas Platen • Qiji Jim Zhu

Scalar and Vector Risk in the General Framework of Portfolio Theory

A Convex Analysis Approach

Stanislaus Maier-Paape
Institute of Mathematics
RWTH Aachen University
Aachen, Germany

Andreas Platen
Düren, Germany

Pedro Júdice
ISCTE Business Research Unit
Lisbon, Portugal

Qiji Jim Zhu
Department of Mathematics
Western Michigan University
Kalamazoo, MI, USA

ISSN 2730-650X ISSN 2730-6518 (electronic)
CMS/CAIMS Books in Mathematics
ISBN 978-3-031-33320-0 ISBN 978-3-031-33321-7 (eBook)
https://doi.org/10.1007/978-3-031-33321-7

Mathematics Subject Classification: 91G10, 91G15, 91G70, 52A41, 90C25

This Springer imprint is published by the registered company Springer Nature Switzerland AG
The registered company address is: Gewerbestrasse 11, 6330 Cham, Switzerland

To the oncology and stem cell transplantation departments at the university clinics in Aachen and Essen, in particular to Prof. Dr. Edgar Jost who escorted me from the very first moment when I was diagnosed with leukemia firstly in 2012. Without the terrific job of Prof. Jost and the teams in Aachen and Essen, this book could have never been written.
Stanislaus Maier-Paape

To my wife Inês for her constant support
Pedro Júdice

To my parents Bianka and Kurt
Andreas Platen

To my parents Dezhao and Shubao
Qiji Jim Zhu

Preface

This monograph is a summary of our recent work on portfolio theory and its applications to bank balance sheet management problems. The book consists of three main blocks. We start with a concise exposition of our recent work on a general framework for portfolio theory extending the framework of Markowitz portfolio theory to involve more general concave utility functions and convex risk measures. This framework encompasses both Markowitz portfolio theory and the growth optimal portfolio theory as special cases. Next, we discuss our new results on portfolio theory allowing multiple types of risks. This line of investigation is motivated by practical financial applications. Bank balance sheet management problems are typical examples. A bank balance sheet usually involves interest risk, credit risk, liquidity risk, and other types of risks. In the course of their activity, banks and regulators prescribe limits on the different risks. Also, some risks are not comparable and thus become very difficult to aggregate in practice (e.g., funding liquidity risk and credit risk). Hence, following the idea of Markowitz to trade-off among rewards and risks, one needs to deal with a multi-objective optimization problem.

We analyze the structure of the Pareto efficient frontier for such a trade-off and its representations and, thus, provide a theoretical foundation for implementing the trade-off between reward and risks in practical problems. Moreover, we illustrate the application of our theory using bank balance sheet management problems of different levels of complexity. Our emphasis here is the general pattern. We begin considering the simple expected return and discuss a linear risk model as well as linear-quadratic risk models with explicit solutions. The role of convex duality is emphasized. Furthermore, we lay out the financial meanings of both primal and dual solutions including conditions characterizing Pareto efficient bank balance sheets. These results are useful in guiding the practitioners in their decision-making.

We also introduce several practically relevant risk functions, for instance, the tracking error and the logarithmic drawdown, as well as utility functions such as the logarithmic terminal wealth relative. Further, we discuss several applied portfolio optimization problems arising from various combinations of these risk and utility functions, in particular, the qualitative structure of their efficient frontier.

This monograph is the result of our industry–academic collaboration in the past several years. It unifies and extends several earlier approaches of the authors in the field of portfolio analysis and illustrates how to apply the theory, in particular, to bank balance sheet optimization problems. Therefore, this monograph might be able to serve as a bridge between applied mathematicians and practitioners in financial industry. In fact, it is our hope that it can attract some researchers of the next generation into such industry–academic collaborations. Therefore, it is written at a level that is accessible to graduate students. Of course, it can also serve researchers in the area of financial mathematics and applied mathematics, as well as practitioners.

In closing this preface, we would like to express our gratitude to Professor Matt Davison, one of the editors of this monograph, whose critical reading of an earlier version of the manuscript and constructive suggestions during the review process helped us to significantly improve the presentation of the material. Furthermore, we are thankful to the editors Angela Stevens and Heinz Bauschke who supported us with encouragement during the formation process of this monograph.

Aachen, Germany — Stanislaus Maier-Paape
Lisbon, Portugal — Pedro Júdice
Düren, Germany — Andreas Platen
Kalamazoo, MI, USA — Qiji Jim Zhu
January, 2023

Contents

Chapter 1
Introduction

1.1 General Motivation

This monograph gives a concise exposition of our recent industry-academic collaboration on a general framework of portfolio theory. In particular, it contains our new results on portfolio problems involving multiple types of risks and a variety of application examples.

Portfolio optimization is one of the fundamental problems of quantitative finance—one that those of us with pension plans or retirement accounts can relate to yet, which is also crucial to many important areas of the financial industry such as the management of hedge funds, mutual funds, enterprise risk management, and bank balance sheet management.

Although portfolio management is pervasive all through financial mathematics, one of the main motivations of our multi-risk approach stems from bank balance sheet management problems. A bank balance sheet includes assets such as loans, corporate credit, bonds and liabilities such as retail deposits, corporate deposits, and shareholders' capital. A bank needs to manage its balance sheet to limit risks while maximizing reward. An important feature of bank balance sheet problems is the involvement of multiple types of risks. This is in contrast to most of the existing literature on portfolio theory [16, 44, 45, 71, 80, 102, 103]. Portfolio optimization problems, in fact, often involve multiple risk factors. However, it is often the practice in portfolio theory to aggregate these risks. As a result, in standard portfolio theory, only one risk (and one utility) function is used.

But that is not acceptable in bank balance sheet management. Three of the main types of risks used there are interest rate risk, credit risk, and liquidity risk.

Interest rate risk is the risk that comes from the impact of interest rate shocks on the balance sheet [10, 14, 61, 97]. Banks typically are funded by short-term liabilities, such as checking and savings deposits, and often invest

S. Maier-Paape et al., *Scalar and Vector Risk in the General Framework of Portfolio Theory*, CMS/CAIMS Books in Mathematics 9,
https://doi.org/10.1007/978-3-031-33321-7_1

in medium- or long-term fixed rate assets such as mortgages or government bonds. This mismatch in maturity profiles may negatively impact the bank's profit and loss [47].

Credit risk is associated with the possibility of credit losses from borrowers who cannot pay the interest or capital outstanding on their loans, [15, 17, 34, 53, 54, 100, 110, 117].

Liquidity risk is associated with sudden withdrawals from depositors or institutional creditors [14, 82, 97]. When such withdrawals arise, the bank needs to have enough liquid assets to sell to make the payments associated with those withdrawals. The institution will face bankruptcy if it does not have enough liquid assets or contingent liquidity arrangements.

There are also other types of risks that banks have to consider, such as legal risk [115] (the risk of lawsuits or fines from regulators) or climate risk [11, 12, 39] (the risk of concentrating the balance sheet on polluting companies, which carries significant potential losses due to the transitional regulatory or tax measures toward a greener economy).

The different nature of the risks and banking regulations makes an aggregation of those risks impractical. Therefore, managers need to consider the problem of how to trade off among the rewards and different types of risks. Thus, the appropriate mathematical model for such a problem is multi-objective optimization.

The importance of our research is also linked to enterprise risk management (ERM). Enterprise risk management is about managing risks in an integrated way throughout the organization [28, 29]. This paradigm avoids business units within institutions managing risks without taking into account the risks of other departments (which is known as a silo approach). Risk is to be ideally managed using a top-down approach, i.e., risk limits should be set at a board level and then allocated internally within business lines. Enterprise risk management involves several steps, such as identifying, reducing, or transferring risks and linking risk management with the bank's strategy. This link is often operationalized via a risk appetite statement.

Multi-objective balance sheet optimization models are today an essential tool to make this much-needed link between the risk appetite of a bank and the strategy it pursues, thus a critical step in implementing enterprise risk management. First, linking risk with strategy can be performed with suitable bank balance sheet optimization models, by setting risk limits that specify the risk budget the bank is willing to take. Second, balance sheet optimization translates risk budgets into decision-making by determining the optimal asset-liability allocations. Since these allocations represent the total amount allocated to each asset class or business line at a bank level, the translation of the risk limits into decisions is done in an integrated way in the spirit of ERM, i.e., at a board level, and not left to the discretion of each business line.

Third, bank balance sheet models assist the risk reduction strategy, a critical aspect of ERM. For ease of exposition, let us assume that a bank has a particular segmentation, by-product (e.g., mortgage, consumer loans, corporate credit, etc.), or geography and that it has an initial asset structure. For example, a very simple bank may have an initial allocation of 50% to mortgages and 50% to Treasury securities. The model determines the desired allocation to each segment, thus identifying the segments where the bank should decrease its assets and, consequently, its risk. Continuing our example, the model may indicate that the bank should allocate 70% to mortgages and 30% to Treasury securities, suggesting that it should reduce the exposure to Treasury securities. Thus, bank balance sheet models help the risk reduction strategy in segments with excessive exposure, given those segments' expected return and risk.

Also, balance sheet models help banks implement reductions in their global risk budgets. Suppose a bank decides the global risk budget is too large for its capital levels. In that case, it can compute the optimal asset-liability structure given a smaller risk budget and thus decide its strategy accordingly. Fourth and finally, the balance sheet optimization models help to improve the efficiency of capital allocation.

The general framework using vector risk measures that we undertake in this book also allows us to guide decision-making in the presence of different types of risk measures and not to be limited to linear- or variance-type measures such as one observes in classical efficient frontiers. Thus, our approach can be used to deal with more sophisticated and state-of-the-art risk measures, such as expected shortfall [1, 13, 46] or conditional value at risk [91, 92]. From the regulatory point of view, this is of utmost importance. For market risk, the Fundamental Review of the Trading Book (FRTB) states that the risk measure to be used in the internal models approach is now expected shortfall, with possibly different liquidity horizons [9]. The problem of optimizing a trading book portfolio, subject to market and liquidity risk limits, is thus a particular case of the general framework we propose.

The most important for guiding the decision-making of bank managers is the Pareto efficient set where the trade-off happens. An important special feature of the bank balance sheet management problem is that the reward is usually captured by an expected utility, which is a concave function. The measures for risks are all convex because diversification reduces risk. Thus, all the functions involved in the multi-objective optimization model are convex or concave. This brings convenience in both qualitatively analyzing and quantitatively (numerically) solving the problem. In fact, there are mature numerical methods and software packages for solving constrained convex optimization problems (see, e.g., [22, 35]).

However, it is important to emphasize that from a practitioner's point of view, finding a numerical solution to a particular optimal portfolio problem is often not that useful. That is because to determine an optimal or efficient portfolio using any such optimization process can only be based on historical data. Yet, optimizing over historical data often falls to the trap of backtest overfitting. For example, Bailey, Borwein, Lopez de Prado, and Zhu [6, 7] have shown that the search for an "optimal" trading strategy often leads to backtest overfitting with high probability. Usually, the shorter the in-sample data and the more searches one does, the better the in-sample performance gets. However, such "over-optimized strategies" often lead to catastrophic performance out of the sample. In fact, the paper [6] includes an example showing that with enough search, one can find an "excellent" trading strategy using randomly generated "historical" data. Portfolios are special cases of trading strategies. Thus, the same kind of overfitting also happens when one searches optimal portfolios. Bailey, Borwein, and Lopez de Prado [5] experimented using historical monthly data of S&P500 components to fit a given target performance by minimizing the sum of squares of the difference between the portfolio time series and the target time series. They considered three targets: steady growth, stair-step growth, and sinusoidal growth. In sample from 1991 through 2005, they can achieve a perfect fit for all the three targets when enough sets of parameters are searched. However, out of the sample (2006–2016), the performances of the fitted portfolios are erratic.

We see that overfitting is a huge problem in portfolio optimization in general and needs to be addressed adequately. This applies, in particular, to bank balance sheet management problems, where in addition to equity the holdings also include bonds, consumer and corporate credit, mortgages, and other assets. For a bank balance sheet problem, the accessible data frequency is often quarterly or even annually. With shorter historical data sequences and a wider range of asset selections, the probability of overfitting is much higher [7]. Therefore, it is of the utmost importance to study the general properties of the Pareto efficient set and its relationship with optimal portfolios.

Such qualitative information can provide insight to asset managers in general and bank balance sheet managers in particular with regard to the construction of their portfolios. We illustrate this with three simple examples:

Firstly, it is known in Markowitz portfolio theory that the Pareto efficient set (efficient frontier) is a bullet-shaped curve called the Markowitz bullet [81]. There is a one-to-one correspondence between a point on the bullet and the corresponding Markowitz efficient portfolio. Although the Markowitz bullet is nonlinear, there is an affine linear relationship among Markowitz efficient portfolios. This leads to the well-known Tobin two-fund theorem [108], which tells us that knowing two distinct Markowitz efficient portfolios, one can generate affine linearly any other Markowitz efficient portfolio using these two. This forms the theoretical foundation for index investing. The logic is that a broad-based index can be regarded approximately as a Markowitz efficient portfolio. Thus, to construct a Markowitz efficient portfolio with desired properties, one only needs to consider, say, an affine linear combination of S&P500 and NASDAQ indices. Hence, index investing is a time-tested sound investment strategy. However, although the awareness of the benefits from (combined) index investing certainly counts as pros of portfolio theory, it is, on the other hand, known that if one constructs Markowitz efficient portfolios using historical data, its out-of-sample performance is in certain cases quite unstable. The obvious reason for that instability is again some kind of overfitting, because in order to attain, for instance, large rewards in sample, often a concentration on only a few assets is necessary. On the other hand, minimum variance portfolios typically work quite well in real applications, presumably a consequence of the good prediction quality of the portfolio variance for out-sample data [55, 114] and an often seen natural diversification for this approach when the portfolio space is rich in low volatility assets.

In conclusion, to overcome overfitting, it is essential to also guarantee diversification which brings us to our next example.

Secondly, diversification in portfolio optimization may be forced by an additional risk constraint. In fact, it is well-known that many fund managers struggle to even beat their benchmark which might be an index or simply the equal-weight portfolio consisting of all the assets of an index with the same weight. Therefore, holding portfolios which are close or at least not too far from a given benchmark is typically beneficial for the performance and may guarantee sufficient diversification. Fortunately, the tracking error constraint does exactly that job, i.e., it forces the portfolios allowed for the optimization to deviate not too far from the benchmark. Since the tracking error is a convex function, it may be viewed as additional risk constraint and –again– leads to a natural application of our multi-risk optimization ansatz where, as a matter of fact, diversification is typically inherited from the benchmark.

We continue with an illustrative example for bank balance sheet management problems:

Thirdly, Júdice and Zhu [65] discussed a bank balance sheet management problem with the reward function being the expected return and which involves only credit and interest rate risks both approximated by linear functions. This leads to a linear programming problem with an explicit solution. However, it turns out that each efficient balance sheet contains only two assets. This kind of solution is certainly unreasonable in terms of a bank balance sheet due to the overconcentration which has to be fixed for practical purposes, for instance, as abovementioned by a constraint guaranteeing diversification. However, the explicit dual solution here reveals that the assets included in the optimal portfolio are characterized by the highest expected return per unit of risk relative to the corresponding risk limit. This is an insight that is very useful to bank managers. In particular, the said ratio for consumer credit is way above all the other asset classes according to recent data. This explains the reason that most of the bank managers intuitively know this advantage and are pursuing it. Advertisements for credit cards flooding our mailboxes might serve as a proof.

On the theoretical side, however, we put particular emphasis on the structure of the efficient frontier. In general, we show in Sect. 3.3 that the efficient frontier is a path-connected set over which the efficient portfolios can be represented by a continuous map. Moreover, this efficient frontier also has several natural representations. It can be represented as a subset of the graph of the concave function maximizing the utility subject to constraints of the varying risk limits. Alternatively, this efficient frontier can also be represented as a subset of a graph of the convex function minimizing one of the risk components subject to constraints of the varying risk limits for the rest of the risk components and the changing lower bound for the utility (cf. Sect. 3.1). In particular, when there is only one risk involved, the efficient frontier can be represented continuously as graphs of a convex or concave function on intervals in rewards and risk space, respectively. This makes the trade-off along the efficient frontier for scalar risk especially convenient (see Sect. 2.3). The structure of the efficient frontier when multiple risks are involved is more complicated. We construct an example showing that the projection of such an efficient frontier to the coordinate hyperplanes may not be convex (see Example 3.25). Also, in the vector risk case, there may be discontinuities of the representing functions of the efficient frontier at least at boundary points (cf. Sect. 3.1.3). Thus, trading-off between different points on the efficient frontier is not trivial and may need to be dealt with using the features related to specific applications in combination with the known topological structure.

In Chap. 4, bank balance sheet management problems of different levels of complexity are used to illustrate the application of the general framework of portfolio theory. The examples here are discussed with a mixture of financial intuition and mathematical rigor. Our emphasis is the general pattern. After a practically oriented introduction to bank balance sheet management problems in Sect. 4.1, we continue considering in Sect. 4.2 the simple expected return and discuss a linear risk model as well as a mix of linear and quadratic risk models in Sect. 4.3. These simple models have explicit solutions. More interestingly, we can explain the financial meaning of the conditions characterizing the optimal balance sheet. Duality plays an important role in the course of solving these problems. Also, the dual solution tells us about the fair price for insuring against the corresponding risk. These insights are useful for guiding the decision of bank managers. The analysis on the bank balance sheet management problems is mostly new, except for a case study published in Júdice and Zhu [65].

In the Markowitz case, all points on the Markowitz bullet correspond to optimal solutions of the underlying trade-off problem and to efficient portfolios alike. However, in general, there is a difference between optimal and efficient portfolios. For instance, on the one hand, a portfolio might assume the best possible utility value and is therefore optimal with respect to the utility. On the other hand, in case there are other portfolios which assume the same utility, but have less risk, then this optimal portfolio is not efficient. We therefore put particular emphasis on the elaboration of the qualitative structure of the efficient frontier in several applied optimization problems concerning the trade-off between a scalar utility and a vector risk function.

In summary, bank balance sheet management motivated us to analyze multi-objective optimization problems with convex data. Our goals are to analyze the Pareto efficient set of such a problem, its relationship with the

Pareto efficient portfolios, and where appropriate to give financial intuitions for the related bank balance sheet problem. Due to the convex nature of the problem, convex duality plays an important role. The special convex structure not only brings about convenience in characterization of the solution but also leads to insightful financial explanations of the solution.

Our industry-academic collaboration is still preliminary. Nevertheless, we find that the mathematical model of bank balance sheet management problems posts a new challenge in the area of multi-objective optimization with convex data. On the other hand, applying our preliminary results in this mathematical model to bank balance sheet problems already yields useful insights for financial practitioners.

Although our model is mainly motivated by bank balance sheet management problems, the results discussed in this book may also be helpful to overcome overfitting and can also be used in other areas of applied mathematics. A few examples are Markowitz portfolio problems with additional tracking error constraint and/or drawdown risk functions, investment funds with illiquid assets, and multiple risks arising in corporations and household financial risks.

We believe that further collaborations between industry practitioners and academic researchers in this important area are desirable. Therefore, it is our hope that this book can serve as a bridge between applied mathematicians and financial practitioners.

This monograph is intended for graduate students and researchers in the area of financial mathematics and applied mathematics, as well as practitioners. The theoretical part (Chaps. 2 and 3) and the application part (Chap. 4) can be read separately. For the convenience of readers, a chapter-by-chapter description of the contents is included in Sect. 1.4. Furthermore, as stated above, we give a short literature review of related portfolio theory in Sect. 1.2 and add some remarks on duality in finance in Sect. 1.3.

1.2 Literature Review on Portfolio Theory

The existing literature on portfolio theory is large. There has been a great deal written on this topic since, and even prior to, the groundbreaking work of Markowitz [80] in the 1950s, that is, 70 years of literature in various different fields, e.g., applied probability, economics, finance, mathematics, and operations research. However, the existing literature does not completely address our key concerns from a practitioner's perspective, and a comprehensive introduction to all relevant facets of the literature would certainly be a book on its own. Therefore, we restrict ourselves to a few directions related to our current work trying to focus on addressing what is missing and not intended to be comprehensive. By and large, existing literature on portfolio theory is evolving in three main directions: Markowitz portfolio theory, growth optimal portfolio theory, and continuous-time portfolio optimization.

(a) Markowitz portfolio theory. Markowitz proposed to determine the best portfolios as a trade-off between expected return and risk measured by the variance [80, 81]. Markowitz' ideal and framework also stimulated related important work of [103, 104, 108]. The textbook by Elton et al. [45] contains an excellent discussion on implementation issues of this theory in investment practice.

Portfolio analysis was and still is deeply rooted in the spirit of trade-off between risk and reward as proposed in Markowitz portfolio theory after close to seven decades. This is also the idea that we follow. Nevertheless, we have to extend the original Markowitz framework.

Firstly, the linear quadratic model, though mathematically attractive, is practically unsuitable due to not counting into the compounding nature of practical portfolio management. Thus, it is necessary to go beyond the linear-quadratic setting. Maier-Paape and Zhu have extended the framework of Markowitz portfolio theory to involve more general concave utility functions and convex risk measures in [78, 79]. Therein, we also emphasized the role of duality both in analyzing the efficient frontier and its corresponding financial meaning in dealing with application problems. This framework was further extended to involve discrete multi-period market models together with Platen in [77, 89]. The multi-period model provides a natural framework to conduct model calibration using historical data and allows a realistic modeling of drawdown risk functions. These publications provided the foundation for our discussion here.

Secondly, in the Markowitz portfolio framework and many of the generalizations that followed (see [103, 108, 109]), the trade-off has always been between one reward and one risk measure. As alluded to above, for bank balance sheet management problems, this is inadequate. We must deal with multiple risks of different nature. Thus, the corresponding mathematical model is inherently one of multi-objective. Several risk functions can be used. Important risk functions are defined by the variance as in [80, 81], the tracking error as in [41], (conditional) value at risk (see, e.g., Rockafellar and Uryasev [91, 92]), the drawdown (see, e.g., Chekhlov, Uryasev and Zabarankin [26], Maier-Paape and Zhu [79]), and (conditional) expected drawdown (see, e.g., Goldberg and Mahmoud [52]). Risk measures specifically related to the bank balance sheet problems will also be discussed in Chap. 4.

We will show later that for such a problem, even if all the rewards and risk measures involved are convex in nature, the projection of the efficient frontiers can still be intrinsically nonconvex. This is very much in contrast to the case of one risk. There the projection of the efficient frontier will always be an interval. However, in Sect. 3.3, we can show that this set is connected, a qualitative behavior that has important financial manifestations. This result tells us that continuously moving from one efficient portfolio to another is possible (cf. Corollary 3.47).

While convexity is a reasonable assumption for risk measures, we need to mention that not all risk measures used in practice are convex. These situations are not discussed in this book. For example, an important part of the classical book of McNeil, Frey, and Embrechts [83] discusses risks associated with extreme events. Value at risk discussed in detail in Jorion's book [62] has been very influential in the practices of both banks and regulatory agencies with recent shift toward the convexified conditional value at risk due to the work of Rockafellar and Uryasev [91, 92].

(b) Growth optimal portfolio theory (GOP). Kelly [70] discussed a special form of the growth optimal portfolio theory as an explanation of Shannon's information rate in the information theory [101]. Thorp first used Kelly's result to the blackjack money management problem [107], and then together with Kassouf, he applied it to stock and warrant trading [67]. Lintner independently proposed the growth optimal portfolio theory [71]. Growth optimal portfolio theory proposes to determine the optimal portfolio by maximizing the log utility function. The name comes from the fact that such an optimal portfolio grows utility in terms of expected logarithmic multiplicative returns (terminal wealth; see Sect. 2.2) faster than any other portfolios. This property of a unique fastest growth optimal portfolio is attractive. On the other hand, the growth optimal portfolio is known to be extremely risky in practice. The book edited by MacLean, Thorp, and Ziemba [76] contains most of the important articles in the area of growth optimal portfolio theory up to 2011. In particular, it contains several papers discussing how to scale back from the growth optimal portfolio to mitigate the risks mostly by ad hoc methods.

Recently, Vince and Zhu [113] as well as Lopez de Prado, Vince, and Zhu [73] took a more systematical approach. They recognize that in practice investment decisions are always about the results in a finite horizon instead of the infinite horizon implicitly assumed in the GOP theory. For example, if we consider the total leverage level of a portfolio, focusing on the total expected return over such a finite horizon, one derives a bell-shaped risk-return curve. This reveals that in addition to the peak (absolute maximum) corresponding to the growth optimal portfolio, two other reasonable portfolio choices will be the reward-risk ratio optimal point and the inflection point, which is where the ratio of reward to risk changes from increasing to decreasing. These two points are considerably more conservative compared to GOP. In applications such as money management for the game of blackjack [113] and bank leverage level [38], those alternative more conservative choices are close to practically deemed reasonable.

However, the GOP theory and its more recent developments essentially use a single utility function, and thus, they are unable to deal with the multiple kinds of risks required in guiding the design of concrete bank balance sheets.

(c) Continuous-time portfolio optimization: Merton [84, 86] pioneered continuous-time portfolio optimization based on stochastic modeling of the equity price processes. Besides optimizing the portfolio, the stochastic process also provided a tool to replicate financial derivatives. In fact, given an appropriate market model, the replicating portfolio of the payoff of a financial derivative can be derived from a backward stochastic differential equation (see [42, 43, 75] and the references therein). The relationship between the stochastic differential equation and the corresponding partial differential equation stipulated by the Feynman-Kac formula [66] provides a solid foundation for pricing financial derivatives. This is, in fact, the foundation for the celebrated Black-Scholes formula [99]. Also, the continuous-time setup of Merton can be used to compute the optimal portfolio using stochastic dynamic programming and the Bellman principle [49]. In this framework, returns follow normal processes, i.e., prices are specified by a geometrical Brownian motion, with the known limitations associated with the normal distributions and the failure to capture extreme events.

As a result, derivative pricing and dynamic portfolio management have grown into the primary tool for the mainstream financial analysis that serves the major financial institutions for whom design, pricing, and trading financial derivatives have become an increasing share of their business. On the other hand, for those familiar with the practice of derivative pricing, the memory of LTCM debacle in 1998 [40] and the 2008 financial crisis are still fresh. These financial disasters are related to the pricing or better mispricing of financial derivatives and the lack of capturing tail events in the modeling. There have been numerous pieces of research with regard to the reasons for those mishaps. One of the consensuses is that adopting stochastic market models with thin tail distributions, such as the log-normal stock price model, is one of the main reasons those models fail in adverse market conditions.

Therefore, it is crucial to generalize the returns to more general distributions and induce skewness and kurtosis. Some generalizations of the Merton model have used other distributions for returns, such as Foque, Sircar, and Zariphopoulou [51], who model the portfolio problem under stochastic volatility, or Aït-Sahalia, Cacho-Diaz, and Hurd [3], who use jump processes. Goldberg and Mahmoud [52] also used cadlag processes to model drawdown risk. However, even so, the Merton-type approach to portfolio optimization is challenging to use in practice. Usually, a portfolio manager or a balance sheet manager has to specify certain risk limits, the so-called risk budget, even in the single risk case. The Merton model does not set a risk limit for the portfolio since the modeling is only based on the utility of consumption or terminal wealth.

The continuous-time setup is also not practical for balance sheet management problems, as many of the observed risk factors, such as default rates, are reported yearly and are not continuous. Thus, the discrete-time framework is much more natural in these problems.

In balance sheet management, the decision-maker has to carefully balance his asset and liability allocation decisions in compliance with multiple risk limits. Such limits are often different by nature, such as funding liquidity risk and credit risk, whose aggregation is extremely difficult if not impossible.

Again, the continuous-time framework of Merton and similar models cannot deal with multiple risks.

Finally, continuous-time models in the Merton framework stem from dynamic programming and the Bellman principle, which can only be implemented for low-dimensional problems in the state variable space. It is widely known that continuous dynamic programs, even if solved numerically, can only deal with a limited number of state variables, except in exceptional situations because the optimization problem explodes with the number of state variables. This is known as the "curse of dimensionality." To the best of our knowledge, there has been one attempt to formulate the balance sheet problem in continuous time by Lipton [72]. Still, the approach, even for a low number of state variables, is complicated to solve numerically, forcing the author to reduce the number of state variables in the problem to a single one in order to solve the problem, thus eliminating most of the features of the initial model.

Thus, using continuous-time stochastic models in bank balance sheet management is not practical. For these reasons, we consider one-period and finite multi-period models (see, e.g., [94] and [105]). The multi-period scenario has been studied for decades in the continuous-time setting (see, e.g., Merton [85]), as well as in the discrete-time setup (see, e.g., Phelps [88], Samuelson [96], and Cariño and Ziemba [24]). More literature can be found, e.g., in Steinbach [106] and Boyd et al. [21]. A portfolio in a multi-period market is defined by a trading strategy which makes the situation more involved in contrast to the one-period market case. In general, in a multi-period model, the information is gradually revealed, modeled mathematically as an associated filtration. In this monograph, the main purpose of introducing the multi-period market is to be able to define other reasonable risk functions. For this, we focus on the fixed fraction trading strategy and define utility functions related to the GOP (see (b) above) and risk functions based on drawdowns (see (a) above). The original idea of fixed fraction betting or investing goes back to Kelly [70] and it was extensively discussed in Vince [112]. In the context of the optimal consumption-investment problem, Merton [84] also showed that under a geometric Brownian motion and a constant relative risk aversion utility, it is optimal to hold fixed proportions in the risky assets.

We would also like to mention a direction of interesting research initiated by Jouini, Meddeb, and Touzi [63] and developed further by Hamel and Heyde [56] as well as Hamel, Heyde, and Rudloff [57] on vector-valued or set-valued coherent risk measures. The research reported in this book and the research of Jouini, Meddeb, Touzi, Hamel, Heyde, and Rudloff are both dealing with vector risk measures. However, the focuses and intellectual roots of these two lines of research are different and complementary.

The focus of the research in [56, 57, 63] is on the question mostly concerned by the regulators: what assets and related quantities can be allowed in a portfolio given concerns for (certain coherent) risks from the regulatory point of view. In other words, given a multitude of risk concerns, they try to determine an acceptable set of assets that can be allowed in a portfolio to ensure the stability of the market as a whole from a regulator's point of view. This is a further development of the idea of coherent risk measures and related coherent acceptance cones pioneered by Artzner et al. [4]. Within this acceptance set, the question of how an individual institution constructs its portfolio to fulfill its investment goal is deliberately left out. This is certainly a reasonable approach for a regulatory agency: regulators do not determine portfolios for institutions they regulate; they only set guidelines.

In contrast, our focus in this monograph is exactly how an individual institute, such as a bank, manages its portfolio or balance sheet in view of the trade-off between a vector-valued risk and reward. In dealing with this problem, we followed the tradition of Markowitz portfolio theory in seeking Pareto efficient portfolios by way of exploring their relationship with Pareto efficient frontiers in the (vector-valued) risk-reward space. To get there, we extensively use methods from convex analysis, in particular the concept of duality. This and some related literature is the subject of the next subsection.

1.3 The Role of Duality

As we have seen in Sect. 1.1, the model we obtain for, e.g., bank balance sheet problems is a multi-objective optimization problem with convex data. Multi-objective optimization has a long history and extensive literature. Keller's recent two-volume book [68, 69] contains a comprehensive discussion of the state of art of the theory and practice of multi-objective optimization. On the theoretical side, the focus there has been the existence of (Pareto) solutions, necessary and sufficient conditions for such solutions so as to identify particular Pareto solutions and numerical methods. Since our problem is convex, these topics are well covered in convex analysis and duality theory.

We therefore give a brief summary of the most relevant results in the appendix for the convenience of the reader. In Appendix A.1, semi-continuity for extended-valued functions (on Banach spaces) is discussed. Afterward, we provide several useful results for extended-valued convex and concave functions in Appendix A.2. Furthermore, in Appendix A.3, a standard convex programming problem is stated and solved with a Lagrange multiplier ansatz. After setting up the basic facts for convex analysis, Appendix A.4 gives a concise description of the duality theory. Classical references in this area are [48, 87, 90]. The emphasis is on Lagrangian duality as it is most directly applicable to the financial problems in this book.

We emphasize the duality and Lagrange multipliers due to their important roles in many financial applications. For example, the success of the classical Markowitz portfolio theory [80, 81] can be largely attributed to examining a portfolio optimization problem from the perspective of its dual problem in the space of risk and rewards. The celebrated Black-Scholes option pricing model

[99] can be understood as the interplay between the dual processes of the price of the option and the cash borrowed for hedging, linked by a dynamic delta hedging [25]. Furthermore, the fundamental theory of asset pricing links the no arbitrage pricing of financial assets to a martingale measure. The prototype of this theorem is first formed in [32, 33] as an alternative to the Black-Scholes option pricing method [99]. More general forms are developed in [36, 58, 59]. From a convex duality point of view, the pricing martingale measure is the dual solution to a constrained optimization problem of maximizing expected utility. For details of those insights and more financial applications of convex duality, we refer to Carr and Zhu [25]. Convex duality also provides useful insights in analyzing bank balance sheet management problems as we will elaborate in the book, in particular in Chap. 4.

What is missing so far, yet crucial for the practice of bank balance sheet management problems, is the topological structure of the Pareto efficient set. This is the central theoretical discussion of this monograph and will be presented in Chaps. 2 and 3 together with several applications in Chap. 4. A more detailed review of the forthcoming is given in Sect. 1.4.

1.4 Preview of the Forthcoming

As our primary motivation for this book was managing a bank balance sheet, we provide insights to do so in Chap. 4. For readers primarily interested in the applied side of this book, it should be noted that –with a little knowledge of convex duality at hand– a lot of Chap. 4 works "on its own," independently of Chaps. 2 and 3. In fact, Chaps. 2 and 3 are more devoted to the discussion of the theoretical side of the general framework of portfolio theory for scalar and vector risk. However, the machinery developed in Chap. 2 and, in particular, in Chap. 3 is also brought to bear its value to the risk management professional within Chap. 4 in several occasions.

In order to get a better understanding of the forthcoming, for convenience of the reader, we now describe the contents of this book step by step:

Chapter 2

In Chap. 2, we start off with a state-of-the-art summary of the general framework of portfolio theory in the scalar risk case. The main ideas here go back to Maier-Paape and Zhu [78], but we generalize that theory in many directions using also ideas from [77, 79, 89].

Accordingly, we construct efficient portfolios as a trade-off between abstract extended-valued (scalar) risk and utility functions, respectively. While the risk functions are always convex, the utility functions are always concave. Section 2.1 is used to lay out the basics for such risk and utility functions on a one-period financial market, while in Sect. 2.2, multi-period financial markets together with the so-called fixed fraction trading ansatz come into play. With these different tools at hand, not only simple risk functions like the variance can be used, but also more complex ones like (logarithmic) drawdown or maximal drawdown can be modeled. Most of these ideas essentially stem from Platen [89]. Commonly used utility functions are expected return with an auxiliary function (see Sect. 2.1.5) and the (logarithmic) terminal wealth relative (TWR) (cf. Sect. 2.2), which is used in growth optimal trading (cf. Vince [111, 112]).

Afterward, in Sect. 2.3, the theory for the efficient frontier is explained; in particular we introduce the Pareto efficient set in the two-dimensional risk-reward space and its representations as continuous graphs over certain intervals in risk or utility space. Moreover, we here also prove new properties like the path-connectedness of the efficient frontier (cf. Corollary 2.73) which was ignored in earlier approaches, yet essential for investors who want to adjust strategies in a continuous manner. At last, in Sect. 2.4, existence and uniqueness of efficient portfolios is discussed, and their relation with, e.g., the maximum utility optimization problem with constraint on the risk is revealed (see Problem 2.77). In fact, in the main theorem (Theorem 2.79), a continuous efficient portfolio map can be constructed which represents the efficient portfolios as a continuous graph on the efficient frontier. Section 2.4 is closed with formulas practically relevant for the calculation of the efficient frontier and with several examples of efficient frontiers occurring often in applications. Furthermore, as an application of the theory, in Example 2.85, solutions for the Markowitz setup are calculated under an additional "close-to-benchmark constraint," also known as tracking error bound (see [41]). This may be used to enforce diversification or simply to urge fund managers to not deviate too far from their benchmark.

Chapter 3

In contrast to Chap. 2, in Chap. 3 we allow the risk function to be vector-valued, while the utility function remains scalar-valued as before. Thus, Chap. 3 is used to develop the general framework of portfolio theory for the vector-valued risk case, which –to the best knowledge of the authors– has not yet been discussed in the literature in this generality. In contrast to Chap. 2, the corresponding maximum utility optimization problem now has a vector-valued risk constraint.

This theoretical extension allows the underlying financial markets to be both one-period and multi-period. But like in Chap. 2, this is completely hidden in the abstract risk and utility function. The proceeding starts in Sect. 3.1 similar as before with a discussion of the efficient frontier, now in a space of dimension larger than two. Although the representing functions related to the efficient frontier need a more involved notation, at first glance several results seem to generalize almost canonically (cf. Proposition 3.12, Theorem 3.13 and Theorem 3.14). However, there are several points, where the theory for the vector-valued risk functions deviates. For instance, in Sect. 3.1.3, we see that the representing functions of the efficient frontier, in general, no longer have to be continuous. All that is left is the continuity in the interior of their domains and a partial continuity on the boundary (cf. Corollary 3.24). Also, the domains of these representing functions no longer need to be convex (see Example 3.25), which is in contrast to the scalar risk case where the representing functions of the efficient frontier in the two-dimensional risk-reward space were defined on intervals (cf. Corollary 2.74 and Theorem 2.76). Fortunately, an important topological property, the path-connectedness of the efficient frontier, could be preserved, although the effort to derive that result was much more involved (see Sect. 3.2 and Theorem 3.43) than for scalar risk, although the known path-connectedness of the efficient frontiers for the scalar risk case could be utilized essentially in a geometrical proof. On the other hand, the efficient frontier for the vector risk case is in general no longer closed (see examples in Sect. 3.1.3). This was different for scalar risk, where the efficient frontier is always a closed subset of the two-dimensional risk-reward space (cf. Remark 2.84).

The last subsection, Sect. 3.3, deals with the efficient portfolios in the vector risk case, especially the uniqueness (Theorem 3.46) in a similar setting as in Sect. 2.4. As a consequence, these efficient portfolios can be parameterized as various graphs, e.g., as graph of a continuous one-to-one efficient portfolio map from the efficient frontier to the portfolio space (Corollary 3.47). In addition, a link to the scalar risk case is given by defining a scalar risk function as a linear combination of the components of the vector-valued risk function. At last, we reconsider the Markowitz setup with additional close-to-benchmark constraint from Example 2.85, but now from a vector risk point of view. In this example, the expected return is the scalar utility, but besides the portfolio variance, as a second risk function, the tracking error is used. It turns out that the qualitative structure of the efficient frontier for this seemingly simple (vector risk, utility) problem is already quite complex (cf. Example 3.50, in particular Fig. 3.13 for two different projections of this efficient frontier).

Chapter 4

In Chap. 4, we discuss several applications of the previous theory, in particular to bank asset-liability management problems. Section 4.1 describes some of the risks which corporations and banks face and motivate the need for a multi-risk framework in practice, stemming from the limits on the different risks set by banks and regulators. In the subsequent subsections, we describe some of the risk measures that banks can use, focusing on linear and quadratic measures that yield tractable solutions.

We then turn to solve different bank balance sheet management problems with increasing levels of complexity. Section 4.2 discusses the linear case, which was developed by Júdice and Zhu [65], where the authors have solved the problem using duality and showed that the dual multipliers correspond to shadow prices of credit and interest rate risk. We describe the calculation of optimal balance sheets using real-world data using this framework.

Section 4.3 tackles the linear-quadratic case, which is new. We use a quadratic interest rate risk constraint and a linear credit risk constraint in this setting.

Enforcing the non-negativity and non-positivity restrictions for assets and liabilities, respectively, makes the problem hard to solve. Furthermore, it leads to many corner solutions which are intractable to deal with in practice. Therefore, we resort again to the close-to-benchmark constraint that forces the solution to be in the feasible region by allowing a maximal volatility-weighted distance to a prescribed benchmark. This formulation of the problem yields higher tractability and an exact solution when the volatility-based neighborhood of the benchmark is inside the credit and interest rate risk constraints.

Moreover, the solution is much easier to implement in practice because it yields diversification of the balance sheet. Again, we apply this example to real-world data.

Section 4.4 contains several further developments. We start with two subsections concerning the general form of a linear as well as a linear-quadratic model involving more than two kinds of risks. Also, we extend the discussion on the linear-quadratic model utilizing a benchmark constraint to guarantee diversification. In the last two subsections, we discuss a minimum drawdown problem with bounded variance constraint and bounded below logarithmic terminal wealth relative (log TWR) utility. In Sect. 4.4.4, we start with some rearrangements of this problem, in particular of the non-differentiable logarithmic drawdown risk function, so that numerical calculations of solutions may be done with standard interior-point algorithms. Finally, in Sect. 4.4.5, we discuss the same problem as above from a qualitative point of view. In particular, the main theorem, Theorem 4.39, finds that in this case, the efficient frontier has more structure than in the general theory worked out in Sect. 3.1. For instance, this efficient frontier is closed and it is furthermore a simply connected bounded two-dimensional surface (cf. Fig. 4.8).

Chapter 5

Finally, in the concluding Chap. 5, we provide a concise roundup of the contents and discuss several potential directions for further investigations.

Chapter 2
Efficient Portfolios for Scalar Risk Functions

In this chapter, we carefully develop a general framework of portfolio theory involving a scalar risk. This will establish a foundation for our central result on a portfolio theory involving multiple risks (or a vector risk) in the next chapter.

Modern portfolio theory was developed in the pioneering work of Markowitz [80, 81]. The principle of trading-off between the standard deviation as risk and the expected return of portfolios proposed in Markowitz portfolio theory has been very influential. The choice of the standard deviation as risk and the expected return as reward in Markowitz portfolio theory, however, is too restrictive in applications. This shortcoming was overcome in the general framework for portfolio theory proposed in Maier–Paape and Zhu [78]. Therein, efficient portfolios were defined as Pareto efficient points in the trade-off between abstract extended-valued convex risk and concave utility functions, $\mathfrak{r}\colon A \to \mathbb{R} \cup \{+\infty\}$ and $\mathfrak{u}\colon A \to \mathbb{R} \cup \{-\infty\}$, respectively, on admissible portfolio sets $A \subset \mathbb{R}^{M+1}$ with one risk-free and M risky assets. In contrast to [78], the setup here is more general in many ways. For instance, in [78] the utility function $\mathfrak{u}$ is restricted to the form of an expected utility of the portfolio payoff (cf. Definition 2.31), which is only a subclass of the here allowed abstract concave utility functions. Accordingly, our approach here is more in the spirit of Platen [89] (cf. also [77]), but without focusing too much on abstract trading strategies and a different approach to the properties of the efficient frontier. In fact, by allowing abstract convex risk and concave utility functions, we can separate the portfolio optimization question (see Problem 2.77) from the construction of those risk and utility functions. The latter is addressed in Sect. 2.1 for an underlying one-period financial market. Similarly, in Sect. 2.2, several new risk and utility functions are provided for multi-period financial markets and a fixed fraction trading ansatz. Nevertheless, it should be noted that all examples of risk and utility functions given in Sects. 2.1 and 2.2 can only touch a small fraction of possible applications of the general framework of portfolio theory.

S. Maier-Paape et al., *Scalar and Vector Risk in the General Framework of Portfolio Theory*, CMS/CAIMS Books in Mathematics 9,
https://doi.org/10.1007/978-3-031-33321-7_2

Afterward, in Sect. 2.3, we resume with our summary and further development of the theory from [78]. Firstly, we construct the efficient frontier in risk-utility space and develop a whole zoo of helpful properties, for instance, the important connectedness property of the efficient frontier, a question which was neglected in [78].

Later on, in Sect. 2.4, existence and uniqueness of optimal portfolios related to abstract convex risk and concave utility functions is discussed (see Theorem 2.78 and 2.79). Furthermore, in Example 2.85, we discuss a linear-quadratic optimization problem and line out how our abstract theory here applies. This example can also be viewed as a connecting link to the vector risk theory we develop in Chap. 3.

As already noted, the theory presented here may serve as a generalization of the results of [78]. As a matter of fact, in particular at the beginning of Sect. 2.1 and to a lesser extent at the beginning of Sect. 2.3, there is a small overlap in notation and simple results with [78]. For convenience of the reader, we restate these simple results here and add citations where appropriate.

2.1 One-Period Financial Markets and Related Risk and Utility Functions

In this section, we provide our basic setup for one-period financial markets and introduce relevant properties for risk and utility functions defined on so-called admissible portfolios.

2.1.1 The Payoff Space

For simplicity, we here start with a model of a one-period financial market S_t on a finite sample space $\Omega = \{\omega_1, \omega_2, \ldots, \omega_N\}$ which represents the states of the economy at time $t = 0$ or 1 and the occurring probabilities $\mathcal{P}(\omega) > 0$ for all $\omega \in \Omega$. This setup for financial markets is in line with the theory presented in [78]. In Sect. 2.2, we pursue an ansatz with multi-period financial markets and on more general sample spaces.

Notation 2.1 (Payoff Space) *Let $\left(\Omega, 2^\Omega, \mathcal{P}\right)$ be a probability space, where 2^Ω is the algebra of all subsets of Ω. The space of random variables $Y\colon \Omega \to \mathbb{R}$ is denoted $RV\left(\Omega, 2^\Omega, \mathcal{P}\right)$. Since Ω is finite, $RV\left(\Omega, 2^\Omega, \mathcal{P}\right)$ is a finite-dimensional vector space:*

$$Y\colon \Omega \to \mathbb{R} \quad \text{is identified with} \quad [Y(\omega)]_{\omega\in\Omega} \in \mathbb{R}^N.$$

Using the inner product

$$\langle Y, Z\rangle := \mathrm{E}[Y \cdot Z] = \sum_{\omega\in\Omega} \mathcal{P}(\omega)\cdot Y(\omega)\cdot Z(\omega),\ Y, Z \in RV\left(\Omega, 2^{\Omega}, \mathcal{P}\right),$$

$RV\left(\Omega, 2^{\Omega}, \mathcal{P}\right)$ becomes a Hilbert space. Let $RV_{\geq 0}\left(\Omega, 2^{\Omega}, \mathcal{P}\right)$ represent the cone of non-negative random variables $Y\colon \Omega \to \mathbb{R}_{\geq 0}$.

Remark 2.2 *Often in financial mathematics, $(\Omega, \Sigma, \mathcal{P})$ is a general probability space with σ-algebra Σ, and the random variables Y and Z lie in $L^2(\Omega; \mathbb{R})$ such that*

$$\langle Y, Z\rangle := \mathrm{E}[Y \cdot Z] = \int_{\Omega} Y(\omega)\cdot Z(\omega)\,\mathrm{d}\mathcal{P}(\omega) \quad \textit{for} \quad Y, Z\colon (\Omega, \Sigma) \to (\mathbb{R}, \mathcal{B})$$

is well defined. But for the moment, we simplify to Ω finite, which is for many applications reasonable, since historical data of returns of a market are discrete and finite anyway.

2.1.2 One-Period Financial Markets

Definition 2.3 (One-Period Financial Market; [78, Definition 1]) *Let Ω be a finite sample space with $|\Omega| = N$ elements and with $\mathcal{P}(\omega) > 0$ for all $\omega \in \Omega$. For $M \in \mathbb{N}$ fixed, we say that $S_t = \left(S_t^0, S_t^1, \dots, S_t^M\right)^\top$, $t \in \{0, 1\}$, is a* financial market *in a one-period economy, provided that $S_0 \in \mathbb{R}_{>0}^{M+1}$ and $S_1 \in (0, \infty) \times \left[RV_{\geq 0}\left(\Omega, 2^{\Omega}, \mathcal{P}\right)\right]^M$. Here, $S_0^0 = 1$, $S_1^0 = \varphi_0 > 0$ represents a* risk-free asset *with a positive return when $\varphi_0 > 1$. The rest of the components S_t^m, $m = 1, \dots, M$, represent the price of the m-th* risky financial asset *at time t.*

Remark 2.4

(a) The risk-free asset can be viewed as a benchmark. Different agents in the market tend to use different assets as their risk-free asset. For example, for individuals often an appropriate risk-free asset is a money market fund or a certificate of deposit. On the other hand, banks and large financial institutions often use deposit accounts or Treasury bills from major economies as their risk-free asset. Furthermore, even bonds are often used as a synonym for risk-free assets, although depending on the market situation they might in fact bear risk.
Nevertheless, in consensus with the literature, we call the risk-free asset often as the "risk-free bond" or simply "bond," although in reality the price process of bonds might be nondeterministic.

(b) The vector $S_0 \in \mathbb{R}_{>0}^{M+1}$ represents a (fixed) initial price vector at $t = 0$, whereas $S_1 \in \mathbb{R}_{\geq 0}^{M+1}$ is a random state at $t = 1$. The component S_1^m with $m = 0$ is deterministic and corresponds to the price of a (risk-free) bond,

i.e., φ_0 *is a (fixed) constant. On the other hand,* S_1^m *with* $m = 1, \ldots, M$ *involve randomness (of risky assets).*

(c) *The* returns *of the i-th asset (over one period) then are*

$$\frac{S_1^i(\omega) - S_0^i}{S_0^i} = \frac{S_1^i(\omega)}{S_0^i} - 1 \quad \text{(depend on } \omega \in \Omega \textit{ in case } i = 1, \ldots, M)$$

and the return R_0 *of the risk-free bond is* $R_0 := \frac{S_1^0 - S_0^0}{S_0^0} = \varphi_0 - 1$, *which is positive under usual market situations. For the mathematics, however, that is not required, i.e.,* $R_0 \in (-1, 0]$ *works as well.*

(d) *A finite sample space is a natural choice because in the real world, only finite quantities of information can be used. This also avoids the distraction of technical difficulties. In this subsection, we restrict to the one-period market model. More complex sample spaces and market models such as multi-period financial models can be applied similarly (cf. Sect. 2.2, or Platen [89]).*

Notation 2.5 (Price Vector of Risky Assets) *Set* $\widehat{S}_t := \left(S_t^1, \ldots, S_t^M\right)^\top$, $t \in \{0, 1\}$, *with* $S_1^m \in RV_{\geq 0}\left(\Omega, 2^\Omega, \mathcal{P}\right)$ *for* $m = 1, \ldots, M$.

2.1.3 Portfolios

Next, we want to introduce portfolios in which investors can hold assets. Moreover, we introduce relevant properties of portfolios, in particular in relation to the financial market.

Definition 2.6 (Portfolio, Wealth, and Payoff) *A* portfolio *is a column vector* $x = (x_0, \ldots, x_M)^\top \in \mathbb{R}^{M+1}$, *whose components* x_m *represent the shares of the m-th asset in the portfolio. Thus,* $S_t^\top x$ *is the* wealth *of the portfolio at time t, where* $S_0^\top x$ *represents the* initial investments *and* $S_1^\top x$ *represents the* payoff. *A portfolio with* $S_0^\top x = \sum_{i=0}^M S_0^i x_i = 1$ *is called* unit initial cost *portfolio.*

Hence, for unit initial cost portfolios, we get

$$\left. \begin{array}{l} S_t^m x_m \text{ is the fraction (or portion) of capital} \\ \text{invested in asset } m \text{ at time } t. \end{array} \right\} \tag{2.1.1}$$

Furthermore, $x_0 = S_0^0 x_0$ corresponds to the investment (fraction) in the risk-free bond. Using $\widehat{x} := (x_1, \ldots, x_M)^\top \in \mathbb{R}^M$ for the risky part, unit initial cost portfolios satisfy

$$1 = S_0^\top x = x_0 + \widehat{S}_0^\top \widehat{x}. \tag{2.1.2}$$

Note that the difference between the wealth at time $t = 1$ and the initial investment, i.e.,

$$S_1^\top x - S_0^\top x = (S_1 - S_0)^\top x \tag{2.1.3}$$

is the *gain* for the portfolio x from time $t = 0$ to $t = 1$.

Definition 2.7 (Admissible Portfolios) *We say that $A \subset \mathbb{R}^{M+1}$ is a set of* admissible portfolios *provided that A is*

- *nonempty,*
- *closed, and*
- *convex.*

In case furthermore

$$A \subset \left\{x \in \mathbb{R}^{M+1} \,:\, S_0^\top x = 1\right\} =: \mathcal{A}_1 \tag{2.1.4}$$

holds, we call A a set of admissible portfolios with unit initial cost.

Notation 2.8 (Risky Parts of Portfolios) *Similar to the above for the risky assets, we define the risky parts of the portfolios as*

$$\widehat{A} := \left\{\widehat{x} \in \mathbb{R}^M \,:\, \exists\, x_0 \in \mathbb{R} \quad \text{such that} \quad \left(x_0, \widehat{x}^\top\right)^\top \in A\right\}. \tag{2.1.5}$$

Remark 2.9 *Thus, as projection of A to $\mathbb{R}^M$, $\widehat{A}$ is also nonempty, closed, and convex, and hence admissible, whenever A is a set of admissible portfolios.*

The following definitions are essential in finance.

Definition 2.10 (Riskless Portfolio) *We say a portfolio $x \in \mathbb{R}^{M+1}$ is* riskless *for the financial market S_t in Definition 2.3 if*

$$(S_1 - \varphi_0 S_0)^\top x \geq 0 \qquad (\mathcal{P}\text{–a.s.}). \tag{2.1.6}$$

In particular, unit initial cost portfolios are riskless if

$$S_1^\top x \geq \varphi_0 \qquad (\mathcal{P}\text{–a.s.}). \tag{2.1.7}$$

A portfolio $x^\top = \left(x_0, \widehat{x}^\top\right)$ is called trivial *if $\widehat{x} = \widehat{0} \in \mathbb{R}^M$. Accordingly, the market has* "no nontrivial riskless portfolio" *if there does not exist a riskless portfolio x with $\widehat{x} \neq \widehat{0}$.*

Note that the *pure bond portfolio*

$$x^* = \left(x_0^*, \widehat{x}^{*\top}\right)^\top := \left(1, \widehat{0}^\top\right)^\top \in \mathbb{R}^{M+1} \tag{2.1.8}$$

is a trivial riskless portfolio with unit initial cost, i.e., everything is invested in the risk-free bond. Using

$$(S_1 - \varphi_0 S_0)^\top x^* = \left(S_1^0 - \varphi_0 S_0^0\right) \cdot x_0^* = 0,$$

in particular $S_1^\top x^* = \varphi_0$, i.e., x "riskless" in (2.1.7) means, that in any case, x performs at least as good as x^*. However, a nontrivial riskless portfolio is not to be expected and we will often use this assumption.

Definition 2.11 (Arbitrage Portfolio) *We say a portfolio $x \in \mathbb{R}^{M+1}$ is an* arbitrage *for the market S_t if it is riskless* and *if*

$$\mathcal{P}\left[(S_1 - \varphi_0 S_0)^\top x > 0\right] > 0. \tag{2.1.9}$$

In particular, since Ω is finite, there must exist some $\omega_0 \in \Omega$ with $\mathcal{P}(\omega_0) > 0$ and

$$(S_1(\omega_0) - \varphi_0 S_0)^\top x > 0.$$

We say the market S_t has no arbitrage *(or is* arbitrage-free*) if an arbitrage portfolio for S_t does not exist.*

An arbitrage is a way to make return above the risk-free rate, without taking any risk of losing money:

$$(S_1 - \varphi_0 S_0)^\top x > 0 \;\Leftrightarrow\; S_1^\top x > \varphi_0 S_0^\top x,$$

where $S_1^\top x$ is the *wealth at* $t = 1$ and $\varphi_0 S_0^\top x$ is the initial investment including interest paid as for bonds. If such an opportunity exists, investors will try to take advantage of it and in this process bid up the price of the risky asset causing the arbitrage opportunity to disappear.

Note that a trivial portfolio cannot be an arbitrage. Hence,

$$\begin{aligned} \left\{x \in \mathbb{R}^{M+1} : x \text{ is arbitrage}\right\} &= \left\{x \in \mathbb{R}^{M+1} : x \text{ is arbitrage and nontrivial}\right\} \\ &\subset \left\{x \in \mathbb{R}^{M+1} : x \text{ is riskless and nontrivial}\right\} \end{aligned} \tag{2.1.10}$$

Therefore, the *no nontrivial riskless portfolio* condition on a market S_t is stronger than the *no arbitrage* condition.

The "difference" between these conditions is given next.

Definition 2.12 (Nontrivial Bond Replicating Portfolio) *We say a portfolio $x \in \mathbb{R}^{M+1}$ is a* nontrivial bond replicating *portfolio if $\widehat{x} \neq \widehat{0}$ and*

$$(S_1 - \varphi_0 S_0)^\top x = 0 \qquad (\mathcal{P}\text{–a.s.}). \tag{2.1.11}$$

A market has "no nontrivial bond replicating portfolio" *if no such portfolio with* (2.1.11) *exists.*

The last three definitions are related as follows.

Proposition 2.13 (Characterization of No Nontrivial Riskless Portfolio) *Consider the financial market S_t from Definition 2.3. Then*

$$\left.\begin{array}{l}\textit{There is no nontrivial}\\ \textit{riskless portfolio in } S_t\end{array}\right\} \Leftrightarrow \left\{\begin{array}{l} S_t \textit{ has no arbitrage portfolio}\\ \textit{and}\\ S_t \textit{ has no nontrivial bond replicating}\\ \textit{portfolio}\end{array}\right.$$

Proof This follows immediately from the definitions. □

Corollary 2.14 *"No nontrivial riskless portfolio" implies "no arbitrage portfolio" but not the other way around.*

In order to get more used to these notations, we want to construct a one-period financial market which has no arbitrage portfolio, but which has a nontrivial riskless portfolio. The following lemma is helpful.

Lemma 2.15 *A one-period financial market as in Definition 2.3 which satisfies*

$$\mathrm{E}\left[\widehat{S}_1 - \varphi_0 \widehat{S}_0\right] = \widehat{0} \in \mathbb{R}^M \tag{2.1.12}$$

has no arbitrage portfolio.

Proof If the market had an arbitrage portfolio $x = \left(x_0, \widehat{x}^\top\right)^\top \in \mathbb{R}^{M+1}$, then due to $S_1^0 = \varphi_0 = \varphi_0 S_0^0$, the inequalities

$$\left(\widehat{S}_1(\omega) - \varphi_0 \widehat{S}_0\right)^\top \widehat{x} \geq 0 \quad \text{for all } \omega \in \Omega$$

and

$$\left(\widehat{S}_1(\omega^*) - \varphi_0 \widehat{S}_0\right)^\top \widehat{x} > 0 \quad \text{for at least one} \quad \omega^* \in \Omega \quad \text{with} \quad \mathcal{P}(\omega^*) > 0$$

would hold. But then $\mathrm{E}\left[\left(\widehat{S}_1 - \varphi_0 \widehat{S}_0\right)^\top \widehat{x}\right] > 0$, which contradicts (2.1.12). □

This leads us to the following example:

Example 2.16 *Consider a one-period financial market with $M = 2$ risky assets, $\varphi_0 = \frac{6}{5}$, $\widehat{S}_0 = (1,1)^\top$, $\Omega = \{\omega_1, \omega_2\}$, $\mathcal{P}(\omega_1) = \mathcal{P}(\omega_2) = \frac{1}{2}$ and*

$$\widehat{S}_1(\omega) = \begin{cases} (\alpha, \beta)^\top, & \textit{for } \omega = \omega_1,\\ \left(\frac{8}{5}, \frac{11}{10}\right)^\top, & \textit{for } \omega = \omega_2 \end{cases} \tag{2.1.13}$$

for some parameters $\alpha, \beta \in \mathbb{R}$. *We ask the question whether or not* $(\alpha, \beta)^\top$ *might be adjusted in a way such that this market has no arbitrage portfolio. This is indeed possible: according to Lemma 2.15, a sufficient condition for no arbitrage portfolio is*

$$\mathrm{E}\left[\widehat{S}_1 - \varphi_0 \widehat{S}_0\right] = (0,0)^\top = \widehat{0} \in \mathbb{R}^2. \tag{2.1.14}$$

Hence, a short calculation reveals that the market in (2.1.13) *with*

$$\begin{pmatrix} \alpha \\ \beta \end{pmatrix} = -\begin{pmatrix} \frac{8}{5} \\ \frac{11}{10} \end{pmatrix} + \frac{12}{5}\begin{pmatrix} 1 \\ 1 \end{pmatrix} = \begin{pmatrix} \frac{4}{5} \\ \frac{13}{10} \end{pmatrix} \tag{2.1.15}$$

has no arbitrage portfolio. Furthermore, with $(\alpha, \beta)^\top$ *from* (2.1.15) *fixed, we obtain*

$$\left(\widehat{S}_1 - \varphi_0 \widehat{S}_0\right)(\omega_1) = -\left(\widehat{S}_1 - \varphi_0 \widehat{S}_0\right)(\omega_2) = \begin{pmatrix} -\frac{2}{5} \\ \frac{1}{10} \end{pmatrix}.$$

Thus, any portfolio $x^* = \left(x_0^*, \frac{1}{4}, 1\right)^\top \in \mathbb{R}^3$ *is a nontrivial bond replicating portfolio, since* $\left(\widehat{S}_1 - \varphi_0 \widehat{S}_0\right)^\top \widehat{x^*} = 0$ *($\mathcal{P}$–a.s.). Clearly, this portfolio is also a nontrivial riskless portfolio. In total, we have constructed a financial market with no arbitrage portfolio, where on the other hand, the no nontrivial riskless portfolio condition fails.*

Next, we give various other statements for "no nontrivial riskless portfolio."

Theorem 2.17 (Further Characterizations of No Nontrivial Riskless Portfolio; [78, Theorem 2]) *Assume a financial market* S_t *in Definition 2.3 (with* $\Omega = \{\omega_1, \ldots, \omega_N\}$ *finite and* $\mathcal{P}(\omega) > 0$ *for all* $\omega \in \Omega$*). Then the following assertions are equivalent:*

(i) There is no nontrivial riskless portfolio.
(ii) For every nontrivial portfolio $x \in \mathbb{R}^{M+1}$ *(with* $\widehat{x} \neq \widehat{0}$*), there exists some* $\omega \in \Omega$ *such that*

$$\left(S_1(\omega) - \varphi_0 S_0\right)^\top x < 0. \tag{2.1.16}$$

(ii) For every risky portfolio* $\widehat{x} \in \mathbb{R}^M$ *with* $\widehat{x} \neq \widehat{0}$*, there exists some* $\omega \in \Omega$ *such that*

$$\left(\widehat{S}_1(\omega) - \varphi_0 \widehat{S}_0\right)^\top \widehat{x} < 0. \tag{2.1.17}$$

(iii) The market has no arbitrage and *the matrix*

$$V := \left[S_1^j(\omega_i) - \varphi_0 S_0^j\right]_{\substack{1\le i\le N\\ 1\le j\le M}}$$
$$= \begin{bmatrix} S_1^1(\omega_1) - \varphi_0 S_0^1 & \cdots & S_1^M(\omega_1) - \varphi_0 S_0^M \\ \vdots & & \vdots \\ S_1^1(\omega_N) - \varphi_0 S_0^1 & \cdots & S_1^M(\omega_N) - \varphi_0 S_0^M \end{bmatrix} \in \mathbb{R}^{N\times M} \tag{2.1.18}$$

has rank M, in particular $N \geq M$.

Proof We give a cyclic proof.

"(i) $\Rightarrow$ (ii)": Assume (ii) fails. Then there exists a nontrivial portfolio x (with $\widehat{x} \neq \widehat{0}$), such that

$$(S_1(\omega) - \varphi_0 S_0)^\top x \geq 0$$

for all $\omega \in \Omega$. By Proposition 2.13, x cannot be bond replicating. Thus, we get

$$\mathcal{P}\left[(S_1 - \varphi_0 S_0)^\top x > 0\right] > 0.$$

Hence, x is an arbitrage. But this is not possible, again due to Proposition 2.13 $\lightning$.

"(ii) $\Rightarrow$ (ii)*": This is obvious, because for $x = \left(x_0, \widehat{x}^\top\right)^\top \in \mathbb{R}^{M+1}$

$$(S_1(\omega) - \varphi_0 S_0)^\top x = \underbrace{(S_1^0 - \varphi_0 S_0^0)}_{=0} \cdot x_0 + \left(\widehat{S}_1(\omega) - \varphi_0 \widehat{S}_0\right)^\top \widehat{x}. \tag{2.1.19}$$

"(ii)* $\Rightarrow$ (iii)": Assume firstly that the market has an arbitrage portfolio $z^* \in \mathbb{R}^{M+1}$.

Then z^* is nontrivial, i.e., $\widehat{z}^* \neq \widehat{0}$ as seen before (see (2.1.10)) and z^* is riskless by definition, i.e.,

$$\left(\widehat{S}_1 - \varphi_0 \widehat{S}_0\right)^\top \widehat{z}^* = (S_1 - \varphi_0 S_0)^\top z^* \overset{(2.1.6)}{\geq} 0 \qquad (\mathcal{P}\text{–a.s.}),$$

contradicting (2.1.17).

Thus, it remains to show that $V \in \mathbb{R}^{N\times M}$ in (2.1.18) has full rank M. If this is not true, then $V\widehat{x} = 0 \in \mathbb{R}^N$ has a nontrivial solution $\mathbb{R}^M \ni \widehat{x} \neq \widehat{0}$. This contradicts again (2.1.17) since

$$V\widehat{x} = \begin{bmatrix} \left(\widehat{S}_1(\omega_1) - \varphi_0 \widehat{S}_0\right)^\top \\ \vdots \\ \left(\widehat{S}_1(\omega_N) - \varphi_0 \widehat{S}_0\right)^\top \end{bmatrix} \cdot \widehat{x} \in \mathbb{R}^N$$

"(iii) $\Rightarrow$ (i)": Due to Proposition 2.13, it remains to show that the market S_t has no nontrivial bond replicating portfolio.

Indirect: Assume there exists a portfolio z^* with $\widehat{z}^* \neq \widehat{0}$ which replicates the bond, i.e.,

$$(S_1 - \varphi_0 S_0)^\top z^* = 0 \qquad (\mathcal{P}\text{–a.s.}).$$

Arguing as in (2.1.19) implies

$$\left(\widehat{S}_1 - \varphi_0 \widehat{S}_0\right)^\top \widehat{z}^* = 0 \qquad (\mathcal{P}\text{–a.s.}).$$

But with $\mathcal{P}(\omega_j) > 0 \quad$ for all $\; j = 1, \ldots, N$ we obtain

$$\left(\widehat{S}_1(\omega_j) - \varphi_0 \widehat{S}_0\right)^\top \widehat{z}^* = 0 \quad \text{for all } \; j = 1, \ldots, N,$$

giving $V\widehat{z}^* = 0$ which contradicts the fact that V has full rank M. □

Remark 2.18 *Whereas the "no arbitrage" condition is very common in finance, the "no nontrivial riskless portfolio" condition appears not so often.*

An application of this yields that the covariance matrix of risky assets must be positive definite.

Corollary 2.19 (Positive Definite Covariance Matrix; [78, Corollary 1]) *Assume the market S_t in Definition 2.3 has no nontrivial riskless portfolio. Then the* covariance matrix *of the risky assets*

$$\begin{aligned}\operatorname{Cov}(\widehat{S}) &:= \mathrm{E}\left[\left(\widehat{S}_1 - \mathrm{E}\left(\widehat{S}_1\right)\right) \cdot \left(\widehat{S}_1 - \mathrm{E}\left(\widehat{S}_1\right)\right)^\top\right] \\ &= \left(\mathrm{E}\left[\left(S_1^i - \mathrm{E}\left(S_1^i\right)\right) \cdot \left(S_1^j - \mathrm{E}\left(S_1^j\right)\right)\right]\right)_{\substack{1 \leq i \leq M \\ 1 \leq j \leq M}} \in \mathbb{R}^{M \times M}\end{aligned}$$

is positive definite.

Proof Note that under the "no nontrivial riskless portfolio" assumption,

$$\widehat{S}_1^\top \cdot \widehat{x} \quad \text{represented as} \quad \left[\widehat{S}_1^\top(\omega) \cdot \widehat{x}\right]_{\omega \in \Omega} \in \mathbb{R}^N$$

cannot be a *constant* random variable for every nontrivial portfolio $\widehat{x} \neq \widehat{0}$.

Otherwise, there exists some $\widehat{x} \in \mathbb{R}^M$ with $\widehat{x} \neq \widehat{0}$ such that $\left(\widehat{S}_1 - \varphi_0 \widehat{S}_0\right)^\top \widehat{x}$ would be constant, which contradicts (2.1.17) used for $\widehat{x}$ and $-\widehat{x}$. Note that $\left(\widehat{S}_1 - \varphi_0 \widehat{S}_0\right)^\top (-\widehat{x}) = -\left(\widehat{S}_1 - \varphi_0 \widehat{S}_0\right)^\top \widehat{x}$. Hence, for every nontrivial risky portfolio $\widehat{x} \neq \widehat{0}$, we get

$$0 < \operatorname{Var}\left(\widehat{S}_1^\top \widehat{x}\right) = \widehat{x}^\top \operatorname{Cov}(\widehat{S})\widehat{x}.$$

Therefore, $\operatorname{Cov}(\widehat{S})$ is positive definite. □

2.1.4 Risk Functions

Recall from Sect. 2.1.3 that $A \subset \mathbb{R}^{M+1}$ is a set of admissible portfolios when it is nonempty, closed, and convex (see Definition 2.7).

Investors are often sensitive to the "risk of a portfolio" which can be gauged by a risk function.

Assumption 2.20 (Risk Functions; [78, Assumption 2])
Consider an extended-valued risk function $\mathfrak{r}\colon A \to \mathbb{R} \cup \{+\infty\}$ *on the admissible portfolios* $A \subset \mathbb{R}^{M+1}$ *(cf. Definition 2.7). We always assume risk functions to be lower semi-continuous (see Appendix A.1). Furthermore, we will often use some of the following assumptions:*

(r1) (Riskless asset contributes no risk) *The risk function* $\mathfrak{r}(x) = \widehat{\mathfrak{r}}(\widehat{x})$ *is a function of only the risky part of the portfolio, where* $x = \left(x_0, \widehat{x}^\top\right)^\top \in A$, *i.e.,* $\widehat{x} \in \widehat{A}$ *(cf. Notation 2.8).*

(r1n) (Non-negativity and normalization) *The risk function is non-negative, i.e.,* $\mathfrak{r}(x) \geq 0$ *for all* $x \in A$ *and there is at least one portfolio of purely bonds in A. Furthermore,* $\mathfrak{r}(x) = 0$ if and only if $x \in A$ *contains only a riskless bond, i.e.,* $x^\top = \left(x_0, \widehat{0}^\top\right)$ *for some* $x_0 \in \mathbb{R}$.

(r2) (Diversification reduces risk) *The risk function* $\mathfrak{r}$ *is proper convex (cf Definition A.10).*

(r2s) (Diversification strictly reduces risk $\mathfrak{r}$) *The risk function* $\mathfrak{r}$ *is strictly convex on* $\operatorname{dom}(\mathfrak{r}) \subset A$.

($\widehat{\text{r2s}}$) (Diversification strictly reduces risk $\widehat{\mathfrak{r}}$) *The risk function* $\widehat{\mathfrak{r}}$ *in* (r1) *is strictly convex (on its convex domain* $\operatorname{dom}(\widehat{\mathfrak{r}}) = \{\widehat{x} \in \widehat{A} \subset \mathbb{R}^M : \widehat{\mathfrak{r}}(\widehat{x}) < \infty\}$*).*

We firstly search for sufficient conditions to ensure (r1), i.e., $\mathfrak{r}(x) = \widehat{\mathfrak{r}}(\widehat{x})$ should depend only on the risky part of the portfolio. In order to get that, we can use a one-period financial market (Definition 2.3), i.e., $S_t = \left(S_t^0, S_t^1, \ldots, S_t^M\right)^\top \in \mathbb{R}^{M+1}$, $t = 0, 1$, with a payoff $S_1^\top x$ at time $t = 1$.

Lemma 2.21 (Conditions to Ensure (r1)**)** *Let* S_t, $t = 0$ *or* 1 *be a one-period financial market. Assume a set of admissible portfolios A (cf. Definition 2.7) and a risk function of the form*

$$\mathfrak{r}(x) := \rho\left(S_1^\top x\right), \quad x \in A \subset \mathbb{R}^{M+1},$$

for a given function $\rho\colon RV\left(\Omega, 2^\Omega, \mathcal{P}\right) \to \mathbb{R} \cup \{+\infty\}$. *Then* (r1) *is satisfied if (a)* $\rho(Y) = \rho(Y+c)$ *for any* $Y \in RV\left(\Omega, 2^{\Omega}, \mathcal{P}\right)$ *and* $c \in \mathbb{R}$

or if

(b) A *consists only of portfolios with unit initial cost.*

Proof *ad(a).* Note that $S_1^\top x = \varphi_0 x_0 + \widehat{S}_1^\top \widehat{x}$ for $x = \left(x_0, \widehat{x}^\top\right)^\top \in A \subset \mathbb{R}^{M+1}$. Thus,

$$\mathfrak{r}(x) = \rho\left(S_1^\top x\right) = \rho\left(\widehat{S}_1^\top \widehat{x}\right) =: \widehat{\mathfrak{r}}\left(\widehat{x}\right). \tag{2.1.20}$$

ad(b). For $x \in A$ with unit initial cost, we have $S_0^\top x = 1$ and therefore $x_0 = 1 - \widehat{S}_0^\top \widehat{x}$. Hence,

$$\mathfrak{r}(x) = \rho\left(S_1^\top x\right) = \rho\left(\varphi_0 x_0 + \widehat{S}_1^\top \widehat{x}\right) = \rho\left(\varphi_0 + \left(\widehat{S}_1 - \varphi_0 \widehat{S}_0\right)^\top \widehat{x}\right) =: \widehat{\mathfrak{r}}\left(\widehat{x}\right), \tag{2.1.21}$$

which completes the proof. □

Note that, for instance, $\rho(Y) = \sqrt{\mathrm{Var}(Y)}$ (standard deviation) satisfies the assumption (a) of Lemma 2.21.

Often risk functions do not depend on the bond component (see (r1)). Then, in order to get (r2s), additionally unit initial cost is helpful.

Lemma 2.22 (Sufficient Conditions for (r2s)**)** *Consider a lower semi-continuous, extended-valued risk function* $\mathfrak{r}\colon A \to \mathbb{R} \cup \{+\infty\}$ *on admissible portfolios* $A \subset \mathbb{R}^{M+1}$ *with* unit (or fixed) initial cost *such that* (r1) *and* $(\widehat{\text{r2s}})$ *of Assumption 2.20 hold. Then* $\mathfrak{r}$ *is also strictly convex, i.e.,* (r2s) *of Assumption 2.20 is satisfied as well.*

Proof Let $\lambda \in (0,1)$, $x^{(0)} \neq x^{(1)} \in \mathrm{dom}(\mathfrak{r}) \subset A$ be given. Since A is with *unit (or fixed) initial cost,* we obtain $\widehat{x}^{(0)} \neq \widehat{x}^{(1)}$. Then, with $x^{(\lambda)} := \lambda x^{(1)} + (1-\lambda)x^{(0)}$ follows

$$\begin{aligned}\mathfrak{r}\left(x^{(\lambda)}\right) \overset{(\text{r1})}{=} \widehat{\mathfrak{r}}\left(\widehat{x}^{(\lambda)}\right) &\overset{(\widehat{\text{r2s}})}{<} \lambda \widehat{\mathfrak{r}}\left(\widehat{x}^{(1)}\right) + (1-\lambda)\widehat{\mathfrak{r}}\left(\widehat{x}^{(0)}\right)\\ &\overset{(\text{r1})}{=} \lambda\, \mathfrak{r}\left(x^{(1)}\right) + (1-\lambda)\, \mathfrak{r}\left(x^{(0)}\right),\end{aligned}$$

i.e., (r2s) holds as well. □

Remark 2.23 *Condition* (r2s) *is particularly helpful to show uniqueness of "efficient portfolios" in Sect. 2.4. Given Lemma 2.22, we learn that* $(\widehat{\text{r2s}})$ *is sufficient as well, when the initial cost is fixed.*

In addition, some risk functions turn out to be positive homogeneous. In this regard, the cone property is sometimes useful.

Definition 2.24 (Positive Scaling Invariance, Cone Property) *Let A be an admissible set of portfolios (cf. Definition 2.7). We say $\widehat{A}$ (see* (2.1.5)*) is* positive scaling invariant, *if*

$$\text{for} \quad t \geq 0 \quad \text{and} \quad \widehat{x} \in \widehat{A}, \quad \text{always} \quad t\widehat{x} \in \widehat{A} \quad \text{holds.} \tag{2.1.22}$$

Note that positive scaling invariance in (2.1.22) *implies that $\widehat{A}$ is a cone since as an admissible set it is in particular convex (see Remark 2.9).*

Assumption 2.25 (Positive Homogeneous Risk Functions) *For a risk function $\mathfrak{r}\colon A \to \mathbb{R} \cup \{+\infty\}$ satisfying (r1) sometimes further assumptions regarding $\widehat{\mathfrak{r}}\colon \widehat{A} \to \mathbb{R} \cup \{+\infty\}$ are in order.*

(r3) (Positive homogeneous) *$\widehat{A}$ is positive scaling invariant, i.e.,* (2.1.22) *holds, and for all $t \geq 0$ and $\widehat{x} \in \widehat{A}$ always $\widehat{\mathfrak{r}}(t\widehat{x}) = t \cdot \widehat{\mathfrak{r}}(\widehat{x})$ follows.*

(r3s) (Diversification strictly reduces risk on level sets) *The risk function $\widehat{\mathfrak{r}}$ satisfies (r3) and, moreover, for all $\widehat{x}, \widehat{y} \in \widehat{A}$ with $\widehat{x} \neq \widehat{y}$ and $\mathfrak{r}(\widehat{x}) = \widehat{\mathfrak{r}}(\widehat{y}) = 1$, the following inequality holds true:*

$$\widehat{\mathfrak{r}}(\alpha\widehat{x} + (1-\alpha)\widehat{y}) < \alpha\widehat{\mathfrak{r}}(\widehat{x}) + (1-\alpha)\widehat{\mathfrak{r}}(\widehat{y}) = 1 \quad \text{for all} \quad \alpha \in (0,1).$$

Apparently, condition (r3) rules out (r2s). Thus, condition (r3s) serves as a replacement for (r2s) when the risk function satisfies (r3). In this case, the following lemma is useful.

Lemma 2.26 ([78, Lemma 1]) *Assume a risk function $\mathfrak{r}\colon A \to \mathbb{R} \cup \{+\infty\}$ satisfies* (r1), (r1n), *and* (r3s), *in particular also* (r3) *holds. Then*

(a) $\mathfrak{r}$ satisfies (r2), *and*
(b) $f(x) = \widehat{f}(\widehat{x}) := [\widehat{\mathfrak{r}}(\widehat{x})]^2$, $\widehat{x} \in \widehat{A}$, satisfies (r1), (r1n) *and* $(\widehat{\text{r2s}})$.

Proof *ad (a).* We only need to show that $\mathfrak{r}$ is convex, since $\mathfrak{r}$ is proper by its definition in (r1n). Let $\alpha \in (0,1)$ and $\widehat{x}, \widehat{y} \in \widehat{A}$ with $\widehat{x} \neq \widehat{y}$ be given.

Case I: $\widehat{x}$ and $\widehat{y}$ lie on the same ray through $\widehat{0}$, say $\widehat{x} = c\widehat{y}$ for some $c \geq 0$. Then $\alpha\widehat{x} + (1-\alpha)\widehat{y} \in \widehat{A}$ since $\widehat{A}$ is convex. Thus,

$$\begin{aligned}\widehat{\mathfrak{r}}(\alpha\widehat{x} + (1-\alpha)\widehat{y}) = \widehat{\mathfrak{r}}([\alpha c + (1-\alpha)]\,\widehat{y}) &\overset{\text{(r3)}}{=} [\alpha c + (1-\alpha)]\,\widehat{\mathfrak{r}}(\widehat{y}) \\ &\overset{\text{(r3)}}{=} \alpha\widehat{\mathfrak{r}}(c\widehat{y}) + (1-\alpha)\widehat{\mathfrak{r}}(\widehat{y}).\end{aligned}$$

Hence, convexity of $\widehat{\mathfrak{r}}$ on that ray follows, giving convexity of $\mathfrak{r}$ by (r1).

Case II: $\widehat{x}$ and $\widehat{y}$ do not lie on the same ray through $\widehat{0}$. Then due to (r1) and (r1n),

$$\widehat{\mathfrak{r}}(\widehat{x}) > 0 \quad \text{and also} \quad \widehat{\mathfrak{r}}(\widehat{y}) > 0 \quad \text{with} \quad \frac{\widehat{x}}{\widehat{\mathfrak{r}}(\widehat{x})} \neq \frac{\widehat{y}}{\widehat{\mathfrak{r}}(\widehat{y})}.$$

Defining

$$\lambda := \frac{\alpha\widehat{\mathfrak{r}}(\widehat{x})}{\alpha\widehat{\mathfrak{r}}(\widehat{x}) + (1-\alpha)\widehat{\mathfrak{r}}(\widehat{y})} \in (0,1) \quad \text{we have} \quad 1-\lambda = \frac{(1-\alpha)\widehat{\mathfrak{r}}(\widehat{y})}{\alpha\widehat{\mathfrak{r}}(\widehat{x}) + (1-\alpha)\widehat{\mathfrak{r}}(\widehat{y})}$$

Furthermore, with (2.1.22) follows $\frac{\widehat{x}}{\widehat{r(x)}}, \frac{\widehat{y}}{\widehat{r(y)}} \in \widehat{A}$. Thus, (r3) implies $\widehat{\mathfrak{r}}\left(\frac{\widehat{x}}{\widehat{\mathfrak{r}(x)}}\right) = 1 = \widehat{\mathfrak{r}}\left(\frac{\widehat{y}}{\widehat{\mathfrak{r}(y)}}\right)$ and from (r3s), we obtain

$$\begin{aligned} 1 &\overset{(\text{r3s})}{>} \widehat{\mathfrak{r}}\left(\lambda\frac{\widehat{x}}{\widehat{\mathfrak{r}}(\widehat{x})} + (1-\lambda)\frac{\widehat{y}}{\widehat{\mathfrak{r}}(\widehat{y})}\right) = \widehat{\mathfrak{r}}\left(\frac{\alpha\widehat{x} + (1-\alpha)\widehat{y}}{\alpha\widehat{\mathfrak{r}}(\widehat{x}) + (1-\alpha)\widehat{\mathfrak{r}}(\widehat{y})}\right) \\ &\overset{(\text{r3})}{=} \frac{1}{\alpha\widehat{\mathfrak{r}}(\widehat{x}) + (1-\alpha)\widehat{\mathfrak{r}}(\widehat{y})} \cdot \widehat{\mathfrak{r}}\left(\alpha\widehat{x} + (1-\alpha)\widehat{y}\right) \end{aligned} \tag{2.1.23}$$

verifying convexity for $\widehat{\mathfrak{r}}$ and due to (r1) also for $\mathfrak{r}$.
ad (b). We leave this as an exercise to the reader. □

Remark 2.27 (Deviation Measure) *A risk function as defined here is in the literature often called* deviation measure *(cf. [93]), if* (r1), (r1n), (r2), *and* (r3) *are satisfied. In particular, the standard deviation*

$$\sigma(\widehat{x}) = \widehat{\mathfrak{r}}(\widehat{x}) := \sqrt{\mathrm{Var}(\widehat{S}_1^\top \widehat{x})} = \sqrt{\widehat{x}^\top \mathrm{Cov}(\widehat{S})\widehat{x}} \tag{2.1.24}$$

of Corollary 2.19 with $\mathrm{Cov}(\widehat{S})$ *positive definite is a deviation measure (again, we leave the verification of* (r2) *to the reader; the remaining claims are obvious). Furthermore, note that assumption* (r1) *excludes the widely used coherent risk measure introduced in [4], which requires that cash reserve reduces risk.*

In the following, we give sufficient conditions for (r2) and (r2s) for a risk function which is similarly defined as in Lemma 2.21.

Lemma 2.28 (Induced Risk Function) *Fix a financial market S_t as in Definition 2.3 and a set of admissible portfolios $A \subset \mathbb{R}^{M+1}$ (cf. Definition 2.7). Suppose after identification $RV\left(\Omega, 2^\Omega, \mathcal{P}\right) \widehat{=} \mathbb{R}^N$ (since $|\Omega| = N < \infty$), and that the function $\rho\colon RV\left(\Omega, 2^\Omega, \mathcal{P}\right) \to \mathbb{R} \cup \{+\infty\}$ is*

(a)
- *lower semi-continuous,*
- *proper convex such that there exists some $x^* \in A$ with $S_1^\top x^* \in \mathrm{dom}(\rho)$, and*
- *positive homogeneous.*

Then the risk function $\mathfrak{r}\colon A \to \mathbb{R} \cup \{+\infty\}$, $x \mapsto \mathfrak{r}(x) := \rho\left(S_1^\top x\right)$, is lower semi-continuous and satisfies (r2) *in Assumption 2.20. If, moreover, A is positive scaling invariant, i.e., $t \geq 0$ and $x \in A$ imply $tx \in A$, then $\mathfrak{r}$ is also positive homogeneous, i.e.,*

$$\mathfrak{r}(tx) = t\mathfrak{r}(x) \quad \textit{for all} \;\; x \in A, \;\; t \geq 0. \tag{2.1.25}$$

Furthermore, in case additionally (a) *of Lemma 2.21 holds, then for* $\widehat{\mathfrak{r}}$ *from* (2.1.20) *also* (r3) *holds.*

(b) If

- S_t *has "no nontrivial riskless portfolio" (Definition 2.10),*
- ρ *is strictly convex, and*
- A *only contains portfolios with unit initial cost,*

then $\widehat{\mathfrak{r}}$ *defined in* (2.1.21) *satisfies* $(\widehat{\text{r2s}})$ *in Assumption 2.20 and thus also* (r2s) *holds.*

Proof *ad (a).* Note that $x \mapsto S_1^\top x$ is a linear mapping in x. Thus, the risk function $\mathfrak{r}$ inherits the properties of ρ giving lower semi-continuity and proper convexity of $\mathfrak{r}$, i.e., (r2) with ($x^* \in \operatorname{dom}(\mathfrak{r})$). Moreover, since ρ is positive homogeneous, positive homogeneity of $\mathfrak{r}$ follows as well for positive scaling invariant A, i.e., (2.1.25) holds. Clearly from (2.1.20) then (r3) follows for $\widehat{\mathfrak{r}}$.

ad (b). We only have to show $(\widehat{\text{r2s}})$ for $\widehat{\mathfrak{r}}$ because then (r2s) follows from Lemma 2.22. Since S_t has no nontrivial riskless portfolio, by Theorem 2.17 (iii), the matrix

$$V = \begin{bmatrix} \left(\widehat{S}_1(\omega_1) - \varphi_0 \widehat{S}_0\right)^\top \\ \vdots \\ \left(\widehat{S}_1(\omega_N) - \varphi_0 \widehat{S}_0\right)^\top \end{bmatrix} \in \mathbb{R}^{N \times M}$$

has full rank $M (N \geq M)$. Thus, for $\widehat{x}, \widehat{y} \in \operatorname{dom}(\widehat{\mathfrak{r}}) \subset \mathbb{R}^M$ with $\widehat{x} \neq \widehat{y}$ follows

$$V\widehat{x} = \left(\widehat{S}_1 - \varphi_0 \widehat{S}_0\right)^\top \widehat{x} \neq \left(\widehat{S}_1 - \varphi_0 \widehat{S}_0\right)^\top \widehat{y} = V\widehat{y}. \tag{2.1.26}$$

Then for $\lambda \in (0,1)$, we obtain

$$\widehat{\mathfrak{r}}(\lambda\widehat{x} + (1-\lambda)\widehat{y}) \overset{(2.1.21)}{=} \rho\left(\varphi_0 + \left(\widehat{S}_1 - \varphi_0 \widehat{S}_0\right)^\top \cdot (\lambda\widehat{x} + (1-\lambda)\widehat{y})\right)$$

since A contains only portfolios with unit initial cost. Hence, using that ρ is strictly convex, we get

$$\begin{aligned} &\widehat{\mathfrak{r}}(\lambda\widehat{x} + (1-\lambda)\,\widehat{y}) \\ &\quad \overset{(2.1.26)}{<} \lambda\rho\left(\varphi_0 + \left(\widehat{S}_1 - \varphi_0 \widehat{S}_0\right)^\top \widehat{x}\right) + (1-\lambda)\rho\left(\varphi_0 + \left(\widehat{S}_1 - \varphi_0 \widehat{S}_0\right)^\top \widehat{y}\right) \\ &\quad \overset{(2.1.21)}{=} \lambda\widehat{\mathfrak{r}}(\widehat{x}) + (1-\lambda)\widehat{\mathfrak{r}}(\widehat{y}), \end{aligned}$$

giving strict convexity of $\widehat{\mathfrak{r}}$, i.e., $(\widehat{\text{r2s}})$. □

2.1.5 Utility Functions

The merit or return of a portfolio is often judged by a utility function.

Assumption 2.29 (Utility Functions) *Consider an extended-valued* utility function $\mathfrak{u}\colon A \to \mathbb{R} \cup \{-\infty\}$ *on the admissible portfolios* A *(cf. Definition 2.7). We will always assume utility functions to be upper semi-continuous. Furthermore, we use some of the following assumptions:*

(u1) (Increasing payoff implies increasing utility) *For distinct portfolios* $x^{(1)} \neq x^{(2)} \in A$ *we have*

$$S_1^\top \left(x^{(2)} - x^{(1)}\right) \geq 0 \ (\mathcal{P}\text{–}a.s.) \quad \Rightarrow \quad \mathfrak{u}\left(x^{(2)}\right) \geq \mathfrak{u}\left(x^{(1)}\right).$$

(u2) (Diminishing marginal utility) *The utility function* $\mathfrak{u}$ *is proper concave.*

(u2s) (Strict diminishing marginal utility) *The utility function* $\mathfrak{u}$ *is strictly concave on its domain* $\operatorname{dom}(\mathfrak{u}) = \{x \in A : \mathfrak{u}(x) > -\infty\}$.

(u3) (Unbounded payoff implies unbounded utility) *For all sequences* $x^{(n)} \in A$, $n \in \mathbb{N}$, *the following holds:*

$$S_1^\top x^{(n)} \to \infty \quad (n \to \infty) \ (\mathcal{P}\text{–}a.s.) \quad \Rightarrow \quad \mathfrak{u}\left(x^{(n)}\right) \to \infty \quad (n \to \infty).$$

Remark 2.30 *Economists often also assume that the Inada conditions hold: e.g., that the slope of the utility function converges to 0 as the input approaches infinity. Although many of the utility functions we use do satisfy this condition, we do not include it in the definition of the utility function. The reason is that we want to allow the identity function as one of the auxiliary functions in the construction of (expected) utility functions in Definition 2.31. This way the Markowitz portfolio theory can be treated as a special case of the general portfolio problem of trading-off between utility and risks which we discuss here.*

A large class of utility functions is constructed as "expected utility."

Definition 2.31 (Expected Utility) *Consider a one-period financial market* S_t *as in Definition 2.3. Using an auxiliary function* $\phi\colon \mathbb{R} \to \mathbb{R}\cup\{-\infty\}$, *the function* $\mathfrak{u}\colon A \to \mathbb{R} \cup \{-\infty\}$, $\mathfrak{u}(x) := \mathrm{E}\left[\phi\left(S_1^\top x\right)\right]$, *defined on the admissible portfolios* A, *is called* expected utility.

The following assumptions on ϕ are sometimes assumed:

Assumption 2.32 (Conditions on Auxiliary Function) *The auxiliary function* $\phi\colon \mathbb{R} \to \mathbb{R} \cup \{-\infty\}$ *is assumed to be upper semi-continuous and to satisfy some of the following properties:*

($\phi 1$) (Profit seeking) *The auxiliary function ϕ is an increasing function.*
($\phi 2$) (Bankruptcy forbidden) *For $t < 0$, $\phi(t) = -\infty$.*
($\phi 3$) (Unlimited) *For $t \to \infty$ we have $\phi(t) \to \infty$.*

Example 2.33 $\phi = \mathrm{id}$ *satisfies* ($\phi 1$), ($\phi 3$) *and is proper concave, while* $\phi = \ln$ *(natural logarithm) is even strictly proper concave and satisfies* ($\phi 1$), ($\phi 2$) *and* ($\phi 3$).

Next, we give sufficient conditions for the expected utility $\mathfrak{u}$ to satisfy (u2) as well as (u2s).

Lemma 2.34 (Properties of Expected Utility) *Let the auxiliary function ϕ used in Definition 2.31 be upper semi-continuous and proper concave, such that, moreover, there exists some $x^* \in A$ with $\mathrm{E}\left[\phi\left(S_1^\top x^*\right)\right] \in \mathbb{R}$. Then the following holds true:*

(a) The expected utility $\mathfrak{u} = \mathrm{E}\left[\phi\left(S_1^\top \cdot\right)\right]$ satisfies (u2) *in Assumption 2.29, i.e., $\mathfrak{u}$ is proper concave.*
(b) Let furthermore ϕ be even strictly concave on its domain

$$\mathrm{dom}(\phi) = \{t \in \mathbb{R} \,:\, \phi(t) > -\infty\} \neq \emptyset$$

and, moreover, let the following hold true:

(i) The financial market has no nontrivial riskless portfolio.
(ii) $A \subset \mathbb{R}^{M+1}$ is a set of admissible portfolios with unit initial cost.

Then the expected utility $x \mapsto \mathfrak{u}(x) = \mathrm{E}\left[\phi\left(S_1^\top x\right)\right]$, $x \in A$, is strictly concave on

$$\mathrm{dom}(\mathfrak{u}) = \{x \in A \,:\, \mathfrak{u}(x) > -\infty\} \neq \emptyset,$$

i.e., (u2s) *holds.*

Proof The claim in *(a)* follows directly from the proper concavity of ϕ since the expectation and $x \mapsto S_1^\top x$ are linear. For the claim in *(b)*, we can use *(i)* and obtain from Theorem 2.17 that

$$V = \left[S_1^j\left(\omega_i\right) - \varphi_0 S_0^j\right]_{\substack{1 \le i \le N \\ 1 \le j \le M}} \in \mathbb{R}^{N \times M}$$

has full rank M. To prove strict concavity of $\mathfrak{u}$, we note that for any two elements from $\mathrm{dom}(\mathfrak{u})$, say $x \neq y$, we get $\widehat{x} \neq \widehat{y} \in \mathbb{R}^M$ by *(ii)* and thus $V\widehat{x} \neq V\widehat{y} \in \mathbb{R}^N$. Using again the unit initial cost condition implies $S_1^\top x \neq S_1^\top y$ on a set with positive measure from which strict concavity of $\mathfrak{u}$ immediately follows since by assumption ϕ is strictly concave. □

2.2 Multi-period Financial Markets and Related Risk and Utility Functions

In Sect. 2.1, some fundamentals with an underlying one-period financial market are given. Now, we use a multi-period financial market model similar to Platen [89] (see also [77]) to introduce some reasonable risk and utility functions. In contrast to Platen [89], and in the spirit of Merton [84], we do not work with general trading strategies, but restrict ourselves to the so-called "fixed fraction" trading strategy, which uses reallocations of the portfolio after each time step to assure that the proportion of the assets in a portfolio stays fixed during all periods [70, 84, 112]. In order to get a nice representation of the related wealth process, it is convenient to let the financial market be given via (geometric) returns as in Brenner [23].

Within this setup, in Sects. 2.2.3–2.2.5, several multi-period utility and risk functions are constructed as needed for the efficient frontier theory presented later on in Sects. 2.3 and 2.4. On a first time reading, however, this section here may be skipped, since the essential properties of risk and utility functions have already been stated in Sect. 2.1 where we only used one-period financial markets.

2.2.1 Geometric Multi-period Financial Market Model

Let $d \in \mathbb{N}$ be fixed. Assume an asset universe with $M+1 \in \mathbb{N}$ assets is given by the *positive* price process over $d+1$ periods:

$$S_t = \left(S_t^0, S_t^1, \ldots, S_t^M\right)^\top \in \mathbb{R}_{>0}^{M+1}, \quad \text{for } t = 0, \ldots, d. \tag{2.2.1}$$

Then the (geometric) return of asset j from period $t-1$ to t is given by

$$G_{j,t} := \frac{S_t^j}{S_{t-1}^j}, \; j = 0, \ldots, M \quad \text{and} \quad t = 1, \ldots, d. \tag{2.2.2}$$

Thus,

$$G = (G_{j,t})_{\substack{0 \le j \le M \\ 1 \le t \le d}} \in \mathbb{R}_{>0}^{(M+1)\times d} \tag{2.2.3}$$

is a return matrix with columns

$$G_t := G_{\cdot,t} = \left(\frac{S_t^0}{S_{t-1}^0}, \frac{S_t^1}{S_{t-1}^1}, \ldots, \frac{S_t^M}{S_{t-1}^M}\right)^\top \in \mathbb{R}_{>0}^{M+1}, \quad \text{for } t = 1, \ldots, d. \tag{2.2.4}$$

As in Sect. 2.1, on the one hand, we assume that asset $j = 0$ is a bond whose returns are deterministic and *risk-free*, i.e.,

$$G_{0,t} \geq 1 \quad \text{for } t = 1, \dots, d. \tag{2.2.5}$$

On the other hand, the assets $j = 1, \dots, M$ are risk affected and modeled by a stochastic process over a filtered probability space $(\Omega, \Sigma, \{\mathcal{F}_t\}_{1 \leq t \leq d}, \mathcal{P})$ with the canonical filtration $\mathcal{F}_t := \sigma(G_s, s \leq t)$. Therefore, G_t may depend on $G_{<t}$, but has no knowledge of $G_{>t}$. We denote the set of all random vectors with values in $\mathbb{R}^M_{>0}$ by

$$\mathcal{L}^0\left(\Omega; \mathbb{R}^M_{>0}\right). \tag{2.2.6}$$

Hence, since the risk-free asset is deterministic, it is

$$G_t \in \mathbb{R}_{\geq 1} \times \mathcal{L}^0\left(\Omega; \mathbb{R}^M_{>0}\right), \quad \text{for } t = 1, \dots, d. \tag{2.2.7}$$

We conclude:

Definition 2.35 (Geometric Return Multi-period Financial Markets; see Platen [89, Definition 2.1.4], Föllmer and Schied [50, Section 5.1]) *Let $M \in \mathbb{N}$ risky assets, indexed by $1, \dots, M$ and one risk-free asset, indexed with zero, be given. The stochastic process*

$$\left\{G_t \colon \Omega \to \mathbb{R}^{M+1}_{>0}\right\}_{1 \leq t \leq d} \tag{2.2.8}$$

defined on the filtered probability space $\left(\Omega, \Sigma, \{\mathcal{F}_t\}_{1 \leq t \leq d}, \mathcal{P}\right)$ with values in $\mathbb{R}^{M+1}_{>0}$ and modeling the (geometric) asset returns with

$$G_t = \left(G_{0,t},\ G_{1,t}, \dots,\ G_{M,t}\right)^\top \in \mathbb{R}_{\geq 1} \times \mathcal{L}^0\left(\Omega; \mathbb{R}^M_{>0}\right) \subset \mathcal{L}^0\left(\Omega; \mathbb{R}^{M+1}_{>0}\right), \tag{2.2.9}$$

*$t = 1, \dots, d$, is called a d-*period financial market model *of size $M + 1$.*

Note that not only the return process G_t is uniquely defined by S_t, but also for a given initial price vector $S_0 \in \mathbb{R}^{M+1}_{>0}$ the price process S_t for $1 \leq t \leq d$ may be derived from G_t by (2.2.4).

2.2.2 Multi-period Wealth Process for the Fixed Fraction Trading Strategy

The idea of the fixed fraction trading strategy is to keep the proportion $x_j \in \mathbb{R}$ to be invested in asset $j = 0, \dots, M$ in our portfolio fixed over all periods. Apparently, since asset prices typically evolve differently, such a fixed proportion can only be achieved if the portfolio is reallocated after each period (see (2.1.1)). Although for a real investment due to transaction costs, this is only reasonable for large time steps, we take that approach here to simplify the mathematics.

Remark 2.36 *The question arises how such an approach can be helpful in a world where trading costs are ubiquitous even for banks. Well, truth is that on a short time scale, such recurring trading costs are not sustainable. Nevertheless, the fixed fraction ansatz can still be useful as an approximation for real-world applications in the sense that resulting optimal portfolios should be monitored as they fluctuate out sample with reallocations executed only when deviations from the in sample calculated optimal portfolio grow too large. In case such a procedure is not acceptable, multi-period financial markets can still be used to construct risk and utility functions, e.g., when time-evolving trading strategies (the simplest and well-known is "buy and hold") are used (see, for instance, Platen [89], or [77]). For the purpose of this monograph, however, this would be way too elaborate.*

So assume for now that our trading generates no transaction costs and the portfolio proportion vector x lies in

$$\mathcal{A}_1 := \left\{ x \in \mathbb{R}^{M+1} : \sum_{j=0}^{M} x_j = 1 \right\}, \tag{2.2.10}$$

where summing up to one means the portfolio is fully invested. Provided we choose w.l.o.g. $S_0 = (1, \dots, 1)^\top \subset \mathbb{R}^{M+1}$, proportion vectors $x \in \mathcal{A}_1$ represent unit initial cost portfolios as already introduced in Definition 2.7. Note further that in theory negative x_j are possible if short positions are allowed. If short positions are not allowed, the proportion vector additionally has to satisfy

$$x_j \geq 0 \qquad \text{for } j = 0, \dots, M. \tag{2.2.11}$$

Given an initial wealth $\mathcal{W}^{(0)} \in \mathbb{R}_{>0}$ to be invested at time $t = 0$, with that approach an equity of $\mathcal{W}^{(0)} \cdot x_j$ is invested in asset j (negative equity for short positions). The wealth of the portfolio $x \in \mathcal{A}_1$ after one period, i.e., at $t = 1$, is then given by

$$\mathcal{W}^{(1)}(x) = \sum_{j=0}^{M} \mathcal{W}^{(0)} \cdot x_j \cdot \frac{S_1^j}{S_0^j} \overset{(2.2.4)}{=} \mathcal{W}^{(0)} \cdot \left(G_1^\top x\right). \tag{2.2.12}$$

Note that $\mathcal{W}^{(1)}(x) > 0$ if and only if $G_1^\top x > 0$, which is a necessary condition for further investments. Similarly, in case the wealth after $t - 1$ time steps remains positive, we get for the wealth of the portfolio $\mathcal{W}^{(t)}(x)$ at time t

$$\mathcal{W}^{(t)}(x) = \mathcal{W}^{(t-1)}(x) \cdot \left(G_t^\top x\right) \quad \text{for } t = 2, \dots, d, \tag{2.2.13}$$

since the portfolio proportion vector $x \in \mathcal{A}_1$ stays fixed due to reallocation.

Remark 2.37 *Since by assumption all geometric returns in* (2.2.4) *are positive, on the one hand, in case no short positions are allowed, i.e., when*

(2.2.11) *holds, the wealth process is always positive. On the other hand, in case short positions are allowed, the factors* $G_t^\top x$ *in* (2.2.13) *may get negative. This represents not only a total loss, but further the wealth process gets negative once that happens the first time. Thus, in this case, the investor ends up with debt and the succeeding wealth process would no longer be well-defined.*

The issue of Remark 2.37 is resolved by the following assumption:

Assumption 2.38 (Intraday Clearing Possible) *In case the wealth process reaches zero (even intraday), we assume that all positions can be cleared immediately and thus the wealth process afterward stays constantly zero.*

With this little trick, the non-negative wealth process

$$\left\{\mathcal{W}^{(t)}(x)\colon \Omega \to \mathbb{R}_{\geq 0}\right\}_{1\leq t\leq d} \tag{2.2.14}$$

is well-defined for arbitrary $x \in \mathcal{A}_1$ and stays positive $\mathcal{P}$–a.s., provided the portfolio is contained in the convex set:

$$\mathcal{A}_{\text{TWR}} := \left\{x \in \mathcal{A}_1 \,:\, G_s^\top x > 0 \quad \mathcal{P}\text{–a.s. for } s = 1, \ldots, d\right\}. \tag{2.2.15}$$

Hence, we obtain

Lemma 2.39 (Fixed Fraction Wealth Process; see Brenner [23, Section 2.2, Page 11]) *We assume that Assumption 2.38 holds. Then for a portfolio with fixed proportion vector* $x \in \mathcal{A}_1$ *and initial wealth* $\mathcal{W}^{(0)} \in \mathbb{R}_{>0}$, *the wealth process* $\left\{\mathcal{W}^{(t)}(x)\colon \Omega \to \mathbb{R}_{\geq 0}\right\}_{1\leq t\leq d}$ *of the portfolio with geometric returns* G_t *as in Definition 2.35 is given as*

$$\mathcal{W}^{(t)}(x) = \begin{cases} \mathcal{W}^{(0)} \cdot \prod_{s=1}^{t} \left(G_s^\top x\right), & \text{if } G_s^\top x > 0 \text{ for } s = 1, \ldots, t, \\ 0, & \text{otherwise.} \end{cases} \tag{2.2.16}$$

Having constructed the wealth process $\mathcal{W}^{(t)}(x)$, we can now define the *terminal wealth relative* after $t \in \{1, \ldots, d\}$ time steps as stochastic process according to Vince [111]

$$\text{TWR}^{(t)}(x) := \frac{\mathcal{W}^{(t)}(x)}{\mathcal{W}^{(0)}} \overset{(2.2.16)}{=} \begin{cases} \prod_{s=1}^{t} \left(G_s^\top x\right), & \text{if } G_s^\top x > 0 \text{ for } s = 1, \ldots, t, \\ 0, & \text{otherwise.} \end{cases} \tag{2.2.17}$$

2.2.3 The Log TWR Utility Function

Based on the terminal wealth relative and given a d-period financial market as in Definition 2.35, we set the $\log \text{TWR}$ *utility function* $\mathfrak{u}_{\ln\text{TWR}}\colon \mathcal{A}_1 \to$

$\mathbb{R} \cup \{-\infty\}$ as

$$x \mapsto \mathfrak{u}_{\ln\text{TWR}}(x) := \frac{1}{d}\,\mathrm{E}\left[\ln \mathrm{TWR}^{(d)}(x)\right], \tag{2.2.18}$$

provided the expectation exists. In Lemma 2.40, we give some integrability conditions on the market which guarantee well-definedness of $\mathfrak{u}_{\ln\text{TWR}}$. Using (2.2.17) and the set $\mathcal{A}_{\text{TWR}}$ from (2.2.15), this is equivalent to

$$\mathfrak{u}_{\ln\text{TWR}}(x) = \begin{cases} \frac{1}{d}\,\mathrm{E}\left[\ln\left(\prod\limits_{s=1}^{d}\left(G_s^\top x\right)\right)\right], & \text{in case } x \in \mathcal{A}_{\text{TWR}}, \\ -\infty, & \text{otherwise.} \end{cases} \tag{2.2.19}$$

Note that for $x \in \mathcal{A}_{\text{TWR}}$, $\mathfrak{u}_{\ln\text{TWR}}$ can also be written as

$$\mathfrak{u}_{\ln\text{TWR}}(x) = \frac{1}{d}\sum_{s=1}^{d} \mathrm{E}\left[\ln\left(G_s^\top x\right)\right], \tag{2.2.20}$$

so that $\mathfrak{u}_{\ln\text{TWR}}$ represents the arithmetic mean of the expected logarithmic portfolio returns.

In the following, we are looking for conditions which guarantee that $\mathfrak{u}_{\ln\text{TWR}}$ is concave or even strictly concave, so that it can be used later on as a utility function in our portfolio theory. Although in the completely different context of "consistency," Brenner [23, Subsection 5.1.1] did discuss similar questions for so-called extended-valued utility functions of log return type, and a lot of his ideas there helped to extract the following results for $\mathfrak{u}_{\ln\text{TWR}}$.

Lemma 2.40 (Log TWR Utility Function 1) *Assume Assumption 2.38 holds. Further, let G_t, $t = 1, \ldots, d$, be the geometric return process as given in Definition 2.35 with additionally*

$$\left[\ln\left(G_{j,t}\right)\right]^+ = \max\left\{0, \ln\left(G_{j,t}\right)\right\} \in \mathcal{L}^1(\Omega;\mathbb{R}) \quad \textit{for } j = 0, \ldots, M \tag{2.2.21}$$

and $t = 1, \ldots, d$. Then $\mathfrak{u}_{\ln\text{TWR}} \colon \mathcal{A}_1 \to \mathbb{R} \cup \{-\infty\}$ given by (2.2.19) is well-defined, proper concave, and $\emptyset \neq \operatorname{dom}\left(\mathfrak{u}_{\ln\text{TWR}}\right) \subset \mathcal{A}_{\text{TWR}}$.

Proof Firstly, $\mathfrak{u}_{\ln\text{TWR}}$ is proper because the pure bond portfolio $x^* := (1, \widehat{0}^\top)^\top \in \mathcal{A}_{\text{TWR}}$ lies in $\operatorname{dom}\left(\mathfrak{u}_{\ln\text{TWR}}\right)$:

$$\mathfrak{u}_{\ln\text{TWR}}\left(x^*\right) \overset{(2.2.20)}{=} \frac{1}{d}\sum_{s=1}^{d} \ln\underbrace{\left(G_{0,s}\right)}_{\geq 1} \overset{(2.2.5)}{\in} \mathbb{R}_{\geq 0}$$

Moreover, $\mathfrak{u}_{\ln\text{TWR}}$ is concave because the natural logarithm is concave and the expectation is linear. It remains to show well-definedness of $\mathfrak{u}_{\ln\text{TWR}}$, i.e., that $\mathrm{E}\left[\ln\left(G_s^\top x\right)\right] \in \mathbb{R} \cup \{-\infty\}$ is well-defined for $x \in \mathcal{A}_{\text{TWR}}$ and for any

$s \in \{1, \ldots, d\}$ (see (2.2.20)). Note that for this, it suffices to show that $\mathrm{E}\left([\ln(G_s^\top x)]^+\right)$ is finite.

So let $x \in \mathcal{A}_{\mathrm{TWR}}$ be given and thus $G_s^\top x > 0 \quad \mathcal{P}$–a.s.. It follows, using the maximum norm $\|\cdot\|_\infty$

$$\begin{aligned} \ln\left(G_s^\top x\right) &\leq \ln\left[(M+1)\,\|G_s\|_\infty \cdot \|x\|_\infty\right] \\ &\overset{(2.2.3)}{\leq} \ln\left[(M+1)\|x\|_\infty\right] + \sum_{j=0}^{M} \left[\ln\left(G_{j,s}\right)\right]^+ . \end{aligned} \tag{2.2.22}$$

By assumption (2.2.21), we have $\mathrm{E}\left([\ln(G_{j,s})]^+\right) < \infty$ giving

$$\mathrm{E}\left([\ln(G_s^\top x)]^+\right) \leq (\ln\left[(M+1)\|x\|_\infty\right])^+ + \sum_{j=0}^{M} \mathrm{E}\left([\ln(G_{j,s})]^+\right) < \infty$$

as claimed and thus $\mathrm{E}\left[\ln\left(G_s^\top x\right)\right] \in \mathbb{R} \cup \{-\infty\}$. Hence, $\mathfrak{u}_{\ln\mathrm{TWR}} \colon \mathcal{A}_1 \to \mathbb{R} \cup \{-\infty\}$ is well-defined and $\mathrm{dom}\left(\mathfrak{u}_{\ln\mathrm{TWR}}\right) \subset \mathcal{A}_{\mathrm{TWR}}$ obviously follows. □

We next want to restrict $\mathfrak{u}_{\ln\mathrm{TWR}}$ to so-called restricted admissible portfolio sets in order to obtain a utility function in terms of Assumption 2.29, in particular (u2) and/or (u2s).

Assumption 2.41 (Restricted Admissible Portfolio Sets) *We say $A \subset \mathcal{A}_1 \subset \mathbb{R}^{M+1}$ is a set of* restricted admissible portfolios *in case A is admissible as in Definition 2.7 (nonempty, closed, and convex) and, moreover, bounded with $A \cap \mathbb{R}^{M+1}_{\geq 0} \neq \emptyset$.*

Remark 2.42 *It is possible to generalize the "no nontrivial riskless portfolio" condition of a one-period financial market (see Definition 2.10) to multi-period markets. Under this condition, Brenner [23, Lemma 2.2.3], shows that $\mathcal{A}_{\mathrm{TWR}}$ in* (2.2.15) *is always bounded. In this case, the boundedness of A in Assumption 2.41 can always be assumed without loss of generality. Because if A were not bounded, we just need to replace A with $\widetilde{A} := A \cap \overline{\mathcal{A}_{\mathrm{TWR}}}$ which is closed and bounded. Then minimizing $\mathfrak{u}_{\ln\mathrm{TWR}|A}$ or minimizing $\mathfrak{u}_{\ln\mathrm{TWR}|\widetilde{A}}$ yield the same solutions because by Lemma 2.40* $\mathrm{dom}\left(\mathfrak{u}_{\ln\mathrm{TWR}}\right) \subset \mathcal{A}_{\mathrm{TWR}}$ *holds.*

Lemma 2.43 (Log TWR Utility Function 2) *Assume Assumption 2.38 holds and, furthermore, let again G_t, $t = 1, \ldots, d$, be the geometric return process from Definition 2.35, but this time with*

$$\ln\left(G_{j,t}\right) \in \mathcal{L}^1(\Omega, \mathbb{R}) \text{ for } j = 0, \ldots, M \text{ and } t = 1, \ldots, d. \tag{2.2.23}$$

Then for any restricted admissible set A of portfolios as in Assumption 2.41, $\mathfrak{u}_{\ln\mathrm{TWR}}$ from (2.2.19) *restricted to A, i.e.,*

$$\mathfrak{u}_{\ln\mathrm{TWR}|A}: A \to \mathbb{R} \cup \{-\infty\}, \tag{2.2.24}$$

is a well-defined utility function in the sense of Assumption 2.29, in particular $\mathfrak{u} = \mathfrak{u}_{\ln\mathrm{TWR}|A}$ *is upper semi-continuous and proper concave, i.e.,* (u2) *therein holds. In fact, we even get*

$$\emptyset \neq A \cap \mathbb{R}^{M+1}_{\geq 0} \subset \mathrm{dom}\left(\mathfrak{u}_{\ln\mathrm{TWR}|A}\right) \subset A \cap \mathcal{A}_{\mathrm{TWR}} \tag{2.2.25}$$

and superlevel sets of $\mathfrak{u}_{\ln\mathrm{TWR}}$*, i.e., sets of the form*

$$\mathcal{B}_A(\mathfrak{u}_{\ln\mathrm{TWR}} \geq \mu) := \{x \in A : \mathfrak{u}_{\ln\mathrm{TWR}}(x) \geq \mu\}, \tag{2.2.26}$$

are compact for $\mu \in \mathbb{R}$ *arbitrary.*

Proof Concavity and well-definedness are already clear from Lemma 2.40. To show properness, we verify (2.2.25). So let an arbitrary $x \in A \cap \mathbb{R}^{M+1}_{\geq 0} \neq \emptyset$ be given. Using $x_j \geq 0$ for $j = 0, \ldots, M$ and $\sum_{j=0}^{M} x_j = 1$, we find at least one $j^* \in \{0, \ldots, M\}$ such that $x_{j^*} > 0$. Moreover, since $G_{j,s}$ are all positive $\mathcal{P}$–a.s., we have $G_{j^*,s} x_{j^*} > 0$ $\mathcal{P}$–a.s. and $G_{j,s} x_j \geq 0$ $\mathcal{P}$–a.s. for all $s = 1, \ldots, d$ and $j = 0, \ldots, M$. Hence, $x \in \mathcal{A}_{\mathrm{TWR}}$ and

$$\begin{aligned}
\mathfrak{u}_{\ln\mathrm{TWR}}(x) &\overset{(2.2.20)}{=} \frac{1}{d}\sum_{s=1}^{d} \mathrm{E}\left[\ln\left(G_s^\top x\right)\right] \geq \frac{1}{d}\sum_{s=1}^{d} \mathrm{E}\left[\ln\left(G_{j^*,s} x_{j^*}\right)\right] \\
&= \ln\left(x_{j^*}\right) + \frac{1}{d}\sum_{s=1}^{d} \mathrm{E}\left[\ln\left(G_{j^*,s}\right)\right] \overset{(2.2.23)}{>} -\infty,
\end{aligned}$$

and (2.2.25) follows.

Upper semi-continuity of $\mathfrak{u}_{\ln\mathrm{TWR}|A}$ follows basically because the function $(g, x) \mapsto \ln\left(g^\top x\right)$ is upper semi-continuous in both arguments $g, x \in \mathbb{R}^{M+1}$ and from the closedness of A. But we refrain from giving details and just refer to Brenner [23, Proposition 5.1.1 (b) and (e)]. At last, the supersets in (2.2.26) are closed, because of the closedness of A and the upper semi-continuity of $\mathfrak{u}_{\ln\mathrm{TWR}|A}: A \to \mathbb{R} \cup \{-\infty\}$ (see Remark A.4). Hence, the sets $\mathcal{B}_A(\mathfrak{u}_{\ln\mathrm{TWR}} \geq \mu)$ are compact because A is bounded. □

In some cases, $\mathfrak{u}_{\ln\mathrm{TWR}|A}$ is even strictly concave. To get this, we need another assumption which goes back to Brenner [23, Assumption 5.1.3].

Assumption 2.44 (Random Injectivity of Geometric Returns) *Let* $A \subset \mathcal{A}_1 \subset \mathbb{R}^{M+1}$ *be a restricted admissible portfolio set as in Assumption 2.41. We further assume that for arbitrary but distinct* $x \neq \widetilde{x} \in A \cap \mathcal{A}_{\mathrm{TWR}}$ *there exists at least one* $s_0 \in \{1, \ldots, d\}$ *such that the geometric returns of* x *and* $\widetilde{x}$ *differ with positive probability, i.e.,*

$$\mathcal{P}\left(G_{s_0}^\top x \neq G_{s_0}^\top \widetilde{x}\right) > 0. \tag{2.2.27}$$

Lemma 2.45 (Log TWR Utility Function 3) *In the situation of Lemma 2.43 for a restricted admissible portfolio set A which furthermore satisfies Assumption 2.44,* $\mathfrak{u}_{\ln\mathrm{TWR}}$ *restricted to A as in* (2.2.24) *is strictly proper concave, i.e.,* (u2s) *in Assumption 2.29 holds.*

Proof Let $x \neq \widetilde{x} \in \operatorname{dom}\left(\mathfrak{u}_{\ln\mathrm{TWR}|A}\right) \subset A \cap \mathcal{A}_{\mathrm{TWR}}$ be given. Then, due to Assumption 2.44, there exists some $s_0 \in \{1, \ldots, d\}$ such that (2.2.27) holds true.

Therefore, since the natural logarithm is strictly concave, we obtain for $\lambda \in (0,1)$

$$\mathcal{P}\Big[\ln\left(G_{s_0}^{\top}\left[\lambda x + (1-\lambda)\widetilde{x}\right]\right) > \lambda \ln\left(G_{s_0}^{\top} x\right) + (1-\lambda)\ln\left(G_{s_0}^{\top}\widetilde{x}\right)\Big] > 0.$$

Hence again by concavity of ln

$$\mathrm{E}\Big[\ln\left(G_{s_0}^{\top}\left[\lambda x + (1-\lambda)\widetilde{x}\right]\right)\Big] > \lambda\, \mathrm{E}\Big[\ln\left(G_{s_0}^{\top} x\right)\Big] + (1-\lambda)\, \mathrm{E}\Big[\ln\left(G_{s_0}^{\top}\widetilde{x}\right)\Big],$$

so that at least one summand of $\mathfrak{u}_{\ln\mathrm{TWR}}$ in (2.2.20) is strictly concave. Since all other summands are already known to be concave, this implies strict concavity of $\mathfrak{u}_{\ln\mathrm{TWR}|A}$. □

2.2.4 Further Multi-period Utility Functions

Without further discussion, we intend to mention that there are other related utility functions for multi-period markets.

Definition 2.46 (Single Period and Minimum Period Log Utility) *With the same setup as given for the definition of* $\mathfrak{u}_{\ln\mathrm{TWR}}\colon \mathcal{A}_1 \to \mathbb{R} \cup \{-\infty\}$ *in* (2.2.18)–(2.2.20) *and the integrability condition* (2.2.21)*, we define the* single period log utility function $\mathfrak{u}_{\mathrm{singlePer}}\colon \mathcal{A}_1 \to \mathbb{R} \cup \{-\infty\}$ *for some fixed period $s^* \in \{1, \ldots, d\}$ by*

$$\mathfrak{u}_{\mathrm{singlePer}}(x) := \begin{cases} \mathrm{E}\left[\ln\left(G_{s^*}^{\top} x\right)\right], & \text{for } x \in \mathcal{A}_{\mathrm{TWR}}, \\ -\infty, & \text{otherwise} \end{cases} \tag{2.2.28}$$

and the minimum period log utility function $\mathfrak{u}_{\mathrm{minPer}}\colon \mathcal{A}_1 \to \mathbb{R} \cup \{-\infty\}$ *by*

$$\mathfrak{u}_{\mathrm{minPer}}(x) := \begin{cases} \mathrm{E}\left[\min\limits_{1 \le s \le d} \ln\left(G_{s}^{\top} x\right)\right], & \text{for } x \in \mathcal{A}_{\mathrm{TWR}}, \\ -\infty, & \text{otherwise.} \end{cases} \tag{2.2.29}$$

Remark 2.47 *We just remark that all properties of* $\mathfrak{u}_{\ln\mathrm{TWR}|A}\colon A \to \mathbb{R} \cup \{-\infty\}$ *of Lemma 2.43 hold similarly for* $\mathfrak{u}_{\mathrm{singlePer}|A}\colon A \to \mathbb{R} \cup \{-\infty\}$ *and* $\mathfrak{u}_{\mathrm{minPer}|A}\colon A \to \mathbb{R} \cup \{-\infty\}$*. Even strict concavity of* $\mathfrak{u}_{\ln\mathrm{TWR}|A}$ *as provided* *in Lemma 2.45 can be picked up similarly if Assumption 2.44 is adapted suitably (for* $\mathfrak{u}_{\text{singlePer}}$ (2.2.27) *has to hold for* $s_0 = s^*$ *and for* $\mathfrak{u}_{\text{minPer}}$ *instead of* (2.2.27)

$$\mathcal{P}\left(\bigcap_{s=1}^{d}\{G_s^\top x \neq G_s^\top \widetilde{x}\}\right) > 0 \tag{2.2.30}$$

is needed).

2.2.5 Multi-period Log Drawdown Risk Functions

In the last subsection, we used the expectation of the stochastic process:

$$\left\{\ln \mathrm{TWR}^{(t)}(x)\colon \Omega \to \mathbb{R} \cup \{-\infty\}\right\}_{1\leq t\leq d}, \tag{2.2.31}$$

where $x \in \mathcal{A}_1$ was a fixed proportion vector, to construct the utility function $\mathfrak{u}_{\ln\mathrm{TWR}}$ (see (2.2.18) and (2.2.19)). With the help of (2.2.17), we get for $x \in \mathcal{A}_{\mathrm{TWR}}$ (see (2.2.15)) and $t = 1, \ldots, d$:

$$\ln \mathrm{TWR}^{(t)}(x) = \ln \prod_{s=1}^{t} \left(G_s^\top x\right) = \sum_{s=1}^{t} \ln\left(G_s^\top x\right). \tag{2.2.32}$$

Therefore, the log TWR process can be interpreted as equity of additive portfolio log returns. We want to apply typical drawdown concepts to this equity.

Definition 2.48 (Log Drawdown Risk Functions) *For a d-period financial market as given in Definition 2.35 based on the additive log returns portfolio equity* (2.2.32)*, we define the following risk functions:*
Expected current log drawdown, $\mathfrak{r}_{\text{drawdown}}\colon \mathcal{A}_1 \to \mathbb{R}_{\geq 0} \cup \{+\infty\}$,

$$\mathfrak{r}_{\text{drawdown}}(x) := \begin{cases} \mathrm{E}\left[\max\left\{0, \max\limits_{1\leq t\leq d}\left[\sum\limits_{s=t}^{d} -\ln\left(G_s^\top x\right)\right]\right\}\right], & x \in \mathcal{A}_{\mathrm{TWR}},\\ +\infty, & \text{otherwise.}\end{cases}$$

Expected maximum log drawdown, $\mathfrak{r}_{\max\mathrm{dd}}\colon \mathcal{A}_1 \to \mathbb{R}_{\geq 0} \cup \{+\infty\}$,

$$\mathfrak{r}_{\max\mathrm{dd}}(x) := \begin{cases} \mathrm{E}\left[\max\left\{0, \max\limits_{1\leq a\leq b\leq d}\left[\sum\limits_{s=a}^{b} -\ln\left(G_s^\top x\right)\right]\right\}\right], & x \in \mathcal{A}_{\mathrm{TWR}},\\ +\infty, & \text{otherwise.}\end{cases}$$

Expected log rundown, $\mathfrak{r}_{\text{rundown}}\colon \mathcal{A}_1 \to \mathbb{R}_{\geq 0} \cup \{+\infty\}$,

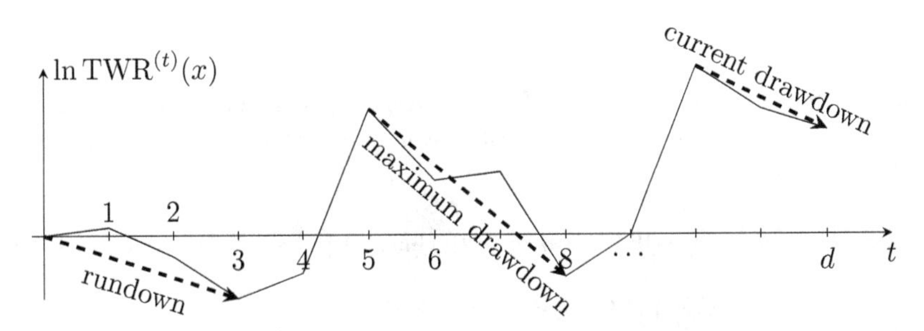

Fig. 2.1 One realization of the equity in (2.2.31)

$$\mathfrak{r}_{\text{rundown}}(x) := \begin{cases} \mathrm{E}\left[\max\left\{0, \max\limits_{1\le t\le d}\left[\sum\limits_{s=1}^{t} -\ln\left(G_s^\top x\right)\right]\right\}\right], & x \in \mathcal{A}_{\text{TWR}}, \\ +\infty, & \text{otherwise.} \end{cases}$$

Remark 2.49 *Platen [89] already introduced an expected* log *drawdown concept comparable to* $\mathfrak{r}_{\text{drawdown}}$, *but it was Brenner [23] who firstly developed the expected current* log *drawdown in the setup of geometric returns. Moreover,* $\mathfrak{r}_{\text{rundown}}$ *was firstly introduced by Henkes [60].*

Well-definedness of the three log drawdown risk functions above is obvious since non-negative arguments of the expectation are always quasi-integrable.

The concepts of these three risk functions are visualized in Fig. 2.1 for one realization $\omega \in \Omega$. It is obvious that the expected maximum log drawdown is an upper bound for the other two, i.e.,

$$\max\left\{\mathfrak{r}_{\text{drawdown}}(x), \mathfrak{r}_{\text{rundown}}(x)\right\} \le \mathfrak{r}_{\text{max dd}}(x) \quad \text{for} \quad x \in \mathcal{A}_1.$$

Note that the equity loss within the drawdowns is measured positive.

Similarly as in Sect. 2.2.3 for $\mathfrak{u}_{\ln\text{TWR}}$, we are now looking for conditions which guarantee that the above defined log drawdown risk functions are convex or even strictly convex, so that they can be used later on as risk functions in our portfolio theory. Again our approach here has gained lots of ideas from Brenner [23, Subsection 5.1.2], who studied so-called extended-valued risk functions of log drawdown type in the context of "consistency."

Lemma 2.50 (Log Drawdown Risk Functions) *Assume that Assumption 2.38 holds and, furthermore, let again* G_t, $t = 1, \dots, d$, *be the geometric return process from Definition 2.35, with*

$$\ln\left(G_{j,t}\right) \in \mathcal{L}^1(\Omega, \mathbb{R}) \text{ for } j = 0, \dots, M \text{ and } t = 1, \dots, d. \tag{2.2.33}$$

Then for any restricted admissible set $A \subset \mathcal{A}_1$ *of portfolios as in Assumption 2.41, any log drawdown risk function* $\mathfrak{r}_{\text{dd}}$ *from Definition 2.48 restricted to* A, *i.e.,*

$$\mathfrak{r}_{\text{dd}|A} := \mathfrak{r}_{\text{rundown}|A},\ \mathfrak{r}_{\max \text{dd}|A},\ \mathfrak{r}_{\text{drawdown}|A} : A \to \mathbb{R}_{\geq 0} \cup \{+\infty\}, \tag{2.2.34}$$

is a risk function in the sense of Assumption 2.20. In particular, these risk functions $\mathfrak{r}_{\text{dd}|A}$ are lower semi-continuous and proper convex, i.e., (r2) *holds. In fact, we even get*

$$\emptyset \neq A \cap \mathbb{R}_{\geq 0}^{M+1} \subset \operatorname{dom}\left(\mathfrak{r}_{\text{dd}|A}\right) = \operatorname{dom}\left(\mathfrak{u}_{\ln\text{TWR}|A}\right) \subset A \cap \mathcal{A}_{\text{TWR}} \tag{2.2.35}$$

and sublevel sets of $\mathfrak{r}_{\text{dd}}$, i.e., sets of the form

$$\mathcal{B}_A(\mathfrak{r}_{\text{dd}} \leq r) := \{x \in A : \mathfrak{r}_{\text{dd}}(x) \leq r\}, \tag{2.2.36}$$

are compact for $r \in \mathbb{R}$ arbitrary.

Proof By definition, it is clear that $\mathfrak{r}_{\text{dd}} \geq 0$. The argument for well-definedness was already given after Definition 2.48. Furthermore, function $x \mapsto -\ln(G_s^\top x)$ is convex. Using the linearity of the expectation and the fact that the maximum of convex functions is still convex, we obtain the convexity of $\mathfrak{r}_{\text{dd}}$. Properness is a consequence of (2.2.35), which we show next. Due to (2.2.25), it suffices to show $\operatorname{dom}\left(\mathfrak{r}_{\text{dd}|A}\right) = \operatorname{dom}\left(\mathfrak{u}_{\ln\text{TWR}|A}\right)$ which in turn follows immediately once

$$0 \leq d \cdot \mathfrak{u}_{\ln\text{TWR}}(x) + \mathfrak{r}_{\text{dd}}(x) \overset{!}{<} \infty \quad \text{for all} \quad x \in A \cap \mathcal{A}_{\text{TWR}} \tag{2.2.37}$$

is shown. We give the argument exemplarily for $\mathfrak{r}_{\text{dd}} = \mathfrak{r}_{\max \text{dd}}$. For $x \in \mathcal{A}_{\text{TWR}}$ we have by (2.2.20)

$$\begin{aligned}
&d \cdot \mathfrak{u}_{\ln\text{TWR}}(x) + \mathfrak{r}_{\max \text{dd}}(x) \\
&= \mathrm{E}\left[\sum_{s=1}^{d} \ln\left(G_s^\top x\right) + \max\left\{0, \max_{1 \leq a \leq b \leq d}\left[\sum_{s=a}^{b} -\ln\left(G_s^\top x\right)\right]\right\}\right] \\
&= \mathrm{E}\left[\max\left\{\sum_{s=1}^{d} \ln\left(G_s^\top x\right), \max_{1 \leq a \leq b \leq d}\left[\sum_{s=1}^{a-1} \ln\left(G_s^\top x\right) + \sum_{s=b+1}^{d} \ln\left(G_s^\top x\right)\right]\right\}\right] \\
&\geq 0,
\end{aligned}$$

where empty sums yield the value zero. Using (2.2.33), we get for arbitrary $s \in \{1, \ldots, d\}$

$$\mathrm{E}\left[\ln\left(G_s^\top x\right)\right] \overset{(2.2.22)}{\leq} \ln\left[(M+1)\|x\|_\infty\right] + \sum_{j=0}^{M} \mathrm{E}\left[\ln\left(G_{j,s}\right)\right]^+ < \infty$$

and (2.2.37) follows.

Furthermore, lower semi-continuity of $\mathfrak{r}_{\text{dd}|A}$ follows from similar arguments as the upper semi-continuity of $\mathfrak{u}_{\ln\text{TWR}}$ (see Lemma 2.43 and Brenner [23,

Proposition 5.1.7 (b) and (f)]), which in turn implies the compactness of the sublevel sets in (2.2.36), since A is bounded (see Remark A.4). □

Remark 2.51 *The question whether or not the log drawdown risk functions of Definition 2.48 are strictly convex is more subtle as the strict concavity problem for* $\mathfrak{u}_{\ln \text{TWR}}$ *in Sect. 2.2.3. For instance, Brenner [23, Proposition 5.1.11 (b)] could show that* $\mathfrak{r}_{\text{drawdown}_{|A}}$ *is strictly convex, provided for* $s = d$ *the two conditions*

$$\mathcal{P}\left(G_s^\top x < 1\right) > 0 \quad \text{for all} \quad x \in A \setminus \left\{(1, \widehat{0}^\top)^\top\right\} \tag{2.2.38}$$

and

$$\mathcal{P}\left(G_s^\top x \neq G_s^\top \widetilde{x} \middle| G_s^\top x < 1\right) > 0 \quad \text{for all} \quad x \neq \widetilde{x} \in A \cap \mathcal{A}_{\text{TWR}} \tag{2.2.39}$$

hold, where (2.2.38) *means that for any portfolio besides the risk-free bond, there occurs a loss with positive probability and by* (2.2.39) *the geometric portfolio returns of any two disjoint portfolios differ with positive probability when one of them is losing. This is needed since different portfolio returns only yield a different drawdown along a finally losing path.*

Without giving any details, $\mathfrak{r}_{\text{rundown}_{|A}}$ *thus is also strictly convex, provided* (2.2.38) *and* (2.2.39) *hold for* $s = 1$. *Suitable conditions for strict convexity of* $\mathfrak{r}_{\max \text{dd}_{|A}}$ *are more subtle and therefore left out here.*

We further note that Platen [89, Theorem 3.1.7] showed that $\mathfrak{r}_{\text{drawdown}}$ *is strictly convex provided the d-period financial market has "no nontrivial risk-free portfolio." But his argument was slightly incorrect as observed by Brenner [23, Remark 5.1.12].*

2.3 Efficient Trade-Off Between Risk and Utility: The Efficient Frontier

This section and the following Sect. 2.4 represent the main part of the generalized framework of portfolio theory for scalar risk functions. The ideas and most of the results presented here go back to Maier-Paape and Zhu [78]. However, the theory here is more general (with both risk and utility modeled as abstract extended-valued functions as introduced in Sect. 2.1) and some arguments are new as well. For instance, in [78] we never discussed whether or not the so-called efficient frontier is a connected set. This is indeed the case and shown in Corollary 2.73. Later on in Chap. 3, a similar theory is developed for vector risk functions.

2.3.1 The Risk-Utility Space

In order to increase the utility of a portfolio, one often has to take on more risk. Therefore, the investment decision of selecting an appropriate portfolio becomes a trade-off between the portfolios' return and risk. To understand this trade-off, we define for a set of admissible portfolios A (Definition 2.3), a risk function $\mathfrak{r}$ and a utility function $\mathfrak{u}$ the *set of valid risk and utility values*

$$\mathcal{G} = \mathcal{G}(\mathfrak{r}, \mathfrak{u}; A) := \left\{(r, \mu) \in \mathbb{R}^2 \, : \, \exists \, x \in A \text{ s.t. } \mathfrak{r}(x) \le r \text{ and } \mathfrak{u}(x) \ge \mu\right\} \quad (2.3.1)$$

as a subset of the two-dimensional *risk-utility space* $\mathbb{R}^2$. Observe that

$$A' \subset A \Rightarrow \mathcal{G}(\mathfrak{r}, \mathfrak{u}; A') \subset \mathcal{G}(\mathfrak{r}, \mathfrak{u}; A). \quad (2.3.2)$$

Here and in the sequel, we work with abstract extended-valued risk functions $\mathfrak{r}\colon A \to \mathbb{R} \cup \{+\infty\}$ which are typically proper convex (see (r2) in Assumption 2.20) and with abstract extended-valued utility functions $\mathfrak{u}\colon A \to \mathbb{R} \cup \{-\infty\}$ which are typically proper concave (see (u2) in Assumption 2.29).

Remark 2.52 *Although the most used risk functions like standard deviation or drawdown have non-negative values so that $\mathfrak{r}\colon A \to \mathbb{R}_{\ge 0} \cup \{+\infty\}$, the theory developed here works for the more general case $\mathfrak{r}\colon A \to \mathbb{R} \cup \{+\infty\}$.*

Definition 2.53 (Risk-Utility Space) *For a risk function $\mathfrak{r}$ and a utility function $\mathfrak{u}$, we define the* sub- *and* superlevel sets *of $\mathfrak{r}$ and $\mathfrak{u}$ for thresholds $r, \mu \in \mathbb{R}$ as*

$$\begin{aligned} \mathcal{B}_A(\mathfrak{r} \le r) &:= \{x \in A\colon \mathfrak{r}(x) \le r\} \subset \operatorname{dom}(\mathfrak{r}) \subset \mathbb{R}^{M+1} \quad \textit{and} \\ \mathcal{B}_A(\mathfrak{u} \ge \mu) &:= \{x \in A\colon \mathfrak{u}(x) \ge \mu\} \subset \operatorname{dom}(\mathfrak{u}) \subset \mathbb{R}^{M+1}, \end{aligned} \quad (2.3.3)$$

respectively, and using its intersection

$$\mathcal{B}_A(\mathfrak{r} \le r; \, \mathfrak{u} \ge \mu) := \mathcal{B}_A(\mathfrak{r} \le r) \cap \mathcal{B}_A(\mathfrak{u} \ge \mu) \subset A, \quad (2.3.4)$$

and the standard preimage notation $f^{-1}(D) := \{z \, : \, f(z) \in D\}$ of a function f gives an alternative representation of the valid risk and utility values:

$$\begin{aligned} \mathcal{G}(\mathfrak{r}, \mathfrak{u}; A) &= \left\{(r, \mu) \in \mathbb{R}^2 \, : \, \mathcal{B}_A(\mathfrak{r} \le r; \, \mathfrak{u} \ge \mu) \neq \emptyset\right\} \\ &= \left\{(r, \mu) \in \mathbb{R}^2 \, : \, \mathfrak{r}^{-1}((-\infty, r]) \cap \mathfrak{u}^{-1}([\mu, \infty)) \cap A \neq \emptyset\right\}. \end{aligned} \quad (2.3.5)$$

Note that the notation used in Definition 2.53 is similar to the one introduced by Platen [89]. The following assumption is sometimes needed in concrete applications. It was first used in [78], but only for the expected utility from Definition 2.31.

Assumption 2.54 (Compact Level Sets 1) *Either (a) for each $\mu \in \mathbb{R}$, the superlevel set $\mathcal{B}_A(\mathfrak{u} \ge \mu) = \mathfrak{u}^{-1}([\mu, \infty)) \cap A \subset \mathbb{R}^{M+1}$ is compact* or *(b)* *for each* $r \in \mathbb{R}$, *the sublevel set* $\mathcal{B}_A(\mathfrak{r} \leq r) = \mathfrak{r}^{-1}((-\infty, r]) \cap A \subset \mathbb{R}^{M+1}$ *is compact.*

Remark 2.55 *The sub- and superlevel sets used above are always closed since* A *is closed and, moreover,* $\mathfrak{u}$ *is upper semi-continuous and* $\mathfrak{r}$ *is lower semi-continuous (cf. Remark A.4). So boundedness is the issue in Assumption 2.54.*

Most of the time, we indeed get along with a weaker compactness assumption, which was firstly introduced in [89] in the context of multi-period markets.

Assumption 2.56 (Compact Level Sets 2) *We assume*

(a) $\mathrm{dom}(\mathfrak{u}) \cap \mathrm{dom}(\mathfrak{r}) \neq \emptyset$ *and*
(b) *for all* $(r, \mu) \in \mathbb{R}^2$, *the sets* $\mathcal{B}_A(\mathfrak{r} \leq r; \mathfrak{u} \geq \mu)$ *from* (2.3.4) *are compact.*

Two obvious possibilities which guarantee the setup in Assumption 2.56 (b) are given below.

Lemma 2.57 *Assumption 2.56 (b) is satisfied in either of the following two cases:*

(i) $A \subset \mathbb{R}^{M+1}$ *is bounded.*
(ii) *Every sequence* $x^{(n)} \in A$, $n \in \mathbb{N}$, *with* $\left\|x^{(n)}\right\| \to \infty \quad (n \to \infty)$, *contains a subsequence* $x^{(n_j)}$, $j \in \mathbb{N}$ *such that either*

$$\mathfrak{r}(x^{(n_j)}) \to \infty \quad or \quad \mathfrak{u}(x^{(n_j)}) \to -\infty \quad (j \to \infty) \text{ holds.}$$

Proposition 2.58 (Properties of $\mathcal{G}(\mathfrak{r}, \mathfrak{u}; A)$) *Assume that* $A \subset \mathbb{R}^{M+1}$ *is a set of admissible portfolios as in Definition 2.7. Moreover, assume the risk function* $\mathfrak{r}$ *satisfies* (r2) *in Assumption 2.20 and assume the utility function* $\mathfrak{u}$ *satisfies* (u2) *in Assumption 2.29. We claim:*

(a) Then the set $\mathcal{G}(\mathfrak{r}, \mathfrak{u}; A)$ *from* (2.3.1) *is convex.*
(b) In addition,

$$(r, \mu) \in \mathcal{G}(\mathfrak{r}, \mathfrak{u}; A) \Rightarrow \begin{cases} (r + k, \mu) \in \mathcal{G}(\mathfrak{r}, \mathfrak{u}; A) & \text{for all } k > 0, \\ (r, \mu - k) \in \mathcal{G}(\mathfrak{r}, \mathfrak{u}; A) & \text{for all } k > 0. \end{cases} \tag{2.3.6}$$

(c) Assume furthermore that Assumption 2.56 (b) holds. Then set $\mathcal{G}(\mathfrak{r}, \mathfrak{u}; A)$ *is closed.*

Proof *ad(a).* Suppose that $(r_0, \mu_0), (r_1, \mu_1) \in \mathcal{G}(\mathfrak{r}, \mathfrak{u}; A)$, and $\alpha \in (0, 1)$. Then by definition, there exist $x^{(0)}, x^{(1)} \in A \subset \mathbb{R}^{M+1}$ such that

$$\mathfrak{r}(x^{(i)}) \leq r_i \quad \text{and} \quad \mathfrak{u}(x^{(i)}) \geq \mu_i \quad \text{for} \quad i \in \{0, 1\}. \tag{2.3.7}$$

The convexity of $\mathfrak{r}$ yields for $x^{(\alpha)} := \alpha x^{(1)} + (1 - \alpha) x^{(0)} \in A$

$$\mathfrak{r}(x^{(\alpha)}) \overset{(r2)}{\leq} \alpha\mathfrak{r}(x^{(1)}) + (1-\alpha)\mathfrak{r}(x^{(0)}) \overset{(2.3.7)}{\leq} \alpha r_1 + (1-\alpha)r_0$$

and concavity of $\mathfrak{u}$ gives analogously

$$\mathfrak{u}(x^{(\alpha)}) \overset{(u2)}{\geq} \alpha\mu_1 + (1-\alpha)\mu_0.$$

Thus, $\alpha\,(r_1,\mu_1) + (1-\alpha)(r_0,\mu_0) \in \mathcal{G}(\mathfrak{r},\mathfrak{u};A)$ and hence $\mathcal{G}(\mathfrak{r},\mathfrak{u};A)$ is convex.
ad(b). This follows directly from the definition of $\mathcal{G}(\mathfrak{r},\mathfrak{u};A)$ in (2.3.1).
ad(c). We leave that part to the reader as an exercise. □

2.3.2 The Efficient Frontier

Now every portfolio $x \in A \subset \mathbb{R}^{M+1}$ may be represented as a point

$$(r_*,\mu_*) := (\mathfrak{r}(x),\mathfrak{u}(x)) \in \mathcal{G}(\mathfrak{r},\mathfrak{u};A) \subset \mathbb{R}^2$$

in the two-dimensional risk-utility space.

Investors prefer portfolios with:

- "lower risk" for a given utility level or with
- "higher utility" for a given risk level.

Portfolios which cannot be improved are called *efficient*.

Definition 2.59 (Efficient Portfolios; [78, Definition 5]) *We say that a portfolio $x^* \in A$ with $\mathfrak{r}\,(x^*) < \infty$ and $\mathfrak{u}\,(x^*) > -\infty$ is* Pareto efficient *(for the risk function $\mathfrak{r}$, the utility function $\mathfrak{u}$, and admissible portfolios A) provided that there does* not exist *any $x' \in A$ such that either*

$$[\mathfrak{r}\,(x') \leq \mathfrak{r}\,(x^*) \quad \textit{and} \quad \mathfrak{u}\,(x') > \mathfrak{u}\,(x^*)] \tag{2.3.8}$$

or

$$[\mathfrak{r}\,(x') < \mathfrak{r}\,(x^*) \quad \textit{and} \quad \mathfrak{u}\,(x') \geq \mathfrak{u}\,(x^*)] \tag{2.3.9}$$

holds.

Definition 2.60 (Efficient Frontier; [78, Definition 5]) *We call the set of images of all efficient portfolios in the two-dimensional "risk-utility" space the* efficient frontier.

Notation: $\mathcal{G}_{\text{eff}} = \mathcal{G}_{\text{eff}}(\mathfrak{r},\mathfrak{u};A) := \{(r,\mu) \in \mathbb{R}^2$: there exists an efficient portfolio $x^* \in A$ with $\mathfrak{r}\,(x^*) = r$ and $\mathfrak{u}(x^*) = \mu\} \subset \mathcal{G}(\mathfrak{r},\mathfrak{u};A)$.

Theorem 2.61 (Efficient Frontier Properties) *Assume again the situation of Proposition 2.58. Then the following holds true:*

(a) Efficient portfolios represented in the two-dimensional risk-utility space all lie on the boundary $\partial\mathcal{G} = \partial\mathcal{G}(\mathfrak{r}, \mathfrak{u}; A)$ of $\mathcal{G} = \mathcal{G}(\mathfrak{r}, \mathfrak{u}; A)$.
(b) $\mathcal{G}_{\text{eff}}(\mathfrak{r}, \mathfrak{u}; A)$ cannot contain vertical or horizontal line segments (of positive length).
(c) In case furthermore Assumption 2.56 (b) holds, then $\mathcal{G}_{\text{eff}}(\mathfrak{r}, \mathfrak{u}; A)$ has the following representation:

$$\mathcal{G}_{\text{eff}}(\mathfrak{r}, \mathfrak{u}; A) = \{(r, \mu) \in \partial\mathcal{G} \,:\, (r-k, \mu), (r, \mu+k) \notin \mathcal{G} \text{ for all } k > 0\}. \tag{2.3.10}$$

Proof *ad(a).* To any interior point $(r, \mu) \in \mathring{\mathcal{G}}(\mathfrak{r}, \mathfrak{u}; A)$ for some $\varepsilon > 0$

$$(r-\varepsilon, \mu) \in \mathcal{G}(\mathfrak{r}, \mathfrak{u}; A) \quad \text{as well as} \quad (r, \mu+\varepsilon) \in \mathcal{G}(\mathfrak{r}, \mathfrak{u}; A)$$

holds. Thus, a portfolio $x \in A$ with $(\mathfrak{r}(x), \mathfrak{u}(x)) = (r, \mu)$ can be improved and therefore cannot be efficient. Hence,

$$\mathcal{G}_{\text{eff}}(\mathfrak{r}, \mathfrak{u}; A) \cap \mathring{\mathcal{G}}(\mathfrak{r}, \mathfrak{u}; A) = \emptyset$$

and $\mathcal{G}_{\text{eff}} \subset \partial\mathcal{G}$ follows.

ad(b). If the boundary $\partial\mathcal{G}(\mathfrak{r}, \mathfrak{u}; A)$ contains any vertical or horizontal line segment, then apparently at most one point on that line segment might correspond to an efficient portfolio.

ad(c). It remains to show (2.3.10). Since any $(r^*, \mu^*) \in \mathcal{G}_{\text{eff}}$ corresponds to an efficient portfolio $x^* \in A$, both $(r^* - k, \mu^*)$ and $(r^*, \mu^* + k)$ cannot be elements of $\mathcal{G}$ for all $k > 0$. This shows "$\subset$" in (2.3.10). In order to show "$\supset$," we can use Proposition 2.58 (c) to get that $\mathcal{G}(\mathfrak{r}, \mathfrak{u};\, A)$ is closed, such that $\partial\mathcal{G} \subset \mathcal{G}$. The remaining argument is left to the reader. □

The following relationship is useful.

Lemma 2.62 (Efficient Frontier of a Subsystem) *Consider sets of admissible portfolios A and $B \subset \mathbb{R}^{M+1}$. Then the following holds:*

$$B \subset A \Rightarrow [\mathcal{G}_{\text{eff}}(\mathfrak{r}, \mathfrak{u}; A) \cap \mathcal{G}(\mathfrak{r}, \mathfrak{u};\, B)] \subset \mathcal{G}_{\text{eff}}(\mathfrak{r}, \mathfrak{u};\, B) \tag{2.3.11}$$

Proof Note that $\mathcal{G}(\mathfrak{r}, \mathfrak{u};\, B) \subset \mathcal{G}(\mathfrak{r}, \mathfrak{u}; A)$. The remaining is an easy exercise. □

Example 2.63 (Empty Efficient Frontier) *Assume that portfolio $x^{(\alpha)} := (\alpha, \widehat{0}^\top)^\top \in A \subset \mathbb{R}^{M+1}$ for all $\alpha \in \mathbb{R}$. Furthermore, the increasing auxiliary function ϕ in Assumption 2.32 (i.e., with (ϕ1)) has no upper bound (i.e., (ϕ3) holds) and the risk function $\mathfrak{r}$ satisfies* (r1) *and* (r1n) *in Assumption 2.20.*

Then $\mathfrak{r}\left(x^{(\alpha)}\right) = 0$ for all $\alpha \in \mathbb{R}$ and

$$\mathfrak{u}\left(x^{(\alpha)}\right) = \mathrm{E}\left[\phi\left(S_1^\top x^{(\alpha)}\right)\right] = \phi(R \cdot \alpha) \to \infty \quad (\alpha \to \infty).$$

Hence, with (2.3.6) *and since* $\mathfrak{r}(x) \geq 0$ *for all* $x \in A$ *due to Assumption 2.20, immediately* $\mathcal{G}(\mathfrak{r}, \mathfrak{u}; A) = [0, \infty) \times \mathbb{R}$ *follows. Thus, with* (2.3.10), $\mathcal{G}_{\text{eff}}(\mathfrak{r}, \mathfrak{u}; A) = \emptyset$.

We conclude that practically meaningful $\mathcal{G}(\mathfrak{r}, \mathfrak{u}; A)$ always correspond to sets of admissible portfolios A whose initial cost $S_0^\top x$ is limited for $x \in A$. Furthermore, if the initial cost has a range and riskless bonds are included in the portfolio, this results in a vertical line segment on the μ-axis contained in $\mathcal{G}(\mathfrak{r}, \mathfrak{u};\ A)$ (with the only possible efficient portfolio of this line segment on its top). Thus, it would suffice to consider sets of portfolios with fixed initial cost (e.g., unit initial cost).

Next, we develop several properties of the efficient frontier under reasonable "standing" assumptions, which are collected below.

Assumption 2.64 (Standing Assumptions for Efficient Trade-Off; Scalar Risk Case) *We assume the following properties:*

(i) $A \subset \mathbb{R}^{M+1}$ *is a set of admissible portfolios according to Definition 2.7.*
(ii) $\mathfrak{r}\colon A \to \mathbb{R} \cup \{+\infty\}$ *is a lower semi-continuous extended-valued risk function satisfying* (r2) *in Assumption 2.20.*
(iii) $\mathfrak{u}\colon A \to \mathbb{R} \cup \{-\infty\}$ *is an upper semi-continuous extended-valued utility function satisfying* (u2) *in Assumption 2.29.*
(iv) Assumption 2.56 concerning compact level sets holds for $A, \mathfrak{r}$ *and* $\mathfrak{u}$.

Remark 2.65 *In Assumption 2.64, there is no need for a fixed or unit initial cost condition, because the compactness assumption in (iv) also rules out pathological examples with empty efficient frontier like in Example 2.63 (see Proposition 2.68 (c)). However, a fixed initial cost conditon is sensible and may very well be incorporated in* A.

In order to represent the efficient frontier as graphs of suitable functions, we use some convex analysis of Appendix A.

Remark 2.66 *Due to Proposition 2.58 (b), which in particular holds when Assumption 2.64 is given, the set* $\mathcal{G}(\mathfrak{r}, \mathfrak{u}; A)$ *defined in* (2.3.1) *may be viewed as*

(i) hypograph *in the "risk-utility" space (see* (A.1.8)*)*

or as

(ii) epigraph *in the "utility-risk" space (see* (A.1.4)*).*

A closed set $\mathcal{G}(\mathfrak{r}, \mathfrak{u}; A)$ (cf. Proposition 2.58 (c)) with the property (2.3.6) naturally defines two functions $\nu\colon \mathbb{R} \to \mathbb{R} \cup \{\pm\infty\}$, $\gamma\colon \mathbb{R} \to \mathbb{R} \cup \{\pm\infty\}$, where

$$\nu(r) := \sup\left\{\mu \,:\, (r, \mu) \in \mathcal{G}(\mathfrak{r}, \mathfrak{u}; A)\right\}, \quad r \in \mathbb{R} \tag{2.3.12}$$

and

$$\gamma(\mu) := \inf\{r \,:\, (r,\mu) \in \mathcal{G}(\mathfrak{r},\mathfrak{u};A)\}, \quad \mu \in \mathbb{R}. \tag{2.3.13}$$

Thus, graph(ν) may be viewed as the boundary of the hypograph in Remark 2.66 (i) and similarly, graph(γ) corresponds to the boundary of the epigraph in Remark 2.66 (ii).

Immediately from the definition of $\mathcal{G}(\mathfrak{r},\mathfrak{u};A)$ (cf. (2.3.1)) follow alternative representations of ν and γ

$$\nu(r) = \sup\{\mathfrak{u}(x) \,:\, \mathfrak{r}(x) \le r,\ x \in A\} \tag{2.3.14}$$

and

$$\gamma(\mu) = \inf\{\mathfrak{r}(x) \,:\, \mathfrak{u}(x) \ge \mu,\ x \in A\}. \tag{2.3.15}$$

Remark 2.67

(a) In Proposition 2.68, we will see that under additional assumptions,

$$\nu(r) < \infty \text{ for all } r \in \mathbb{R} \quad \text{and} \quad \gamma(\mu) > -\infty \text{ for all } \mu \in \mathbb{R} \tag{2.3.16}$$

can be ensured. Then by Propositions A.5 and A.7, $\nu\colon \mathbb{R} \to \mathbb{R} \cup \{-\infty\}$ and $\gamma\colon \mathbb{R} \to \mathbb{R} \cup \{+\infty\}$ are both well-defined and even upper and lower semi-continuous, respectively.

(b) If the risk function $\mathfrak{r}$ has only non-negative values, i.e., $\mathfrak{r}\colon A \to \mathbb{R}_{\ge 0} \cup \{+\infty\}$ (e.g., when (r1n) *in Assumption 2.20 holds), then $\mathcal{G}(\mathfrak{r},\mathfrak{u};A) \cap \{(r,\mu) \,:\, r < 0\} = \emptyset$ and thus $\nu_{|(-\infty,0)} = -\infty$ and $\gamma(\mu) \ge 0 > -\infty$ for $\mu \in \mathbb{R}$ holds. Therefore, in this case, $\gamma\colon \mathbb{R} \to \mathbb{R} \cup \{+\infty\}$ is already well-defined.*

Proposition 2.68 (Representation of the Efficient Frontier 1; Scalar Risk) *Assume for a set $A \subset \mathbb{R}^{M+1}$ of admissible portfolios as well as for extended-valued risk and utility functions $\mathfrak{r}$ and $\mathfrak{u}$ that Assumption 2.64 is given. Then the following holds true:*

(a) $\mathcal{G}(\mathfrak{r},\mathfrak{u};A) \neq \emptyset$.

(b) The function

$$\begin{aligned}\gamma\colon \mathbb{R} &\to \mathbb{R} \cup \{+\infty\}\\ \mu &\mapsto \gamma(\mu) = \inf\{r \,:\, (r,\mu) \in \mathcal{G}(\mathfrak{r},\mathfrak{u};A)\}\end{aligned}$$

is well-defined, increasing, lower semi-continuous, and convex, and

$$\begin{aligned}\nu\colon \mathbb{R} &\to \mathbb{R} \cup \{-\infty\}\\ r &\mapsto \nu(r) = \sup\{\mu \,:\, (r,\mu) \in \mathcal{G}(\mathfrak{r},\mathfrak{u};A)\}\end{aligned}$$

is well-defined, increasing, upper semi-continuous, and concave.

(c) $\mathcal{G}_{\text{eff}}(\mathfrak{r},\mathfrak{u};A) \neq \emptyset$.

Proof *ad(a).* By Assumption 2.64 (iv), in particular Assumption 2.56 (a), $\text{dom}(\mathfrak{u}) \cap \text{dom}(\mathfrak{r}) \neq \emptyset$, holds. Hence $\mathcal{G}(\mathfrak{r}, \mathfrak{u}; A) \neq \emptyset$.

ad(b). Furthermore, we need to show that $\gamma \colon \mathbb{R} \to \mathbb{R} \cup \{+\infty\}$ and $\nu \colon \mathbb{R} \to \mathbb{R} \cup \{-\infty\}$ in (2.3.12) and (2.3.13) are well-defined, i.e., that (2.3.16) holds. To see this, we use a standard compactness argument and assume on the contrary that (2.3.16) were false. So, for instance, assume there exists some $\mu^* \in \mathbb{R}$ with $\gamma(\mu^*) = -\infty$. Then by (2.3.15), there exists a sequence $\{x^{(n)}\}_{n \in \mathbb{N}} \subset A$ with $\mathfrak{u}(x^{(n)}) \geq \mu^*$ and $\mathfrak{r}(x^{(n)}) \to -\infty$ as $n \to \infty$, i.e., $x^{(n)} \in \text{dom}(\mathfrak{r}) \cap \text{dom}(\mathfrak{u})$ and

$$x^{(n)} \in \mathcal{B}_A(\mathfrak{r} \leq 0; \mathfrak{u} \geq \mu^*) \quad \text{for sufficient large } n,$$

where the latter set is compact according to Assumption 2.64 (iv). Hence, $x^{(n)}$ has a convergent subsequence, w.l.o.g. $x^{(n)} \to x^* \in A$ as $n \to \infty$. Finally, lower semi-continuity of $\mathfrak{r}$ implies

$$\liminf_{n \to \infty} \mathfrak{r}\left(x^{(n)}\right) \geq \mathfrak{r}(x^*) > -\infty,$$

a contradiction. The argument for well-definedness of $\nu \colon \mathbb{R} \to \mathbb{R} \cup \{-\infty\}$ is similar and therefore omitted.

The remaining assertions follow immediately: ν and γ are both increasing by the representation in (2.3.14) and (2.3.15), respectively. Moreover, by Proposition 2.58 (a) and (c), $\mathcal{G}(\mathfrak{r}, \mathfrak{u}; A)$ is a closed and convex subset of $\mathbb{R}^2$. Using the well-definedness of $\gamma \colon \mathbb{R} \to \mathbb{R} \cup \{+\infty\}$ and Proposition 2.58 (b), we obtain from Proposition A.5 that γ is lower semi-continuous and $\text{epi}(\gamma) = \mathcal{G}(\mathfrak{r}, \mathfrak{u}; A)$ in the utility-risk space. Thus, γ is in particular also convex (see Lemma A.13). The properties of $r \mapsto \nu(r)$ follow similarly with Proposition A.7 and $\text{hypo}(\nu) = \mathcal{G}(\mathfrak{r}, \mathfrak{u}; A)$ in the risk-utility space.

ad(c). According to (a), we have $\mathcal{G} = \mathcal{G}(\mathfrak{r}, \mathfrak{u}; A) \neq \emptyset$. Starting in any point $(\overline{r}, \overline{\mu}) \in \mathcal{G}$, by lowering r and increasing μ while staying in $\mathcal{G}$, one finally reaches a point on $\partial\mathcal{G} \subset \mathcal{G}$ which cannot be improved and thus lies on $\mathcal{G}_{\text{eff}} = \mathcal{G}_{\text{eff}}(\mathfrak{r}, \mathfrak{u}; A)$. This is a consequence of $\nu(r) < \infty$ for all $r \in \mathbb{R}$ and $\gamma(\mu) > -\infty$ for all $\mu \in \mathbb{R}$ which are well-defined according to (b) (see also (2.3.16)). Thus, $\mathcal{G}_{\text{eff}} \neq \emptyset$. □

The next technical lemma is quite helpful in the succeeding theory.

Lemma 2.69 (Properties of dom(γ) and dom(ν)) *Under the conditions of Proposition 2.68, for any $(r_0, \mu_0) \in \mathcal{G}_{\text{eff}}(\mathfrak{r}, \mathfrak{u}; A)$ the following holds:*

(a) $(-\infty, \mu_0] \subset \text{dom}(\gamma) = \{\mu \in \mathbb{R} \colon \gamma(\mu) < \infty\}$,
(b) $[r_0, \infty) \subset \text{dom}(\nu) = \{r \in \mathbb{R} \colon \nu(r) > -\infty\}$.

Proof Let $(r_0, \mu_0) \in \mathcal{G}_{\text{eff}}(\mathfrak{r}, \mathfrak{u}; A) \subset \mathcal{G}(\mathfrak{r}, \mathfrak{u}; A) \subset \mathbb{R}^2$ be given. By Definition 2.60, there exists an efficient portfolio $x^* \in A$ with $\mathfrak{r}(x^*) = r_0$ and $\mathfrak{u}(x^*) = \mu_0$. Hence, by (2.3.14) and (2.3.15) follows $\gamma(\mu_0) \leq r_0$ and

$\nu(r_0) \geq \mu_0$. On the other hand, since x^* is efficient, necessarily $\gamma(\mu_0) = r_0$ and $\nu(r_0) = \mu_0$ holds. In order to prove (a), assume $\mu \leq \mu_0$. Since γ is increasing according to Proposition 2.68 (b), we get $\gamma(\mu) \leq \gamma(\mu_0) = r_0$. Thus, $\mu \in \mathrm{dom}(\gamma)$. The proof of part (b) is similar. □

Below we need the "exchange operator"

$$\begin{aligned} \widehat{P}: \quad \widehat{\mathbb{R}}^2 &\to \widehat{\mathbb{R}}^2 \\ (\mu, r) &\mapsto (r, \mu), \end{aligned} \tag{2.3.17}$$

where we set $\widehat{\mathbb{R}} := \mathbb{R} \cup \{\pm\infty\}$.

Theorem 2.70 (Representation of the Efficient Frontier 2; Scalar Risk) *Assume for a set $A \subset \mathbb{R}^{M+1}$ of admissible portfolios as well as for extended-valued risk and utility functions $\mathfrak{r}$ and $\mathfrak{u}$ that Assumption 2.64 holds.*

(a) Then, the efficient frontier has the following representation:

$$\mathcal{G}_{\text{eff}} = \mathcal{G}_{\text{eff}}(\mathfrak{r}, \mathfrak{u}; A) = \widehat{P}\,[\mathrm{graph}(\gamma)] \cap \mathrm{graph}(\nu), \tag{2.3.18}$$

where $\mathrm{graph}(\gamma) = \{(\mu, \gamma(\mu)) \,:\, \mu \in \mathbb{R}\} \subset \mathbb{R} \times (\mathbb{R} \cup \{+\infty\})$
and $\mathrm{graph}(\nu) = \{(r, \nu(r)) \,:\, r \in \mathbb{R}\} \subset \mathbb{R} \times (\mathbb{R} \cup \{-\infty\})$,
with ν and γ defined in (2.3.12) *and* (2.3.13), *respectively.*

(b) Furthermore, the right-hand side in (2.3.18) *can be written as*

$$\widehat{P}\,[\mathrm{graph}(\gamma)] \cap \mathrm{graph}(\nu) = \{(\gamma(\mu), \mu) \,:\, \mu \in \mathrm{dom}(\gamma) \subset \mathbb{R}\} \cap \{(r, \nu(r)) \,:\, r \in \mathrm{dom}(\nu) \subset \mathbb{R}\}. \tag{2.3.19}$$

Note in particular that *not* all points in the graphs of ν and γ correspond to efficient portfolios.

Proof (of Theorem 2.70) *ad(b).* We firstly show (2.3.19):
"$\supset$" is clear by definition. In order to show "$\subset$" in (2.3.19), let

$$(r_0, \mu_0) \in \widehat{P}\,[\mathrm{graph}(\gamma)] \cap \mathrm{graph}(\nu). \tag{2.3.20}$$

Since

$$\widehat{P}\,[\mathrm{graph}(\gamma)] = \{(\gamma(\mu), \mu) : \mu \in \mathbb{R}\} \subset (\mathbb{R} \cup \{+\infty\}) \times \mathbb{R},$$

necessarily $\mu_0 \in \mathbb{R}$ and by the definition of $\mathrm{graph}(\nu)$ also $r_0 \in \mathbb{R}$. Hence, $(r_0, \mu_0) \in \mathbb{R}^2$ and therefore by (2.3.20)

$$r_0 = \gamma(\mu_0) \quad \text{yielding} \quad \mu_0 \in \mathrm{dom}(\gamma) \tag{2.3.21}$$

and

$$\mu_0 = \nu(r_0) \quad \text{yielding} \quad r_0 \in \mathrm{dom}(\nu). \tag{2.3.22}$$

Thus, (2.3.19) follows.

ad(a). Next we want to show (2.3.18). In order to see "$\supset$," let again (r_0, μ_0) satisfy (2.3.20). We have to show that

$$(r_0, \mu_0) \overset{!}{\in} \mathcal{G}_{\mathrm{eff}}. \tag{2.3.23}$$

Reasoning as above, we get $(r_0, \mu_0) \in \mathbb{R}^2$ and from (2.3.21) and (2.3.22)

$$r_0 = \gamma(\mu_0) \overset{(2.3.15)}{=} \inf\{\mathfrak{r}(x) : \mathfrak{u}(x) \geq \mu_0,\ x \in A\} \tag{2.3.24}$$

and

$$\mu_0 = \nu(r_0) \overset{(2.3.14)}{=} \sup\{\mathfrak{u}(x) : \mathfrak{r}(x) \leq r_0,\ x \in A\}. \tag{2.3.25}$$

By (2.3.25) we can select a sequence $x^{(n)} \in A,\ n \in \mathbb{N}$, such that

$$\mathfrak{r}\left(x^{(n)}\right) \leq r_0 \quad \text{and} \quad \mathfrak{u}\left(x^{(n)}\right) \nearrow \mu_0 \quad \text{as} \quad n \to \infty. \tag{2.3.26}$$

Note that by Assumption 2.56 (b)

$$\mathcal{B}_A(\mathfrak{r} \leq r_0; \mathfrak{u} \geq \mu_0 - 1) \overset{(2.3.4)}{=} \{x \in A : \mathfrak{r}(x) \leq r_0 \text{ and } \mathfrak{u}(x) \geq \mu_0 - 1\}$$

is compact. Hence, there exists a subsequence (w.l.o.g. the whole sequence) which is convergent to some $x^* \in A$, since A is closed, i.e., $x^{(n)} \to x^*$ $(n \to \infty)$. Lower semi-continuity of $\mathfrak{r}$ and upper semi-continuity of $\mathfrak{u}$ immediately imply

$$\mathfrak{r}(x^*) \leq r_0 \quad \text{and} \quad \mathfrak{u}(x^*) \geq \mu_0. \tag{2.3.27}$$

So in particular, $(r_0, \mu_0) \in \mathcal{G} = \mathcal{G}(\mathfrak{r}, \mathfrak{u}; A)$ (cf. (2.3.1)). In fact, using the representations (2.3.24) and (2.3.25), we even get

$$\mathfrak{r}(x^*) = r_0 \quad \text{and} \quad \mathfrak{u}(x^*) = \mu_0. \tag{2.3.28}$$

Finally, we show that (r_0, μ_0) is on the efficient frontier. Consider $(r_1, \mu_1) \in \mathcal{G}$ arbitrary. Efficiency of x^* represented by (r_0, μ_0) means that the following two cases are not allowed to occur (see Definition 2.59):
Case I: $\mu_1 > \mu_0$ and $r_1 \leq r_0$.
Case II: $\mu_1 \geq \mu_0$ and $r_1 < r_0$.

We argue for both cases indirectly. Assuming firstly $(r_1, \mu_1) \in \mathcal{G}$ satisfies *Case I* yields

$$\nu(r_1) \overset{(2.3.12)}{=} \sup\{\mu : (r_1, \mu) \in \mathcal{G}(\mathfrak{r}, \mathfrak{u}; A)\} \geq \mu_1 > \mu_0 \overset{(2.3.25)}{=} \nu(r_0),$$

in contradiction to the fact that ν is increasing (see Proposition 2.68 (b)). Analogously, assuming $(r_1, \mu_1) \in \mathcal{G}$ satisfies *Case II* gives

$$\gamma(\mu_1) \overset{(2.3.13)}{=} \inf\{r : (r, \mu_1) \in \mathcal{G}(\mathfrak{r}, \mathfrak{u}; A)\} \le r_1 < r_0 \overset{(2.3.24)}{=} \gamma(\mu_0),$$

contradicting that γ is increasing due to Proposition 2.68 (b).

Hence, $(r_0, \mu_0) \in \mathcal{G}_{\text{eff}}(\mathfrak{r}, \mathfrak{u}; A)$ as claimed in (2.3.23). The remaining claim "$\subset$" in (2.3.18) is simple and therefore left to the reader as an exercise. □

2.3.3 Connectedness of the Efficient Frontier

Although (2.3.18) already is a very nice representation of the efficient frontier, so far we are not able to state any topological properties of $\mathcal{G}_{\text{eff}}$. In particular, it is not yet clear whether or not $\mathcal{G}_{\text{eff}}$ locally is a continuous curve, or even globally path-connected. Both properties in fact hold true, as will be shown in the following.

To start with, in the following corollary, we mark out those subsets of $\text{dom}(\nu)$ or $\text{dom}(\gamma)$ on which the efficient frontier eventually may be parameterized as graph of ν or γ (see Theorem 2.76):

Corollary 2.71 (Projection of $\mathcal{G}_{\text{eff}}$ to the Axes) *In the situation of Theorem 2.70, we define the* projections *of $\mathcal{G}_{\text{eff}} = \mathcal{G}_{\text{eff}}(\mathfrak{r}, \mathfrak{u}; A)$ to the axes, i.e.,*

$$I = I(\mathfrak{r}, \mathfrak{u}; A) := \{r \in \mathbb{R} : \exists\, \mu \in \mathbb{R} \text{ with } (r, \mu) \in \mathcal{G}_{\text{eff}}(\mathfrak{r}, \mathfrak{u}; A)\} \tag{2.3.29}$$

and

$$J = J(\mathfrak{r}, \mathfrak{u}; A) := \{\mu \in \mathbb{R} : \exists\, r \in \mathbb{R} \text{ with } (r, \mu) \in \mathcal{G}_{\text{eff}}(\mathfrak{r}, \mathfrak{u}; A)\}. \tag{2.3.30}$$

Then the following holds true:

(a) $I = \text{dom}(\nu) \cap \text{range}(\gamma) \subset \mathbb{R}$
(b) $J = \text{dom}(\gamma) \cap \text{range}(\nu) \subset \mathbb{R}$.

Proof Using (2.3.18) and (2.3.19), we have

$$\mathcal{G}_{\text{eff}}(\mathfrak{r}, \mathfrak{u}; A) = \{(\gamma(\mu), \mu) : \mu \in \text{dom}(\gamma)\} \cap \{(r, \nu(r)) : r \in \text{dom}(\nu)\}. \tag{2.3.31}$$

Hence, by definition of I as projection of $\mathcal{G}_{\text{eff}}$ to the r-axis, we get $I = \text{range}(\gamma) \cap \text{dom}(\nu)$, i.e., (a) holds in this case. Similarly (b), i.e., $J = \text{dom}(\gamma) \cap \text{range}(\nu)$, holds true, since J is the projection of $\mathcal{G}_{\text{eff}}$ to the μ-axis. □

A typical situation of (2.3.31) may be viewed in Fig. 2.2.

The following key lemma in consequence yields that the efficient frontier is a path-connected continuous curve.

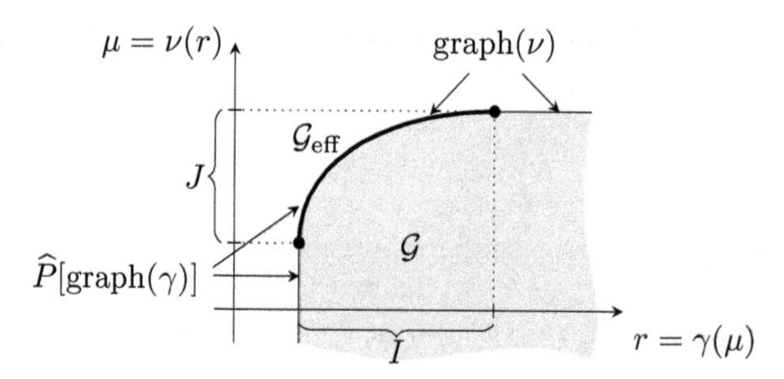

Fig. 2.2 $\mathcal{G}_{\text{eff}}$ as graph of γ and ν

Lemma 2.72 (Path-Connectedness of $\mathcal{G}_{\text{eff}}$) *In the situation of Theorem 2.70, let two different, but arbitrary, points $(r_i, \mu_i) \in \mathcal{G}_{\text{eff}} = \mathcal{G}_{\text{eff}}(\mathfrak{r}, \mathfrak{u}; A)$, $i \in \{1, 2\}$, be given, such that w.l.o.g. $r_1 < r_2$ and $\mu_1 < \mu_2$. Then the following holds true for ν and γ defined in* (2.3.12) *and* (2.3.13) *and its restrictions:*

(a) $\gamma_{|[\mu_1,\mu_2]} \colon [\mu_1, \mu_2] \to \mathbb{R}$ *is continuous and* $\operatorname{range}\left(\gamma_{|[\mu_1,\mu_2]}\right) = [r_1, r_2]$.
(b) $\nu_{|[r_1,r_2]} \colon [r_1, r_2] \to \mathbb{R}$ *is continuous and* $\operatorname{range}\left(\nu_{|[r_1,r_2]}\right) = [\mu_1, \mu_2]$.
(c) $\widehat{P}\left[\operatorname{graph}\left(\gamma_{|[\mu_1,\mu_2]}\right)\right] = \operatorname{graph}\left(\nu_{|[r_1,r_2]}\right) \subset \mathcal{G}_{\text{eff}}$. *In particular,* $[\mu_1, \mu_2] \subset J$ *and* $[r_1, r_2] \subset I$.

Proof Firstly, note that any two arbitrary points (r_1, μ_1) and (r_2, μ_2) on the efficient frontier can be ordered w.l.o.g. in such a way that $r_1 < r_2$ and $\mu_1 < \mu_2$, because any other constellation would immediately contradict the properties of efficient portfolios (see Definition 2.59).

ad(a) and *(b)*. By Proposition 2.68 (b), $\gamma \colon \mathbb{R} \to \mathbb{R} \cup \{+\infty\}$ is convex and $\nu \colon \mathbb{R} \to \mathbb{R} \cup \{-\infty\}$ is concave. So $-\nu$ is convex as well. Hence, by Theorem A.14 (b), both ν and γ are continuous in the interior of $\operatorname{dom}(\nu) = \{r \in \mathbb{R} \colon \nu(r) > -\infty\}$ and $\operatorname{dom}(\gamma) = \{\mu \in \mathbb{R} \colon \gamma(\mu) < +\infty\}$, respectively. Both domains are convex sets (see Remark A.11 (a)) and thus intervals with positive length, since $\mathcal{G}_{\text{eff}}$ contains by assumption the two different points (r_1, μ_1) and (r_2, μ_2) with $r_1 < r_2$ and $\mu_1 < \mu_2$. (see (2.3.31)).

In particular, the assumption $(r_i, \mu_i) \in \mathcal{G}_{\text{eff}}$, $i \in \{1, 2\}$, yields with (2.3.31)

$$\mu_i = \nu\left(r_i\right) \in J \quad \text{and} \quad r_i = \gamma\left(\mu_i\right) \in I, \quad i \in \{1, 2\}. \tag{2.3.32}$$

Moreover, Lemma 2.69 gives

$$[r_1, \infty) \subset \operatorname{dom}(\nu) \quad \text{and} \quad (-\infty, \mu_2] \subset \operatorname{dom}(\gamma) \tag{2.3.33}$$

and thus, $\gamma_{|[\mu_1,\mu_2)}$ and $\nu_{|(r_1,r_2]}$ are continuous (see Theorem A.14) with a priori possible discontinuities at the endpoints. But this does not happen:
Claim: ν is right continuous at the left endpoint r_1.

On the one hand, according to Proposition 2.68 (b), ν is increasing, and thus the limit

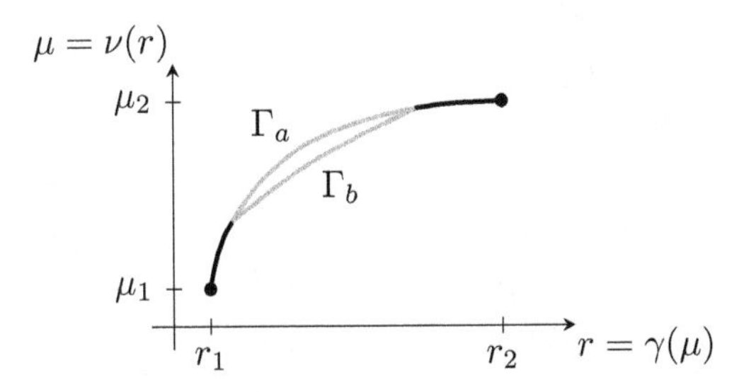

Fig. 2.3 Γ_a and Γ_b in the proof of Lemma 2.72

$$\mu^* := \lim_{r \searrow r_1} \nu(r) \geq \mu_1 = \nu(r_1) \tag{2.3.34}$$

must exist. On the other hand, again by Proposition 2.68 (b), ν is upper semi-continuous yielding

$$\limsup_{r \to r_1} \nu(r) \leq \nu(r_1) = \mu_1. \tag{2.3.35}$$

However, (2.3.35) together with (2.3.34) implies $\mu^* = \lim_{r \searrow r_1} \nu(r) = \mu_1$ and right continuity at r_1 follows.

The left continuity of γ at its right endpoint μ_2 can be shown similarly. Moreover, since γ is increasing, the continuity of γ also implies range $\left(\gamma_{|[\mu_1,\mu_2]}\right) = [r_1, r_2]$. Similarly, range $\left(\nu_{|[r_1,r_2]}\right) = [\mu_1, \mu_2]$ follows. Hence, claims (a) and (b) are proved.

ad (c). We only have to show that

$$\Gamma_a := \widehat{P}\left[\text{graph}\left(\gamma_{|[\mu_1,\mu_2]}\right)\right] \stackrel{!}{=} \text{graph}\left(\nu_{|[r_1,r_2]}\right) =: \Gamma_b \tag{2.3.36}$$

because then with (2.3.31) already follows that all those points lie on $\mathcal{G}_{\text{eff}}$. By (a) and (b), Γ_a and Γ_b are continuous curves with equal endpoints $(r_i, \mu_i) \in \mathcal{G}_{\text{eff}}$, $i \in \{1, 2\}$. If (2.3.36) were not true, i.e., $\Gamma_a \neq \Gamma_b$, then a part of Γ_a must lie below Γ_b, or vice versa (see Fig. 2.3).

But this is impossible, since both finite points on graph(ν) and finite points on $\widehat{P}$[graph(γ)] lie on $\partial\mathcal{G} = \partial\mathcal{G}(\mathfrak{r}, \mathfrak{u}; A)$ by definition (see (2.3.12) and (2.3.13)). This, however, is in contradiction with (2.3.6) on the lower part of the above two curves. □

Corollary 2.73 (Topological Structure of $\mathcal{G}_{\text{eff}}$) *Assume for a set $A \subset \mathbb{R}^{M+1}$ of admissible portfolios as well as for extended-valued risk and utility functions $\mathfrak{r}$ and $\mathfrak{u}$ that Assumption 2.64 holds. Then, the efficient frontier $\mathcal{G}_{\text{eff}} = \mathcal{G}_{\text{eff}}(\mathfrak{r}, \mathfrak{u}; A)$ is a single point or a path-connected continuous curve (one-sided continuous at finite endpoint(s)).*

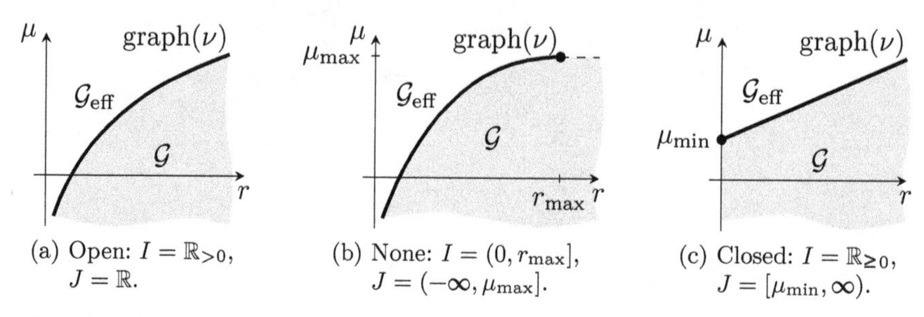

(a) Open: $I = \mathbb{R}_{>0}$, $J = \mathbb{R}$.

(b) None: $I = (0, r_{\max}]$, $J = (-\infty, \mu_{\max}]$.

(c) Closed: $I = \mathbb{R}_{\geq 0}$, $J = [\mu_{\min}, \infty)$.

Fig. 2.4 $\mathcal{G}_{\text{eff}}$ examples for a risk function $\mathfrak{r}$ with non-negative values

Proof All assertions are an immediate consequence of Lemma 2.72, with a connecting continuous path inside $\mathcal{G}_{\text{eff}}$ between arbitrary points of $\mathcal{G}_{\text{eff}}$ constructed in part (c) of that lemma. □

Corollary 2.74 (I and J Are Intervals) *In the situation of Corollary 2.73, the following holds true:*

(a) The projections of the efficient frontier $\mathcal{G}_{\text{eff}} = \mathcal{G}_{\text{eff}}(\mathfrak{r}, \mathfrak{u}; A) \subset \mathbb{R}^2$ to the axes, $I = I(\mathfrak{r}, \mathfrak{u}; A)$, as well as $J = J(\mathfrak{r}, \mathfrak{u}; A)$ defined in (2.3.29) *and* (2.3.30) *are intervals.*

(b) $\mathcal{G}_{\text{eff}}$, I, and J are all just single points, or otherwise, I and J are both nondegenerate intervals, i.e., with positive length.

Proof *ad(a).* This is an immediate consequence of the path-connectedness of $\mathcal{G}_{\text{eff}}$ (see Lemma 2.72 and Corollary 2.73).

ad(b). If $\mathcal{G}_{\text{eff}}$ is not a single point, but either I or J was just a point, then $\mathcal{G}_{\text{eff}}$ would be a straight line parallel to one axis, contradicting Theorem 2.61 (b). □

Remark 2.75 *The intervals I and J may be open, closed, or none of both. See Fig. 2.4.*

Theorem 2.76 (Parameterization of $\mathcal{G}_{\text{eff}}$ as Graphs on I and J) *Assume for a set $A \subset \mathbb{R}^{M+1}$ of admissible portfolios as well as for extended-valued risk and utility functions $\mathfrak{r}$ and $\mathfrak{u}$ that Assumption 2.64 holds. Then for I and J defined in* (2.3.29) *and* (2.3.30) *as well as ν and γ defined in* (2.3.12) *and* (2.3.13), *respectively, the following holds true:*

(a) $\gamma\colon J \to \mathbb{R}$ and $\nu\colon I \to \mathbb{R}$ are continuous (one-sided continuous at finite endpoints).

(b) Furthermore, $\gamma\colon J \to I$ and $\nu\colon I \to J$ are strictly increasing, bijective, and inverse to each other, i.e.,

$$(\gamma \circ \nu)(r) = r \quad \text{for all} \quad r \in I \text{ and } (\nu \circ \gamma)(\mu) = \mu \quad \text{for all} \quad \mu \in J. \tag{2.3.37}$$

(c) $\mathcal{G}_{\text{eff}}(\mathfrak{r}, \mathfrak{u}; A)$ *has the* representation

$$\mathcal{G}_{\text{eff}}(\mathfrak{r}, \mathfrak{u}; A) = \widehat{P}\left[\text{graph}\left(\gamma_{|J}\right)\right] = \text{graph}\left(\nu_{|I}\right). \tag{2.3.38}$$

Note that in general, γ and ν are only semi-continuous (cf. Proposition 2.68 (b)). See again Fig. 2.2. Continuity above is only claimed on J and I, respectively.

Proof (of Theorem 2.76) *ad (a).* In case $\mathcal{G}_{\text{eff}}$ and thus I and J are just single points, there is nothing to show. Otherwise, the intervals I and J have obvious representations:

$$I = \bigcup_{\substack{r_1, r_2 \in I \\ r_1 < r_2}} [r_1, r_2] \quad \text{and} \quad J = \bigcup_{\substack{\mu_1, \mu_2 \in J \\ \mu_1 < \mu_2}} [\mu_1, \mu_2]. \tag{2.3.39}$$

Thus, the continuity of $\gamma_{|[\mu_1, \mu_2]}$ and $\nu_{|[r_1, r_2]}$ according to Lemma 2.72 (a) and (b) yields the claimed continuity of $\gamma\colon J \to \mathbb{R}$ and $\nu\colon I \to \mathbb{R}$, since continuity is a local property.

ad (b). Given some arbitrary $\mu \in J$, by definition in (2.3.30), there exists some $r \in \mathbb{R}$ such that $(r, \mu) \in \mathcal{G}_{\text{eff}}$. Hence, $r \in I$ by (2.3.29). Using (2.3.31), we moreover get

$$r = \gamma(\mu) \quad \text{and} \quad \mu = \nu(r). \tag{2.3.40}$$

This proves that $\gamma\colon J \to I$ is well-defined. A similar argument with J and I replaced yields surjectivity of $\gamma\colon J \to I$ as well. Furthermore, by (2.3.40) $(\nu \circ \gamma)(\mu) = \mu$.

It remains to show that $\gamma\colon J \to I$ is strictly increasing, since this will also give injectivity. The increasing property of $\gamma\colon \mathbb{R} \to \mathbb{R} \cup \{+\infty\}$ was already proved in Proposition 2.68 (b). So assume for the contrary, there exist $\mu_1, \mu_2 \in J,\ \mu_1 < \mu_2$ with $\gamma(\mu_1) = \gamma(\mu_2)$.

Hence, $\gamma_{|[\mu_1, \mu_2]} = \text{const} =: r^* \in \mathbb{R}$, because γ is increasing. This means $(\{r^*\} \times [\mu_1, \mu_2]) \subset \partial\mathcal{G} \subset \mathcal{G}$, which contradicts $\mu_2 \in J$, in particular $(r^*, \mu_2) \in \mathcal{G}_{\text{eff}}$ (see Theorem 2.61 (c)).

The arguments for $\nu\colon I \to J$ are similar.

ad (c). Here we only have to show (2.3.38). If $\mathcal{G}_{\text{eff}}$ is just a single point, this is obvious. Otherwise, using Lemma 2.72 (c) once more and (2.3.39), we immediately obtain

$$\widehat{P}\left[\text{graph}(\gamma_{|J})\right] \subset \mathcal{G}_{\text{eff}}.$$

To show "$\supset$," assume some arbitrary $(r, \mu) \in \mathcal{G}_{\text{eff}}$. Then, as argued in (b), $r \in I, \mu \in J$ and $\gamma(\mu) = r$ follows, i.e., $(r, \mu) \in \widehat{P}\left[\text{graph}\left(\gamma_{|J}\right)\right]$. Again, the argument to show $\text{graph}\left(\nu_{|I}\right) = \mathcal{G}_{\text{eff}}$ is similar. □

2.4 Efficient Portfolios

So far we have seen in Sect. 2.3 that an efficient trade-off between risk and utility can be represented by graphs of γ and ν, defined in (2.3.12) and (2.3.13), where $\mu \mapsto \gamma(\mu)$ relates the level of utility μ to a minimum risk, while $r \mapsto \nu(r)$ relates the level of risk r to a maximal possible utility. Next, we want to investigate how the corresponding efficient portfolios behave and give our main existence and uniqueness results for efficient portfolios in the scalar risk case. As in Sect. 2.3, the presented ideas and results essentially go back to [78], but the theory given here is slightly more general. See also [77] and [89] for a similar approach working directly on multi-period financial markets, which of course is included here as well.

2.4.1 Existence of Efficient Portfolios

As before for the properties of the efficient frontier $\mathcal{G}_{\text{eff}} = \mathcal{G}_{\text{eff}}(\mathfrak{r}, \mathfrak{u}; A)$ in Sect. 2.3, we again rely on our standing assumptions, Assumption 2.64.

Problem 2.77 (Trade-Off Optimization) *For a set of admissible portfolios $A \subset \mathbb{R}^{M+1}$ (see Definition 2.7), an extended-valued risk function $\mathfrak{r}\colon A \to \mathbb{R} \cup \{+\infty\}$ and an extended-valued utility function $\mathfrak{u}\colon A \to \mathbb{R} \cup \{-\infty\}$, we consider the following problems:*

(a) For $\mu \in \mathbb{R}$, the minimum risk optimization problem *reads*

$$\min_{x \in A} \mathfrak{r}(x) \quad \textit{subject to} \quad \mathfrak{u}(x) \geq \mu. \tag{2.4.1}$$

(b) For $r \in \mathbb{R}$, the maximum utility optimization problem *reads*

$$\max_{x \in A} \mathfrak{u}(x) \quad \textit{subject to} \quad \mathfrak{r}(x) \leq r. \tag{2.4.2}$$

Theorem 2.78 (Existence for Problem 2.77) *Assume Assumption 2.64 within particular a set $A \subset \mathbb{R}^{M+1}$ of admissible portfolios, as well as extended-valued risk and utility functions $\mathfrak{r}\colon A \to \mathbb{R} \cup \{+\infty\}$ and $\mathfrak{u}\colon A \to \mathbb{R} \cup \{-\infty\}$. Furthermore, consider the intervals $I = I(\mathfrak{r}, \mathfrak{u}; A) \subset \mathbb{R}$ and $J = J(\mathfrak{r}, \mathfrak{u}; A) \subset \mathbb{R}$ defined in (2.3.29) and (2.3.30), respectively. Then the following holds true:*

(a) For arbitrary but fixed $\mu \in J$, the minimum risk problem (2.4.1) has a solution $\bar{x} = \bar{x}_\mu \in A$. Moreover, $x \in A$ solves (2.4.1) if and only if x is an efficient portfolio for $\mathfrak{r}, \mathfrak{u}$, and A according to Definition 2.59 with $\mathfrak{u}(x) = \mu$.

(b) For arbitrary but fixed $r \in I$, the maximum utility problem (2.4.2) *has a solution $\bar{y} = \bar{y}_r \in A$. Moreover, $y \in A$ solves* (2.4.2) *if and only if y is an efficient portfolio for $\mathfrak{r}, \mathfrak{u}$, and A with $\mathfrak{r}(y) = r$.*

Proof *ad(a).* Let $\mu \in J$ arbitrary be given. Then by Theorem 2.76 (c) and (b), $(\gamma(\mu), \mu) \in \mathcal{G}_{\text{eff}}$ and $r := \gamma(\mu) \in I$. Thus, there exists an efficient portfolio $\bar{x} = \bar{x}_\mu \in A$ with $\mathfrak{r}(\bar{x}) = r$ and $\mathfrak{u}(\bar{x}) = \mu$. In particular,

$$r = \gamma(\mu) \overset{(2.3.15)}{=} \inf \{\mathfrak{r}(x) : \mathfrak{u}(x) \geq \mu, \quad x \in A\}, \tag{2.4.3}$$

i.e., $\bar{x}$ is a solution of Problem 2.77 (a). Furthermore, each efficient portfolio with utility value μ is a solution of Problem 2.77 (a) as well (cf. Definition 2.59). Thus, it remains to show that any solution of Problem 2.77 (a) is efficient and necessarily has utility value μ. So let $\widetilde{x} = \widetilde{x}_\mu \in A$ be an arbitrary solution of Problem 2.77 (a). Then with (2.4.3)

$$\mathfrak{r}(\widetilde{x}) = \gamma(\mu) = \mathfrak{r}(\bar{x}) \quad \text{and} \quad \mathfrak{u}(\widetilde{x}) \geq \mu \tag{2.4.4}$$

follow. But since $\bar{x} = \bar{x}_\mu \in A$ is efficient, $\mathfrak{u}(\widetilde{x})$ cannot be strictly larger than μ. Thus, $\mathfrak{u}(\widetilde{x}) = \mu$ and consequently $\widetilde{x} = \widetilde{x}_\mu$ are efficient as well.

ad(b). The arguments leading to (b) are similar and therefore left to the reader. □

2.4.2 Uniqueness of Efficient Portfolios

Under additional assumptions of some kind of "strictness" of risk or utility function, we even obtain unique efficient portfolios for $\mathfrak{r}, \mathfrak{u}$, and A according to Definition 2.59.

Theorem 2.79 (Uniqueness for Problem 2.77 and Continuous Efficient Portfolio Map) *Assume for a set $A \subset \mathbb{R}^{M+1}$ of admissible portfolios as well as for extended-valued risk and utility functions $\mathfrak{r}$ and $\mathfrak{u}$ that Assumption 2.64 holds. Suppose additionally that either $\mathfrak{u}$ is strictly concave on* $\text{dom}(\mathfrak{u}) \subset A$, *i.e.,* (u2s) *in Assumption 2.29 holds, or $\mathfrak{r}$ is strictly convex on* $\text{dom}(\mathfrak{r}) \subset A$, *i.e.,* (r2s) *in Assumption 2.20 holds. Then the following holds true:*

(a) Each point $(r, \mu) \in \mathcal{G}_{\text{eff}} = \mathcal{G}_{\text{eff}}(\mathfrak{r}, \mathfrak{u}; A)$ corresponds to a unique efficient portfolio $X(r, \mu) \in A$ realizing the risk-utility values (r, μ).

(b) The mapping

$$X\colon\ \mathcal{G}_{\text{eff}}(\mathfrak{r}, \mathfrak{u}; A) \to A \subset \mathbb{R}^{M+1}, \qquad (r, \mu) \mapsto X(r, \mu), \tag{2.4.5}$$

is injective and continuous (one-sided continuous at finite endpoint(s) of $\mathcal{G}_{\text{eff}}$). Therefore, the efficient portfolios lie on a continuous curve with no self-intersections.

(c) Moreover, efficient portfolios have the continuous representations

$$x^*: r \mapsto x^*(r) := X(r, \nu(r)), \quad r \in I, \quad \textit{and} \tag{2.4.6}$$
$$y^*: \mu \mapsto y^*(\mu) := X(\gamma(\mu), \mu), \quad \mu \in J, \tag{2.4.7}$$

on intervals I and J defined in (2.3.29) *and* (2.3.30)*, respectively.*

Proof *ad(a).* Let an arbitrary $(r, \mu) \in \mathcal{G}_{\text{eff}}$ be given. Arguing as at the beginning of the proof of Theorem 2.78, we obtain an efficient portfolio $\bar{x} = \bar{x}_\mu \in A$ realizing $(r, \mu) = (\gamma(\mu), \mu)$ as risk-utility value. We only have to show that $\bar{x}$ is unique with this property. Assume for the contrary that two different portfolios $x^{(1)}, x^{(2)} \in A$ corresponding to the risk-utility value (r, μ) exist, i.e.,

$$\mathfrak{r}\left(x^{(i)}\right) = r \quad \text{and} \quad \mathfrak{u}\left(x^{(i)}\right) = \mu, \quad i = 1, 2. \tag{2.4.8}$$

Similar as in (2.4.3), then

$$r \in I, \ \mu \in J \quad \text{and} \quad \gamma(\mu) = r. \tag{2.4.9}$$

The remaining argumentation is similar for (r2s) and (u2s). We give, for instance, (r2s):

Case (r2s): Since A is convex, $x^\star := \left(x^{(1)} + x^{(2)}\right)/2 \in A$. Now concavity of $\mathfrak{u}$ (cf. (u2) in Assumption 2.29) implies

$$\mathfrak{u}(x^\star) \geq \frac{1}{2} \sum_{i=1}^{2} \mathfrak{u}\left(x^{(i)}\right) = \mu$$

and strict convexity of $\mathfrak{r}$ (by (r2s)) gives

$$\mathfrak{r}(x^\star) < \frac{1}{2}\mathfrak{r}\left(x^{(1)}\right) + \frac{1}{2}\mathfrak{r}\left(x^{(2)}\right) = r = \gamma(\mu).$$

But this is a contradiction to the definition of γ (see (2.3.15)). Thus, the efficient portfolio corresponding to (r, μ) is unique. Therefore, the "efficient portfolio" mapping X in (2.4.5) is well-defined. As said before, the remaining case (u2s) works similar and is therefore left to the reader.

ad(b). We start with the injectivity of the efficient portfolio mapping. Let $(r_i, \mu_i) \in \mathcal{G}_{\text{eff}}$, $i = 1, 2$, be given with $\bar{x} := X(r_1, \mu_1) = X(r_2, \mu_2)$. Then by construction,

$$(r_1, \mu_1) = (\mathfrak{r}(\bar{x}), \ \mathfrak{u}(\bar{x})) = (r_2, \mu_2)$$

yielding injectivity and also that the curve $\{X(r,\mu)\colon (r,\mu) \in \mathcal{G}_{\text{eff}}\} \subset \mathbb{R}^{M+1}$ cannot have any self-intersections.

Next, we show continuity of $X\colon \mathcal{G}_{\text{eff}} \to A$. In case $\mathcal{G}_{\text{eff}}$ is just a single point, there is nothing to show. If this is not the case, by Corollary 2.73, $\mathcal{G}_{\text{eff}}$ is a path-connected continuous curve in $\mathbb{R}^2$. Consider $(\bar{r},\bar{\mu}) \in \mathcal{G}_{\text{eff}}$ arbitrary and a sequence $(r_n,\mu_n) \in \mathcal{G}_{\text{eff}}$, $n \in \mathbb{N}$, converging to $(\bar{r},\bar{\mu})$. Let $x^{(n)} := X(r_n,\mu_n) \in A$ be the unique efficient portfolio realizing the risk-utility value (r_n,μ_n) and similarly $\bar{x} := X(\bar{r},\bar{\mu})$. It only remains to show that $x^{(n)} \to \bar{x}$ $(n \to \infty)$. To see this, note that for sufficiently large n

$$x^{(n)} \in \{x \in A : \mathfrak{r}(x) \leq \bar{r}+1,\ \mathfrak{u}(x) \geq \bar{\mu}-1\} \overset{(2.3.4)}{=} \mathcal{B}_A(\mathfrak{r} \leq \bar{r}+1; \mathfrak{u} \geq \bar{\mu}-1)$$

which is a compact set according to Assumption 2.64 (iv). Hence, a subsequence $x^{(n_k)}$, $k \in \mathbb{N}$, is convergent to say $\widetilde{x} := \lim_{k\to\infty} x^{(n_k)} \in A$, since A is closed. Moreover, according to Assumption 2.64 (ii) and (iii), $\mathfrak{r}$ is lower semi-continuous and $\mathfrak{u}$ is upper semi-continuous which gives

$$\mathfrak{r}(\widetilde{x}) \leq \liminf_{k\to\infty} \mathfrak{r}\left(x^{(n_k)}\right) = \liminf_{k\to\infty} r_{n_k} = \bar{r} \tag{2.4.10}$$

and

$$\mathfrak{u}(\widetilde{x}) \geq \limsup_{k\to\infty} \mathfrak{u}\left(x^{(n_k)}\right) = \limsup_{k\to\infty} \mu_{n_k} = \bar{\mu}. \tag{2.4.11}$$

But since $(\bar{r},\bar{\mu}) \in \mathcal{G}_{\text{eff}}$, $\mathfrak{r}(\widetilde{x}) = \bar{r}$ and $\mathfrak{u}(\widetilde{x}) = \bar{\mu}$ follows (see (2.3.10)). Hence, $\widetilde{x}$ is an efficient portfolio realizing the risk-utility value $(\bar{r},\bar{\mu})$, so that by uniqueness

$$\widetilde{x} = X(\bar{r},\bar{\mu}) = \bar{x}.$$

The same argument actually proves via an indirect argument that the whole sequence $x^{(n)}$ necessarily converges to $\bar{x}$ as well. This finishes the proof of (b).

ad(c). The remaining claims immediately follow from the continuity of $\nu_{|I}\colon I \to J$ and $\gamma_{|J}\colon J \to I$ (see Theorem 2.76 (a) and (b)), since concatenation of continuous functions yields again continuous functions. □

Remark 2.80 (Leverage Space)

(a) The classical two-fund separation theorem [108] asserts that all the efficient trade-offs in the Markowitz portfolio theory framework [81] can be derived as the linear combination of a market portfolio *consisting only of risky assets and a risk-free bond. Thus, the trade-off between the expected return (see Sect. 2.1.5, in particular, Definition 2.31 with $\phi = id$) and the risk measured by the covariance (see Remark 2.27) is realized by portfolios lying on a straight line. This result establishes the theoretical foundation for index investing. Of course, the linear structure of the efficient frontier in that case*

crucially relies on the specific setting that the utility function $\mathfrak{u}$ *here is linear and the risk measure* $\mathfrak{r}$ *is the covariance and thus quadratic. Theorem 2.79 shows that although, in general, the linear structure of the efficient frontier is no longer to be expected, the continuous trade-off between general utility and risk functions along a one parameter curve is still possible.*

(b) Trade-off between utility and risk is thus implemented by portfolios $x^* = x^*(r), r \in I$, *and* $y^* = y^*(\mu), \mu \in J$, *from* (2.4.6) *and* (2.4.7), *respectively, which trace out the same curve in portfolio space* $\mathbb{R}^{M+1}$ *(sometimes called "leverage space," cf. Vince [112]). Note that* x^* *and* y^* *both depend on* A *and the risk function* $\mathfrak{r}$ *as well as on the utility function* $\mathfrak{u}$. *This provides a method for systematically selecting portfolios to reduce risk exposure or to increase utility.*

2.4.3 Relationship with the Efficient Frontier

Since the intervals I and J are relevant not only as projections of the efficient frontier to the axes but also in the representation of the efficient portfolios in (2.4.6) and (2.4.7), it is reasonable to ask how I and J can be determined in the applications. Thus, firstly, we define possible endpoints which may or may not be included in I or J. Here and in the remaining section, we will always assume Assumption 2.64.

Definition 2.81 ("Endpoints" of I and J) *For the intervals* $I = I(\mathfrak{r}, \mathfrak{u}; A)$ *and* $J = J(\mathfrak{r}, \mathfrak{u}; A)$ *defined in* (2.3.29) *and* (2.3.30), *as projections of* $\mathcal{G}_{\text{eff}} = \mathcal{G}_{\text{eff}}(\mathfrak{r}, \mathfrak{u}; A)$ *to the axes, we set:*

$$r_{\min} := \inf I = \inf_{(r,\mu)\in\mathcal{G}_{\text{eff}}} \{r\} = \inf\left[\operatorname{dom}(\nu) \cap \operatorname{range}(\gamma)\right], \tag{2.4.12}$$

$$r_{\max} := \sup I = \sup_{(r,\mu)\in\mathcal{G}_{\text{eff}}} \{r\} = \sup\left[\operatorname{dom}(\nu) \cap \operatorname{range}(\gamma)\right], \tag{2.4.13}$$

$$\mu_{\min} := \inf J = \inf_{(r,\mu)\in\mathcal{G}_{\text{eff}}} \{\mu\} = \inf\left[\operatorname{dom}(\gamma) \cap \operatorname{range}(\nu)\right], \tag{2.4.14}$$

$$\mu_{\max} := \sup J = \sup_{(r,\mu)\in\mathcal{G}_{\text{eff}}} \{\mu\} = \sup\left[\operatorname{dom}(\gamma) \cap \operatorname{range}(\nu)\right]. \tag{2.4.15}$$

Note that the representations above in terms of domains and ranges of ν and γ are a consequence of Corollary 2.71. The following proposition connects the above endpoints to the risk and utility function.

Proposition 2.82 (Characterization of "Endpoints" of I and J in Terms of Risk and Utility) *Assume for a set* $A \subset \mathbb{R}^{M+1}$ *of admissible portfolios as well as extended-valued risk and utility functions* $\mathfrak{r}$ *and* $\mathfrak{u}$ *that Assumption 2.64 holds. Then using* $\mathcal{B} := \operatorname{dom}(\mathfrak{u}) \cap \operatorname{dom}(\mathfrak{r}) \subset A$, *the following holds true:*

$$r_{\min} = \inf_{x\in\mathcal{B}} \{\mathfrak{r}(x)\} = \inf\{\mathfrak{r}(x) \,:\, \mathfrak{u}(x) > -\infty,\ x \in A\} < +\infty, \tag{2.4.16}$$

$$\mu_{\max} = \sup_{x\in\mathcal{B}} \{\mathfrak{u}(x)\} = \sup\{\mathfrak{u}(x) \,:\, \mathfrak{r}(x) < +\infty,\ x \in A\} > -\infty, \tag{2.4.17}$$

$$\mu_{\min} = \lim_{r\searrow r_{\min}} \sup\{\mathfrak{u}(x) \,:\, \mathfrak{r}(x) \le r,\ x \in A\} \le \mu_{\max}, \tag{2.4.18}$$

$$r_{\max} = \lim_{\mu\nearrow \mu_{\max}} \inf\{\mathfrak{r}(x) \,:\, \mathfrak{u}(x) \ge \mu,\ x \in A\} \ge r_{\min}. \tag{2.4.19}$$

Proof Firstly, note that $\mathcal{B} = \mathrm{dom}(\mathfrak{u}) \cap \mathrm{dom}(\mathfrak{r})$ is non-empty according to Assumption 2.64 (iv). Moreover, due to Proposition 2.68 (c), $\mathcal{G}_{\mathrm{eff}}$ is non-empty and equals the non-vertical and non-horizontal part of $\partial\mathcal{G} = \partial\mathcal{G}(\mathfrak{r}, \mathfrak{u}; A)$ (cf. Theorem 2.61), in particular $\mathcal{G}_{\mathrm{eff}} \subset \partial\mathcal{G}$.

ad $r_{\min}$. Using (2.3.6) and the convexity of $\mathcal{G}$ due to Proposition 2.58 (a), the horizontal part of $\partial\mathcal{G}$, if it exists, is unbounded to the right. Thus, such horizontal parts of $\partial\mathcal{G}$ are located to the right of $\mathcal{G}_{\mathrm{eff}}$ (see (2.3.10)). Moreover, the vertical part of $\partial\mathcal{G}$ does not change the infimum in the definition of $\mathfrak{r}_{\min}$ such that

$$r_{\min} \overset{(2.4.12)}{=} \inf_{(r,\mu)\in\mathcal{G}_{\mathrm{eff}}} \{r\} = \inf_{(r,\mu)\in\partial\mathcal{G}} \{r\} \overset{\mathcal{G}_{\mathrm{eff}}\neq\emptyset}{<} +\infty \tag{2.4.20}$$

follows. Furthermore, $\mathcal{G} = \mathcal{G}(\mathfrak{r}, \mathfrak{u}; A)$ is not only convex and satisfies (2.3.6), but it is also closed by Proposition 2.58 (c), so that (2.4.16) follows from

$$r_{\min} \overset{(2.4.20)}{=} \inf_{(r,\mu)\in\partial\mathcal{G}} \{r\} = \inf_{(r,\mu)\in\mathcal{G}} \{r\} = \inf_{x\in\mathcal{B}} \{\mathfrak{r}(x)\} < +\infty.$$

The arguments for $\mu_{\max}$ in (2.4.17) are completely symmetric and therefore left to the reader.

ad $\mu_{\min}$. Here we have

$$\mu_{\min} \overset{(2.4.14)}{=} \inf_{(r,\mu)\in\mathcal{G}_{\mathrm{eff}}} \{\mu\} \overset{(2.3.38)}{=} \inf_{r\in I} \{\nu(r)\}.$$

On the other hand, ν is increasing by Proposition 2.68 (b) and $r_{\min} = \inf I$ (cf. (2.4.12)) so that

$$\mu_{\min} = \inf_{r\in I} \{\nu(r)\} = \lim_{r\searrow r_{\min}} \nu(r) \overset{(2.3.14)}{=} \lim_{r\searrow r_{\min}} \sup\{\mathfrak{u}(x) \,:\, \mathfrak{r}(x) \le r,\ x \in A\},$$

which gives (2.4.18). Again, the arguments for proving (2.4.19) are similar and left out. □

Corollary 2.83 ("Endpoint" Properties) *In the situation of Proposition 2.82 with $I = I(\mathfrak{r}, \mathfrak{u}; A)$ and $J = J(\mathfrak{r}, \mathfrak{u}; A)$ defined in (2.3.29) and (2.3.30), we have:*

(a) $r_{\min} \in I$ *if and only if* $\mu_{\min} \in J$.
(b) If $r_{\min} \in I$ *then* $\mu_{\min} = \nu(r_{\min})$ *and* $\gamma(\mu_{\min}) = r_{\min}$.
(c) $r_{\max} \in I$ *if and only if* $\mu_{\max} \in J$.
(d) If $\mu_{\max} \in J$ *then* $r_{\max} = \gamma(\mu_{\max})$ *and* $\nu(r_{\max}) = \mu_{\max}$.
(e) *(i) If* $r_{\min} \in I$ *and* $\mu_{\max} \in J$ *then* $I = [r_{\min}, r_{\max}]$ *and* $J = [\mu_{\min}, \mu_{\max}]$.
(ii) If $r_{\min} \in I$ *and* $\mu_{\max} \notin J$ *then* $I = [r_{\min}, \infty)$ *and* $J = [\mu_{\min}, \mu_{\max})$.
(iii) If $r_{\min} \notin I$ *and* $\mu_{\max} \in J$ *then* $I = (r_{\min}, r_{\max}]$ *and* $J = (-\infty, \mu_{\max}]$.
(iv) If $r_{\min} \notin I$ *and* $\mu_{\max} \notin J$ *then* $I = (r_{\min}, \infty)$ *and* $J = (-\infty, \mu_{\max})$.

Proof *ad (a) and (b).* "$\Rightarrow$" Assume $r_{\min} \in I$. Then due to (2.3.29), there exists some $\bar{\mu} \in \mathbb{R}$ such that $(r_{\min}, \bar{\mu}) \in \mathcal{G}_{\text{eff}} = \mathcal{G}_{\text{eff}}(\mathfrak{r}, \mathfrak{u}; A)$. Hence, with (2.3.30) and Theorem 2.76 (c),

$$\bar{\mu} \in J \quad \text{and} \quad r_{\min} = \gamma(\bar{\mu}) \tag{2.4.21}$$

follow. Since γ is strictly increasing on J due to Theorem 2.76 (b), from $r_{\min} = \inf I = \min I$ we infer $\bar{\mu} = \min J$. Therefore, according to Definition 2.81, $\mu_{\min} = \inf J = \bar{\mu}$ holds. In particular, $\mu_{\min} \in J$ follows as claimed.

Note that this also shows (b), because $r_{\min} = \gamma(\mu_{\min})$ holds by (2.4.21) and with (2.3.37) also $\mu_{\min} = \nu(r_{\min})$ follows.

"$\Leftarrow$" The converse in (a) follows similarly, such that (a) and (b) are proved.

ad (c) and (d). Can be shown as (a) and (b). We therefore leave it to the reader as an exercise.

ad (e). By Definition 2.81 and, since I and J are intervals due to Corollary 2.74 (a), in any case

$$(r_{\min}, r_{\max}) \subset I \quad \text{and} \quad (\mu_{\min}, \mu_{\max}) \subset J \tag{2.4.22}$$

hold. On the other hand, since $r_{\min} = \inf I$ and $r_{\max} = \sup I$, the interval I cannot be larger than the closure of the interval $(r_{\min}, r_{\max})$. Similarly, the interval J cannot be larger than the closure of $(\mu_{\min}, \mu_{\max})$. Thus, in order to show *(e.i)* to *(e.iv)*, it suffices to investigate the possible endpoints.

ad (e.i). Here by assumption $r_{\min} \in I$ and $\mu_{\max} \in J$. Using (a) and (c), we moreover get $\mu_{\min} \in J$ and $r_{\max} \in I$, respectively. Therefore, with the above preliminaries,

$$I = [r_{\min}, r_{\max}] \quad \text{and} \quad J = [\mu_{\min}, \mu_{\max}] \tag{2.4.23}$$

follow.

ad (e.ii). Assume $r_{\min} \in I$ and $\mu_{\max} \notin J$. Then with (a) and (c), $\mu_{\min} \in J$ and $r_{\max} \notin I$. Thus, the above preliminaries imply

$$I = [r_{\min}, r_{\max}) \quad \text{and} \quad J = [\mu_{\min}, \mu_{\max}). \tag{2.4.24}$$

It therefore remains to show that

$$r_{\max} \stackrel{!}{=} \infty. \tag{2.4.25}$$

Assume for the contrary that $r_{\max} < \infty$. We intend to construct an efficient portfolio $\widetilde{x}$ with risk-utility value $(r_{\max}, \mu_{\max}) \in \mathbb{R}^2$ such that $\mu_{\max}$ would be in J, which is false by assumption.

Firstly, note that $r_{\max} \in \mathbb{R}$ since $r_{\max} = \sup I$ and $I \neq \emptyset$ as projection of $\mathcal{G}_{\text{eff}} \neq \emptyset$ to the r-axis (cf. (2.3.29) and Proposition 2.68 (c)). Similarly, $J \neq \emptyset$ and thus $\mu_{\min} < \mu_{\max}$ since $\mu_{\max} \notin J$. Hence, with $\mu_{\max} = \sup J \in \mathbb{R} \cup \{+\infty\}$, we can choose a sequence $(\mu_n)_{n \in \mathbb{N}} \subset J$ with $\mu_n \nearrow \mu_{\max}$ as $n \to \infty$. Setting $r_n := \gamma(\mu_n) \in I$, we get from Theorem 2.76 (c) that $(r_n, \mu_n) \in \mathcal{G}_{\text{eff}}$. Hence, there exist $x^{(n)} \in \operatorname{dom}(\mathfrak{r}) \cap \operatorname{dom}(\mathfrak{u}) \subset A$, $n \in \mathbb{N}$, with

$$\mathfrak{r}\left(x^{(n)}\right) = r_n \leq r_{\max} \quad \text{and} \quad \mathfrak{u}\left(x^{(n)}\right) = \mu_n \geq \mu_{\min}. \tag{2.4.26}$$

Since $r_{\max}$ and $\mu_{\min}$ are finite, this means $x^{(n)} \in \mathcal{B}_A(\mathfrak{r} \leq r_{\max}; \mathfrak{u} \geq \mu_{\min})$ for all $n \in \mathbb{N}$ (see (2.3.4)). The latter sub- and superlevel set, however, is compact due to Assumption 2.64 (iv).

Hence, a subsequence (w.l.o.g. the sequence itself) must converge, say $x^{(n)} \to \widetilde{x} \in A$ as $n \to \infty$, since A is closed by Assumption 2.64 (i). Moreover, according to Assumption 2.64 (ii) and (iii), $\mathfrak{r}\colon A \to \mathbb{R} \cup \{+\infty\}$ is lower semi-continuous and $\mathfrak{u}\colon A \to \mathbb{R} \cup \{-\infty\}$ is upper semi-continuous giving

$$\mathfrak{r}(\widetilde{x}) \leq \liminf_{n \to \infty} \mathfrak{r}\left(x^{(n)}\right) \overset{(2.4.26)}{\leq} r_{\max} < +\infty \tag{2.4.27}$$

and

$$\mathfrak{u}(\widetilde{x}) \geq \limsup_{n \to \infty} \mathfrak{u}\left(x^{(n)}\right) \overset{(2.4.26)}{=} \lim_{n \to \infty} \mu_n = \mu_{\max}. \tag{2.4.28}$$

But since by (2.4.17),

$$\mu_{\max} = \sup \{\mathfrak{u}(x) \,:\, \mathfrak{r}(x) < \infty,\ x \in A\}, \tag{2.4.29}$$

necessarily $\mathfrak{u}(\widetilde{x}) = \mu_{\max}$ must hold, in particular $\mu_{\max} < \infty$. At last, we claim that $\widetilde{x}$ is indeed an efficient portfolio realizing the risk-utility value:

$$(r_{\max}, \mu_{\max}) \overset{!}{\in} \mathcal{G}_{\text{eff}}. \tag{2.4.30}$$

Then as outlined before, $\mu_{\max} \in J$, which is a contradiction. So all that is left to show is (2.4.30), since then $\mathfrak{r}(\widetilde{x}) = r_{\max}$ immediately follows. To see (2.4.30), note that by (2.4.29), there cannot exist any $x \in A$ with

$$\mathfrak{u}(x) > \mu_{\max} \quad \text{and} \quad \mathfrak{r}(x) \leq r_{\max} < \infty. \tag{2.4.31}$$

If on the other hand $(r_{\max}, \mu_{\max})$ were not on the efficient frontier, then due to (2.4.31) there had to exist some $\bar{x} \in A$ with

$$\mathfrak{u}(\bar{x}) = \mu_{\max} \quad \text{and} \quad \mathfrak{r}(\bar{x}) < r_{\max}, \tag{2.4.32}$$

which in turn yields

$$\gamma(\mu_{\max}) \overset{(2.3.15)}{=} \inf\{\mathfrak{r}(x) : \mathfrak{u}(x) \geq \mu_{\max},\ x \in A\} \overset{(2.4.32)}{<} r_{\max} \overset{(2.4.19)}{=} \lim_{\mu \nearrow \mu_{\max}} \gamma(\mu) \overset{\gamma \text{ increasing}}{\leq} \gamma(\mu_{\max}),$$

a contradiction! This concludes the proof of *(e.ii)*. The remaining claims *(e.iii)* and *(e.iv)* follow similarly and are therefore left to the reader. □

Remark 2.84 ("Endpoints" of Efficient Frontier and Closedness) *In particular, it follows from Corollary 2.83 that the efficient frontier is either (one-sided) unbounded, or, if it is (one-sided) bounded, then the endpoint is included in* $\mathcal{G}_{\text{eff}}$. *Thus, as graph of a continuous function,* $\mathcal{G}_{\text{eff}}$ *is a closed subset of* $\mathbb{R}^2$.

2.4.4 Examples of Efficient Frontiers for Scalar Risk

Typically, in applications, the risk function is non-negative, i.e., $\mathfrak{r}(x) \geq 0$ for all $x \in A$. Then clearly $r_{\min} \geq 0$ is finite. $\mu_{\min}$, however, may also be finite or attain $-\infty$. For instance, finite $\mu_{\min}$ and $r_{\min} > 0$ is attained by the minimum variance portfolio which arises as Markowitz efficient portfolio, where $\mathfrak{r}(x) = \widehat{\mathfrak{r}}(\widehat{x}) = \sigma^2\left(\widehat{S}_1^\top \widehat{x}\right)$ and $\mathfrak{u}(x) = \widehat{\mathfrak{u}}(\widehat{x}) = \mathrm{E}\left(\widehat{S}_1^\top \widehat{x}\right)$, $x \in A$, for a one-period financial market as in Definition 2.3 (see Fig. 2.5a). Details can be found in Maier-Paape and Zhu [78, Section 4].

In that paper, also examples for $\mu_{\min} = -\infty$ as well as for $\mu_{\max}$ finite and not attained are constructed (cf. Fig. 2.5b). Details can be found in [78, Examples 1 and 2].

Another important example occurs for the expected utility (cf. Sect. 2.1.5):

$$\mathfrak{u}(x) = \mathrm{E}\left[\phi\left(S_1^\top x\right)\right], \quad x \in A,$$

where the admissible portfolios A have unit initial cost and the auxiliary function $\phi\colon \mathbb{R} \to \mathbb{R} \cup \{-\infty\}$ satisfies $\phi(\varphi_0) > -\infty$. Here, φ_0 is the return of the risk-free bond of a one-period financial market (cf. again Definition 2.3). If $\mathfrak{u}$, A, and a risk function $\mathfrak{r}$ satisfy the conditions of Theorem 2.79 (yielding unique efficient portfolios) and additionally (r1n) in Assumption 2.20 holds, then one can easily verify that

$$r_{\min} = 0, \ \mu_{\min} = \phi(\varphi_0) \ \text{and} \ x_{\min} := X(r_{\min}, \mu_{\min}) = (1, \widehat{0}^\top)^\top \in A.$$

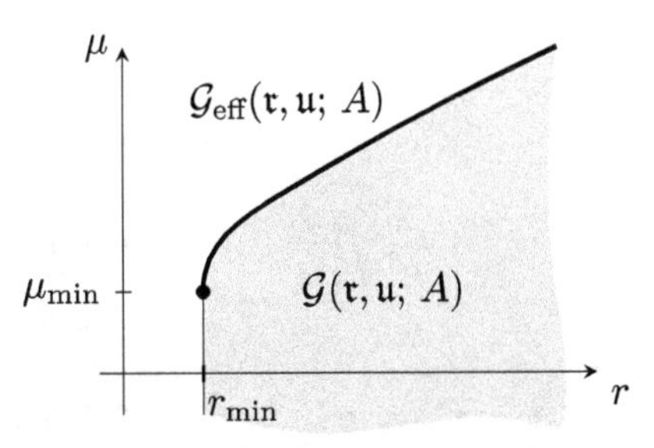

(a) Efficient frontier where both $r_{\min}$ and $\mu_{\min}$ are finite and attained.

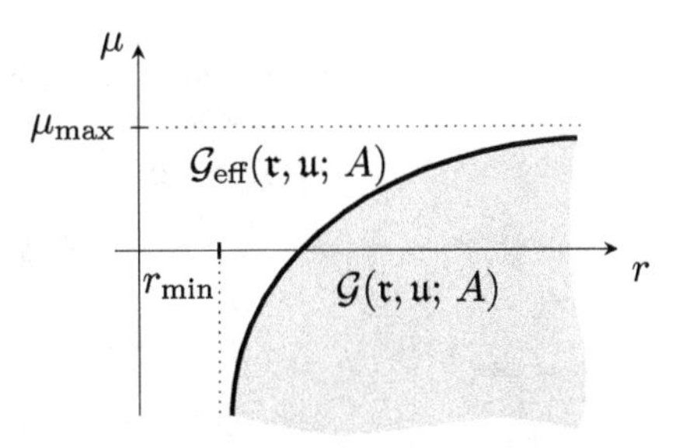

(b) Efficient frontier with $\mu_{\min} = -\infty$ and $\mu_{\max}$ finite but not attained.

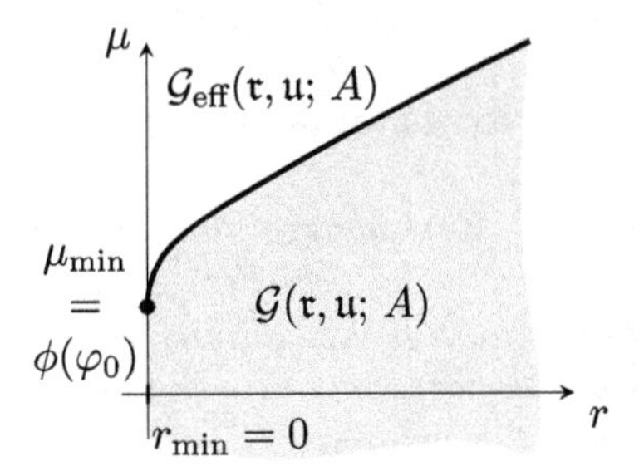

(c) Efficient frontier with $(1, \hat{0}^\top)^\top \in A$.

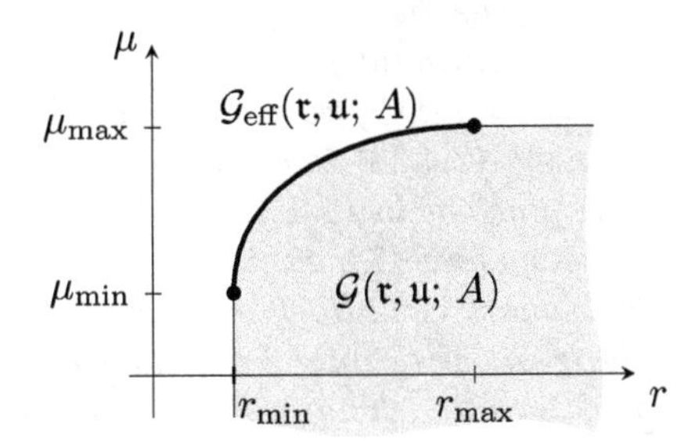

(d) Efficient frontier when $r_{\min} > 0$ and $\mu_{\max}$ is finite and attained as maximum.

Fig. 2.5 Examples for efficient frontiers

The efficient frontier in this case looks like in Fig. 2.5c. Such a situation happens in the capital asset pricing model, where the efficient frontier is even a ray starting in $(\mathfrak{r}_{\min}, \mu_{\min}) = (0, \phi(\varphi_0))$ (see [78, Theorem 8]).

Finally, an example with $\mu_{\max} < \infty$ finite and attained by an unique efficient portfolio $x_{\max} := X\Big(\gamma(\mu_{\max}), \mu_{\max}\Big) \in A$ happens in growth optimal portfolio theory, where

$$\mathfrak{u}(x) = \mathrm{E}\Big[\ln\left(S_1^\top x\right)\Big], \quad x \in A,$$

is used as utility function. Here, $x_{\max} \in A$ solves the optimization problem:

$$\mathrm{E}\Big[\ln\left(S_1^\top x\right)\Big] \stackrel{!}{=} \max, \quad x \in \mathrm{dom}(\mathfrak{r}) \subset A.$$

For details, confer Section 6 in [78] (see Fig. 2.5d for an illustration).

Example 2.85 (Markowitz Portfolios with the Close-to-Benchmark Constraint; Tracking Error Control) *We discuss the linear-quadratic optimization problem:*

$$\frac{1}{2}\sigma^2 = \min_{\widehat{x}\in\mathbb{R}^M} \frac{1}{2}\widehat{x}^\top \Sigma \widehat{x} \tag{2.4.33}$$

subject to

$$\lambda_1 : \quad \mathrm{E}\left[\widehat{S}_1^\top \widehat{x}\right] \geq \mu \tag{2.4.34}$$

$$\lambda_2 : \quad \widehat{S}_0^\top \widehat{x} = 1 \tag{2.4.35}$$

$$\lambda_3 : \quad \frac{1}{2}\left(\widehat{x} - \widehat{x}^*\right)^\top \Sigma \left(\widehat{x} - \widehat{x}^*\right) \leq \varrho, \tag{2.4.36}$$

where $\Sigma \in \mathbb{R}^{M\times M}$ is the positive definite covariant matrix for a financial market $S_\bullet$ as in Definition 2.3 and $\mu \in \mathbb{R}$ is a parameter. Furthermore, let a fixed constant $\varrho > 0$ and a fixed benchmark portfolio $\widehat{x}^* \in \mathbb{R}^M_{>0}$ (e.g., $\widehat{x}^* = \frac{1}{M}(1,\ldots,1)^\top$) be given, such that the intersection of the sets described by (2.4.34)–(2.4.36) *is not empty.*

The problem in (2.4.33)–(2.4.35) *is the usual Markowitz minimum variance setup (see [78, Section 4]) combined with* (2.4.36) *which is a "close-to-benchmark condition" to guarantee diversification. Such solutions are particularly of importance for fund managers whose performance often is seen related to a benchmark. Note that the standard deviation of $\widehat{x} - \widehat{x}^*$, i.e., $\sqrt{(\widehat{x}-\widehat{x}^*)^\top \Sigma (\widehat{x}-\widehat{x}^*)}$ (which is obviously related to* (2.4.36)*), is sometimes called "tracking error" in the literature (see, e.g., [41]).*

The portfolio set related to (2.4.33)–(2.4.36)*, namely,*

$$A = A_\varrho := \left\{ \left(0, \widehat{x}^\top\right)^\top \in \mathbb{R}^{M+1} : \widehat{S}_0^\top \widehat{x} = 1 \text{ and } \frac{1}{2}\left(\widehat{x} - \widehat{x}^*\right)^\top \Sigma \left(\widehat{x} - \widehat{x}^*\right) \leq \varrho \right\}, \tag{2.4.37}$$

is closed, non-empty, and convex, i.e., admissible according to Definition 2.7. It is moreover not hard to see that all conditions of the main existence and uniqueness theorem (Theorem 2.79) are satisfied which already yields (for $\varrho > 0$ and $\widehat{x}^* \in \mathbb{R}^M_{>0}$ fixed) unique solutions $\widehat{x} = \widehat{x}_{\text{opt}}(\mu)$ of (2.4.33)–(2.4.36) *for all $\mu \in J$. The goal here, however, is to get explicit formula for these solutions.*

To get there, we apply the Lagrange multiplier theorem (Theorem A.22). The Lagrangian for (2.4.33)–(2.4.36) *is*

$$L(\widehat{x}, \lambda) = \frac{1}{2}\widehat{x}^\top \Sigma \widehat{x} + \lambda_1 \left(\mu - \widehat{x}^\top \mathrm{E}\left[\widehat{S}_1\right]\right) + \lambda_2 \left(1 - \widehat{x}^\top \widehat{S}_0\right) + \lambda_3 \left(\frac{1}{2}\left(\widehat{x} - \widehat{x}^*\right)^\top \Sigma \left(\widehat{x} - \widehat{x}^*\right) - \varrho\right),$$

where $\lambda_2 \in \mathbb{R}$, $\lambda_1 \geq 0$, and $\lambda_3 \geq 0$ are the Lagrange multipliers. At the optimal solution $\widehat{x} = \widehat{x}_{\text{opt}}$ of (2.4.33)–(2.4.36)*, we must have $0 = \nabla_{\widehat{x}} L$, which gives*

$$\widehat{x}_{\text{opt}} = \widehat{x}_{\text{opt}}(\mu) = \frac{1}{1+\lambda_3}\Sigma^{-1}\left(\lambda_1 \operatorname{E}\left[\widehat{S}_1\right] + \lambda_2 \widehat{S}_0 + \lambda_3 \Sigma \widehat{x}^*\right). \tag{2.4.38}$$

The dependence on μ stems from the fact that $(\lambda_1, \lambda_2, \lambda_3)$ depend implicitly on μ. In the following, we want to assume that

$$\lambda_1 > 0 \quad \textit{and} \quad \lambda_3 > 0, \tag{2.4.39}$$

i.e., (2.4.34) and (2.4.36) are binding for $\widehat{x}_{\text{opt}}$. We firstly claim that in this case

$$\sigma^2 = \sigma^2(\mu) = \widehat{x}_{\text{opt}}^\top \Sigma \widehat{x}_{\text{opt}} \stackrel{!}{=} \frac{2}{2+\lambda_3}\left[\lambda_1 \mu + \lambda_2 + \frac{\lambda_3}{2}\left(\widehat{x}^*\right)^\top \Sigma \widehat{x}^* - \lambda_3 \varrho\right] \tag{2.4.40}$$

holds. Binding means that (2.4.34) and (2.4.36) are satisfied by $\widehat{x}_{\text{opt}}(\mu)$ with equality. Together with (2.4.38), we hence obtain

$$\sigma^2(\mu) = \widehat{x}_{\text{opt}}^\top \Sigma \widehat{x}_{\text{opt}} = \frac{1}{1+\lambda_3} \cdot \left[\lambda_1 \mu + \lambda_2 + \lambda_3 \widehat{x}_{\text{opt}}^\top \Sigma \widehat{x}^*\right]. \tag{2.4.41}$$

From (2.4.36) binding, we get with $\|\widehat{x}\|_\Sigma := \sqrt{\widehat{x}^\top \Sigma \widehat{x}}$ (which is a norm on $\mathbb{R}^M$):

$$\widehat{x}_{\text{opt}}^\top \Sigma \widehat{x}^* = \frac{1}{2}\left\|\widehat{x}_{\text{opt}}\right\|_\Sigma^2 + \frac{1}{2}\left\|\widehat{x}^*\right\|_\Sigma^2 - \varrho. \tag{2.4.42}$$

Putting this into (2.4.41) yields (2.4.40). Next, we want to calculate equations for λ_1, λ_2, and λ_3. For this, it is convenient to use constants α, β, and γ defined by

$$\alpha := \operatorname{E}\left[\widehat{S}_1\right]^\top \Sigma^{-1} \operatorname{E}\left[\widehat{S}_1\right] \in \mathbb{R}^+ \tag{2.4.43}$$

$$\beta := \operatorname{E}\left[\widehat{S}_1\right]^\top \Sigma^{-1} \widehat{S}_0 \in \mathbb{R} \tag{2.4.44}$$

$$\gamma := \widehat{S}_0^\top \Sigma^{-1} \widehat{S}_0 \in \mathbb{R}^+. \tag{2.4.45}$$

Multiplying (2.4.38) from the left with $\operatorname{E}\left[\widehat{S}_1^\top\right]$ and $\widehat{S}_0^\top$, respectively, we obtain two linear equations:

$$\mu = \lambda_1 \alpha + \lambda_2 \beta + \lambda_3 \left(\operatorname{E}\left[\widehat{S}_1^\top\right] \widehat{x}^* - \mu\right), \tag{2.4.46}$$

$$1 = \lambda_1 \beta + \lambda_2 \gamma + \lambda_3 \left(\widehat{S}_0^\top \widehat{x}^* - 1\right). \tag{2.4.47}$$

Therefore, we obtain from (2.4.46) and (2.4.47) that

$$\begin{bmatrix}\lambda_1\\ \lambda_2\end{bmatrix} = C^{-1}\begin{bmatrix}\mu + \lambda_3\mu - \lambda_3 \mathrm{E}\left[\widehat{S}_1^\top\right]\widehat{x}^*\\ 1+\lambda_3-\lambda_3\widehat{S}_0^\top\widehat{x}^*\end{bmatrix} =: \begin{bmatrix}a_1+b_1\lambda_3\\ a_2+b_2\lambda_3\end{bmatrix}. \tag{2.4.48}$$

where the matrix $C := \begin{bmatrix}\alpha & \beta\\ \beta & \gamma\end{bmatrix} \in \mathbb{R}^{2\times 2}$ *is invertible in case* $\mathrm{E}\left[\widehat{S}_1\right]$ *and* $\widehat{S}_0$ *are linearly independent, since* Σ^{-1} *is positive definite as well. With this additional assumption, we see that* λ_1 *and* λ_2 *in fact depend affine linearly on* λ_3.

Using that (2.4.36) *is binding, we can finally derive a quadratic equation for* λ_3. *In fact, using* (2.4.38) *once more, together with the constants from* (2.4.43)–(2.4.45), *we obtain after some calculation*

$$\begin{aligned}2\left(1+\lambda_3\right)^2\varrho = \lambda_1^2\alpha + \lambda_2^2\gamma + \left(\widehat{x}^*\right)^\top \Sigma\widehat{x}^*\\ + 2\lambda_1\lambda_2\beta - 2\lambda_1\left(\widehat{x}^*\right)^\top \mathrm{E}\left[\widehat{S}_1\right] - 2\lambda_2\left(\widehat{x}^*\right)^\top\widehat{S}_0.\end{aligned} \tag{2.4.49}$$

Due to (2.4.48), *we can replace* λ_1 *and* λ_2 *by* $a_i + b_i\lambda_3$, $i \in \{1,2\}$, *and get a quadratic equation for* λ_3, *which is indeed solvable, since all other data are known. Note that the existence of a solution was already a priori clear from Theorem 2.79. One only has to check whether the assumption made in* (2.4.39) *holds true for the calculated* λ_1 *and* λ_3. *With* λ_1, λ_2, *and* λ_3 *determined, at last an explicit formula for* $\widehat{x} = \widehat{x}_{\mathrm{opt}}(\mu)$ *follows from* (2.4.38).

Remark 2.86 *In case the ansatz with* (2.4.34) *and* (2.4.36) *binding does not reveal a suitable solution, at least one of the inequalities in* (2.4.34) *and* (2.4.36) *is irrelevant and can be ignored. This problem is even simpler to solve and therefore left to the reader. In fact, if, for instance, the benchmark constraint* (2.4.36) *is not binding, and thus ignored, then the remaining problem is the standard Markowitz problem, i.e.,* (2.4.33)–(2.4.35) *(cf. Section 4 in Maier-Paape and Zhu [78]). According to [78, Theorem 6], in case* $\widehat{S}_0$ *and* $\mathrm{E}\left[\widehat{S}_1^\top\right]$ *are linearly independent, the Markowitz efficient portfolios lie on a ray* $Y \subset \{\widehat{S}_0^\top\widehat{x} = 1\} \subset \mathbb{R}^M$ *which starts at the minimum variance portfolio* $\widehat{x}_{\mathrm{minVar}} \in \mathbb{R}^M$ *which realizes a minimal possible risk in* $\{\widehat{S}_0^\top\widehat{x} = 1\}$, *given by*

$$\sigma_{\mathrm{minVar}} := \sigma(\widehat{S}_1^\top\widehat{x}_{\mathrm{minVar}}) = \sqrt{\widehat{x}_{\mathrm{minVar}}^\top\Sigma\widehat{x}_{\mathrm{minVar}}} > 0 \tag{2.4.50}$$

and

$$\mu_{\mathrm{minVar}} := \mathrm{E}\left[\widehat{S}_1^\top\widehat{x}_{\mathrm{minVar}}\right] \in \mathbb{R}. \tag{2.4.51}$$

is the realized expected return. So if, for instance,

$$\widehat{x}_{\mathrm{minVar}} \in D_\varrho := \left\{\widehat{x}\in\mathbb{R}^M : \frac{1}{2}\left(\widehat{x}-\widehat{x}^*\right)^\top\Sigma\left(\widehat{x}-\widehat{x}^*\right) \le \varrho\right\}, \tag{2.4.52}$$

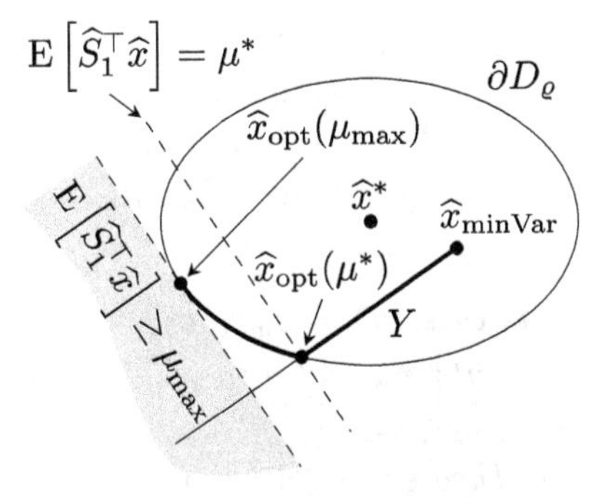

(a) minVar portfolio and ray of Markowitz efficient portfolios projected to the hyperplane $\{\widehat{S}_0^\top \widehat{x} = 1\}$.

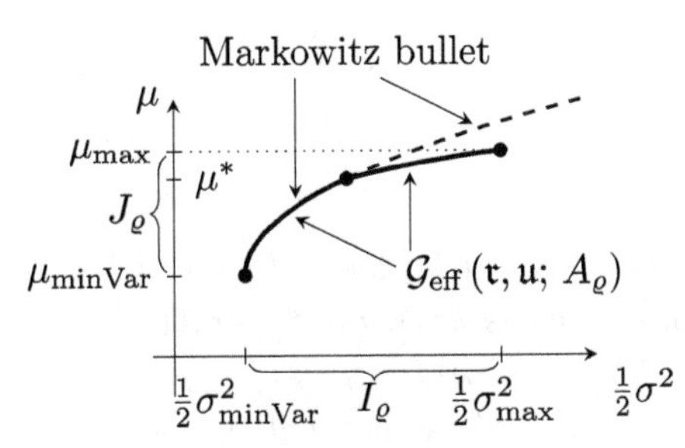

(b) Markowitz bullet and part which is efficient for (2.4.33) – (2.4.36) for $\mathfrak{r}\colon \widehat{x} \mapsto \frac{1}{2}\sigma^2\left(\widehat{S}_1^\top \widehat{x}\right)$, and $\mathfrak{u}\colon \widehat{x} \mapsto \mathrm{E}\left[\widehat{S}_1^\top \widehat{x}\right]$.

Fig. 2.6 Illustrations for Remark 2.86

then this ray Y crosses ∂D_ϱ in exactly one point which turns out to be a solution of (2.4.33)–(2.4.36) *with* (2.4.34) *and* (2.4.36) *binding, i.e.,* $\widehat{x}_{\text{opt}}(\mu^*)$ *for some suitable unique $\mu^* = \mu^*(\varrho)$ (see Fig. 2.6a). Since points on the line segment connecting $\widehat{x}_{\text{minVar}}$ and $\widehat{x}_{\text{opt}}(\mu^*)$ all represent Markowitz efficient portfolios for $\mu \in [\mu_{\text{minVar}}, \mu^*]$, these portfolios are also efficient for* (2.4.33)–(2.4.36), *because all these portfolios are contained in*

$$A = A_\varrho = \left\{ \left(0, \widehat{x}^\top\right)^\top \in R^{M+1} \,:\, \widehat{S}_0^\top \widehat{x} = 1,\; \widehat{x} \in D_\varrho \right\} \tag{2.4.53}$$

from (2.4.37) *as well (see Lemma 2.62). In Fig. 2.6b, we see the Markowitz bullet, i.e., the efficient frontier for* (2.4.33)–(2.4.35) *when* (2.4.36) *is ignored. As argued above, the bold part of the Markowitz bullet which is parameterized by $\mu \in [\mu_{\text{minVar}}, \mu^*]$ coincides with the efficient frontier of* (2.4.33)–(2.4.36).

Furthermore, for $\mu < \mu_{\text{minVar}}$ and $\sigma < \sigma_{\text{minVar}}$, there cannot be any efficient portfolios for (2.4.33)–(2.4.36) *since by [78, Theorem 6] there has to be exactly one portfolio in the set*

$$\left\{ \widehat{x} \in \mathbb{R}^M \,:\, \widehat{S}_0^\top \widehat{x} = 1 \quad \text{and} \quad \frac{1}{2}\widehat{x}^\top \Sigma \widehat{x} \leq \frac{1}{2}\sigma^2_{\text{minVar}} \right\}, \tag{2.4.54}$$

which is $\widehat{x}_{\text{minVar}}$. Otherwise, $\widehat{x}_{\text{minVar}}$ were not the unique minimum variance solution. One the other hand, setting as in (2.4.17)

$$\mu_{\max} = \mu_{\max}(\varrho) := \sup \left\{ \mathrm{E}\left[\widehat{S}_1^\top \widehat{x}\right] \,:\, (0, \widehat{x}^\top)^\top \in A_\varrho \right\} \in \mathbb{R}, \tag{2.4.55}$$

we obtain that $\widehat{x}_{\text{opt}} = \widehat{x}_{\text{opt}}(\mu)$, $\mu \in (\mu^*, \mu_{\max}]$ give other efficient portfolios for (2.4.33)–(2.4.36), *all off which have* (2.4.34) *and* (2.4.36) *binding, but com-*

pared to the Markowitz efficient portfolios for that μ, *they clearly must have higher risk (see again Fig. 2.6). Finally, we obtain* $J = J_\varrho = [\mu_{\text{minVar}}, \mu_{\max}]$ *for* (2.4.33)–(2.4.36) *since both endpoints of the efficient frontier are attained by efficient portfolios. Similarly, setting* $\sigma_{\max} = \sigma_{\max}(\varrho) := \sigma(\widehat{S}_1^\top \widehat{x}_{\text{opt}}(\mu_{\max}))$, *we have* $I = I_\varrho = \left[\frac{1}{2}\sigma^2_{\text{minVar}}, \frac{1}{2}\sigma^2_{\max}\right]$.

It should furthermore be noted that in some cases $\mu^* = \mu_{\max}$ *is possible. This happens when the hyperplane* $\{\mathrm{E}\left[\widehat{S}_1^\top \widehat{x}\right] = \mu^*\}$ *restricted to* $\{\widehat{S}_0^\top \widehat{x} = 1\}$ *is tangent to* ∂D_ϱ. *In this case, the efficient frontier* $\mathcal{G}_{\text{eff}}(\mathfrak{r}, \mathfrak{u}; A_\varrho)$ *is just a part of the Markowitz bullet. Moreover, from Theorem 2.76, we obtain that* $\mathcal{G}_{\text{eff}}(\mathfrak{r}, \mathfrak{u}; A_\varrho)$ *may be represented as graphs of strictly increasing functions* $\gamma_\varrho\colon J_\varrho \to I_\varrho$ *and* $\nu_\varrho\colon I_\varrho \to J_\varrho$, *where, for instance,*

$$\gamma_\varrho(\mu) := \inf\left\{\frac{1}{2}\widehat{x}^\top \Sigma \widehat{x} \,:\, \mathrm{E}\left[\widehat{S}_1^\top \widehat{x}\right] \geq \mu,\ \left(0, \widehat{x}^\top\right)^\top \in A_\varrho\right\}. \tag{2.4.56}$$

Remark 2.87 *Though not originally intended for, the ellipsoidal benchmark constraint (tracking error constraint) could not only be included into the set* A *(as in* (2.4.37)*), but it can as well be considered as a second risk constraint. While in the first case the theory for scalar risk functions as developed in this chapter applies, in the second case we have to deal with multiple risks as will be discussed in the next chapter. For this vector-risk application, the set*

$$A := \left\{\left(0, \widehat{x}^\top\right)^\top \in \mathbb{R}^{M+1} \,:\, \widehat{S}_0^\top \widehat{x} = 1\right\}$$

even is simpler as in (2.4.37). *However,* (2.4.36) *establishes a new risk constraint with a true parameter* $\varrho > 0$. *In fact,* (2.4.33)–(2.4.36) *now involves two risk functions:*

$$\begin{aligned} \mathfrak{r}_1(x) &= \mathfrak{r}_{\text{Var}}(x) = \widehat{\mathfrak{r}}_{\text{Var}}(\widehat{x}) := \frac{1}{2}\widehat{x}^\top \Sigma \widehat{x} \qquad \text{and} \\ \mathfrak{r}_2(x) &= \mathfrak{r}_{\text{Track}}(x) = \widehat{\mathfrak{r}}_{\text{Track}}(\widehat{x}) := \frac{1}{2}\left(\widehat{x} - \widehat{x}^*\right)^\top \Sigma \left(\widehat{x} - \widehat{x}^*\right). \end{aligned} \tag{2.4.57}$$

Thus, from this viewpoint, (2.4.33)–(2.4.36) *is a two-parameter problem for given* (μ, ϱ). *Although the calculations in Example 2.85 now also depend on* ϱ *(e.g.,* $\widehat{x}_{\text{opt}} = \widehat{x}_{\text{opt}}(\mu, \varrho)$*), the formulas given there are still valid. However, the question of efficiency for* $\widehat{x}_{\text{opt}}(\mu, \varrho)$ *is now due to the two parameters more complicated as we will see later on. Thus, in essence, this example is a natural link between Chaps. 2 and 3 (cf. Example 3.50).*

Chapter 3
Efficient Portfolios for Vector Risk Functions

In the last chapter, we presented a thorough outline of the general framework of portfolio theory for abstract, scalar-valued risk and utility functions. In many applications, however, it is essential to control numerous different kinds of risk. Motivating examples in this regard are presented in Sect. 1.1 and related solutions using methods developed here will be discussed later in Chap. 4.

Accordingly, we develop in this chapter the general framework for abstract, but vector-valued risk functions as well. Apart from that, as in Chap. 2, we allow as underlying financial market S_t the one-period case (cf. Definition 2.3) as well as the multi-period case (cf. Definition 2.35).

We furthermore assume that risk and utility are given as abstract extended-valued functions defined on a set of admissible portfolios $A \subset \mathbb{R}^{M+1}$ (see Definition 2.7). As before, portfolios $x = \left(x_0, \widehat{x}^\top\right)^\top \in A$ contain a risk-free component $x_0 \in \mathbb{R}$ and $M \in \mathbb{N}$ risky asset components $\widehat{x} = (x_1, \ldots, x_M)^\top \in \mathbb{R}^M$. Similarly as in Chap. 2, we sometimes need portfolios $x \in A$ with unit initial cost, i.e., we then have

$$A \subset \mathcal{A}_1 = \left\{x \in \mathbb{R}^{M+1} \,:\, S_0^\top x = 1\right\} \tag{3.0.1}$$

(see Definition 2.6 and Definition 2.7).

We here develop the theory of the efficient frontier (Sect. 3.1) and efficient portfolios (Sect. 3.3) for the vector risk case. Particular emphasis is put on proving the connectedness of the efficient frontier (see Sect. 3.2), which turns out to be far harder than the proof in the scalar risk case (cf. Lemma 2.72).

S. Maier-Paape et al., *Scalar and Vector Risk in the General Framework of Portfolio Theory*, CMS/CAIMS Books in Mathematics 9,
https://doi.org/10.1007/978-3-031-33321-7_3

3.1 Efficient Frontier for the Vector Risk-Utility Trade-Off

The efficient frontier constructed and discussed for the scalar risk case in Sect. 2.3 is the set of risk-utility values of all efficient portfolios (see Definition 2.64). This is true for vector-valued risk functions as well. Accordingly, we here develop similar concepts for the vector-valued risk case.

Thus, we search for portfolios which guarantee an efficient trade-off between a scalar utility and a vector risk function.

3.1.1 Technical Assumptions

The scalar utility function $\mathfrak{u}\colon A \to \mathbb{R} \cup \{-\infty\}$ is defined as in Chap. 2 (see especially Sect. 2.1.5). On the other hand, as already said, the risk function now is vector-valued. However, we can use very similar assumptions as in Sect. 2.1.4.

Assumption 3.1 (Conditions on the Vector Risk Function; see Assumptions 2.20) *Consider a vector risk function* $\mathfrak{r} = (\mathfrak{r}_1, \ldots, \mathfrak{r}_K) : A \to (\mathbb{R} \cup \{+\infty\})^K$, $K \geq 2$, *with lower semi-continuous components* $\mathfrak{r}_k$, *where* $A \subset \mathbb{R}^{M+1}$ *is a set of admissible portfolios according to Definition 2.7.*

We say that $\mathfrak{r}$ *satisfies one (or more) of the conditions* (r1), (r1n), *or* (r2) *if* each *component* $\mathfrak{r}_k$, $k = 1, \ldots, K$, *satisfies* (r1), (r1n), *or* (r2) *from Assumptions 2.20, respectively.*

We say that $\mathfrak{r}$ *satisfies* (r2s) *if* $\mathfrak{r}_k$ *for at* least one $k \in \{1, \ldots, K\}$ *satisfies* (r2s) *from Assumptions 2.20.*

Since these properties of a vector risk function can be broken down to its components, the results in Sect. 2.1.4 still can be used.

To simplify the notation, we use $\leq$ and $\geq$ componentwise, i.e., for two vectors $v, w \in \mathbb{R}^K$, we have $v \leq w$ if and only if $v_i \leq w_i$ for all $i = 1, \ldots, K$ (analogously for the other operator). Hence, the risk-utility space

$$\mathcal{G} = \mathcal{G}^{(Kd)}(\mathfrak{r}, \mathfrak{u}; A) := \{(r, \mu) \in \mathbb{R}^{K+1} : \exists\, x \in A \text{ s.t. } \mathfrak{r}(x) \leq r \text{ and } \mathfrak{u}(x) \geq \mu\} \tag{3.1.2}$$

is written in a similar fashion to (2.3.1) but now it is $K+1$ dimensional.

Notation 3.2 *In order to emphasize the fact that the risk-utility space now corresponds to a vector-valued risk function* $\mathfrak{r} : A \to (\mathbb{R} \cup \{+\infty\})^K$, *we sometimes use the notation* $\mathcal{G} = \mathcal{G}^{(K-dim)}(\mathfrak{r}, \mathfrak{u}; A)$ *or for short* $\mathcal{G} = \mathcal{G}^{(Kd)}(\mathfrak{r}, \mathfrak{u}; A)$ *as in* (3.1.2).

Remark 3.3 (Risk-Utility as Row Vectors) *Here and in the sequel, it is convenient to have risk vectors* $r \in \mathbb{R}^K$ *as row vectors. Similarly, the* *risk-utility vectors* $(r, \mu) \in \mathbb{R}^{K+1}$ *are assumed to be row vectors which is in contrast to our portfolio vectors* $x \in A$ *which are always column vectors.*

The sublevel sets of $\mathfrak{r}$ for a vector $r \in \mathbb{R}^K$ can be written as in Definition 2.53 by

$$\mathcal{B}_A^{(Kd)}(\mathfrak{r} \le r) := \{x \in A : \mathfrak{r}(x) \le r\} = \mathfrak{r}^{-1}((-\infty, r]) \cap A \subset \mathrm{dom}(\mathfrak{r}) \subset \mathbb{R}^{M+1}.$$

Using the superlevel sets $\mathcal{B}_A(\mathfrak{u} \ge \mu)$ per Definition 2.53 of the scalar utility function $\mathfrak{u}$ and its intersection with $\mathcal{B}_A^{(Kd)}(\mathfrak{r} \le r)$, namely,

$$\mathcal{B}_A^{(Kd)}(\mathfrak{r} \le r;\, \mathfrak{u} \ge \mu) := \mathcal{B}_A^{(Kd)}(\mathfrak{r} \le r) \cap \mathcal{B}_A(\mathfrak{u} \ge \mu) \subset A, \tag{3.1.3}$$

gives us the alternative representation of the valid risk and utility values similar to (2.3.5),

$$\mathcal{G} = \mathcal{G}^{(Kd)}(\mathfrak{r}, \mathfrak{u};\, A) = \left\{(r, \mu) \in \mathbb{R}^{K+1} \,:\, \mathcal{B}_A^{(Kd)}(\mathfrak{r} \le r;\, \mathfrak{u} \ge \mu) \ne \emptyset\right\}. \tag{3.1.4}$$

The following compactness assumption will be needed in concrete applications. It is the vector risk analog of Assumption 2.56.

Assumption 3.4 (Compact Level Sets; Vector Risk Case) *We assume*

(a) $\mathrm{dom}(\mathfrak{u}) \cap \mathrm{dom}(\mathfrak{r}) \ne \emptyset$ *and*
(b) for all $(r, \mu) \in \mathbb{R}^{K+1}$, *the sets* $\mathcal{B}_A^{(Kd)}(\mathfrak{r} \le r;\, \mathfrak{u} \ge \mu)$ *from* (3.1.3) *are compact.*

In case only one component of the risk vector is discussed, similarly to Assumption 2.56 sometimes the following assumption is useful.

Assumption 3.5 (Further Compact Level Sets 1; $k \in \{1, \dots, K\}$ fixed) *We assume*

(a) $\mathrm{dom}(\mathfrak{u}) \cap \mathrm{dom}(\mathfrak{r}_k) \ne \emptyset$ *and*
(b) for all $(s, \mu) \in \mathbb{R}^2$, *the sets* $\mathcal{B}_A(\mathfrak{r}_k \le s;\, \mathfrak{u} \ge \mu)$ *from* (2.3.4) *are compact.*

Since Assumption 3.5 depends on $k \in \{1, \dots, K\}$, it in fact stands for K different assumptions as k may vary. Clearly, Assumption 3.5 (b) is stronger than Assumption 3.4 (b), but it is only needed for the discussion of some further (partial) continuity results in Sect. 3.1.3.

Proposition 3.6 (Properties of $\mathcal{G} = \mathcal{G}^{(Kd)}(\mathfrak{r}, \mathfrak{u};\, A)$; see Proposition 2.58) *Assume that* $A \subset \mathbb{R}^{M+1}$ *is a set of admissible portfolios as in Definition 2.7. In addition, assume that the vector risk function* $\mathfrak{r}$ *satisfies (r2) in Assumption 3.1 and the utility function* $\mathfrak{u}$ *satisfies (u2) in Assumption 2.29. We claim:*

(a) Then set $\mathcal{G}^{(Kd)}(\mathfrak{r}, \mathfrak{u};\, A) \subset \mathbb{R}^{K+1}$ *is convex.*

(b) $(r,\mu) \in \mathcal{G}^{(Kd)}(\mathfrak{r},\mathfrak{u};\ A)$ *implies that, for any* $(k,k_0) \in \mathbb{R}^K_{\geq 0} \times \mathbb{R}_{\geq 0}$, $(r+k,\mu) \in \mathcal{G}^{(Kd)}(\mathfrak{r},\mathfrak{u};\ A)$ *and* $(r,\mu-k_0) \in \mathcal{G}^{(Kd)}(\mathfrak{r},\mathfrak{u};\ A)$.

(c) Assume furthermore that Assumption 3.4 (b) holds. Then $\mathcal{G}^{(Kd)}(\mathfrak{r},\mathfrak{u};\ A)$ *is closed.*

Proof *ad(a).* Similar to the proof of Proposition 2.58 (a) and so omitted.

ad(b). The property $(r,\mu) \in \mathcal{G}^{(Kd)}(\mathfrak{r},\mathfrak{u};\ A)$ implies that, for any $(k,k_0) \in \mathbb{R}^K_{\geq 0} \times \mathbb{R}_{\geq 0}$, $(r+k,\mu) \in \mathcal{G}^{(Kd)}(\mathfrak{r},\mathfrak{u};\ A)$ and $(r,\mu-k_0) \in \mathcal{G}^{(Kd)}(\mathfrak{r},\mathfrak{u};\ A)$, which follows directly from the definition of $\mathcal{G}^{(Kd)}(\mathfrak{r},\mathfrak{u};\ A)$ in (3.1.2).

ad(c). Suppose that $(r_n,\mu_n) \to (r,\mu)$ as $n \to \infty$, for a sequence in $\mathcal{G}^{(Kd)}(\mathfrak{r},\mathfrak{u};\ A) \subset \mathbb{R}^{K+1}$. Then there exists a sequence $x^{(n)} \in A$ such that

$$r_n \geq \mathfrak{r}(x^{(n)}) \text{ and } \mu_n \leq \mathfrak{u}(x^{(n)}). \tag{3.1.5}$$

By Assumption 3.4 (b), a subsequence of $x^{(n)}$ (denoted again by $x^{(n)}$) converges to, say, $\bar{x} \in A$, because A is closed. Taking limits in (3.1.5), we arrive at

$$r \geq \mathfrak{r}(\bar{x}) \text{ and } \mu \leq \mathfrak{u}(\bar{x}). \tag{3.1.6}$$

Thus, $(r,\mu) \in \mathcal{G}^{(Kd)}(\mathfrak{r},\mathfrak{u};\ A)$ and hence $\mathcal{G}^{(Kd)}(\mathfrak{r},\mathfrak{u};\ A)$ is a closed set. □

3.1.2 Representation of the Efficient Frontier

This subsection is related to Sect. 2.3.2 where we now discuss the case of the vector risk function. Again, we will consider a set of admissible portfolios $A \subset \mathbb{R}^{M+1}$.

We can represent a portfolio $x \in A$ as a point, $(\mathfrak{r}(x),\mathfrak{u}(x)) \in \mathcal{G}^{(Kd)}(\mathfrak{r},\mathfrak{u};\ A)$ in the $K+1$-dimensional risk-utility space. As in the scalar risk case, here again investors prefer portfolios with lower risk if the utility is the same or with higher utility given the same level of risk. Accordingly, the efficiency concept is now adapted.

Definition 3.7 (Efficient Portfolio; Vector Risk Case, see Definition 2.59) *We say that a portfolio* $x^* \in A$ *with finite risk and utility values is* Pareto efficient *(for vector risk* $\mathfrak{r}$*, scalar utility* $\mathfrak{u}$*, and admissible portfolios* A*) provided that there does not exist any portfolio* $x' \in A$ *such that either*

$$\left[\mathfrak{r}(x') \leq \mathfrak{r}(x^*) \text{ and } \mathfrak{u}(x') > \mathfrak{u}(x^*)\right]$$

or

$$\left[\mathfrak{r}(x') \leq \mathfrak{r}(x^*),\ \mathfrak{r}(x') \neq \mathfrak{r}(x^*) \text{ and } \mathfrak{u}(x') \geq \mathfrak{u}(x^*)\right]$$

holds.

Definition 3.8 (Efficient Frontier; Vector Risk Case, see Definition 2.60) *We call the set of images of all efficient portfolios in the $K+1$-dimensional risk-utility space the* efficient frontier *and denote it by* $\mathcal{G}^{(Kd)}_{\text{eff}} = \mathcal{G}^{(Kd)}_{\text{eff}}(\mathfrak{r}, \mathfrak{u};\, A)$.

The next theorem characterizes efficient portfolios in the risk-utility space similar to Theorem 2.61.

Theorem 3.9 (Efficient Frontier Properties; Vector Risk Case) *Efficient portfolios represented in the $K+1$-dimensional risk-utility space are all located on the boundary $\partial\mathcal{G} = \partial\mathcal{G}^{(Kd)}(\mathfrak{r}, \mathfrak{u};\, A)$ of the set $\mathcal{G}^{(Kd)}(\mathfrak{r}, \mathfrak{u};\, A)$. In case the boundary of $\mathcal{G}^{(Kd)}(\mathfrak{r}, \mathfrak{u};\, A)$ contains a line segment parallel to any of the $K+1$ coordinate axes, on such a line segment lies at most one efficient portfolio.*

Proof Assume an efficient portfolio $x \in A$ is represented in the $K+1$-dimensional risk-utility space as a point $(r, \mu) \in \mathbb{R}^{K+1}$ which is not on the boundary of the set $\mathcal{G}^{(Kd)}(\mathfrak{r}, \mathfrak{u};\, A)$. Then for $\varepsilon > 0$ small enough, we have both $(r, \mu + \varepsilon) \in \mathcal{G}^{(Kd)}(\mathfrak{r}, \mathfrak{u};\, A)$ and $(r - \varepsilon e^k, \mu) \in \mathcal{G}^{(Kd)}(\mathfrak{r}, \mathfrak{u};\, A)$ for some k where $e^k \in \mathbb{R}^K$ is the unit vector with the kth component equals 1. This means x can be improved and thus cannot be efficient.

In case the boundary of $\mathcal{G}^{(Kd)}(\mathfrak{r}, \mathfrak{u};\, A)$ contains a line segment parallel to any of the $K+1$ coordinate axis, then a portfolio x represented by (r, μ) lying on that segment can be similarly improved unless r is minimal (μ is maximal, respectively) among all risk-utility values on that segment. Hence, there is at most one efficient portfolio on each such segment. □

The following relationship is straightforward and similar to Lemma 2.62.

Lemma 3.10 (Efficient Frontier of Subsystem) *Consider sets of admissible portfolios A, B. If $B \subset A$ then $\mathcal{G}^{(Kd)}_{\text{eff}}(\mathfrak{r}, \mathfrak{u};\, A) \cap \mathcal{G}^{(Kd)}(\mathfrak{r}, \mathfrak{u};\, B) \subset \mathcal{G}^{(Kd)}_{\text{eff}}(\mathfrak{r}, \mathfrak{u};\, B)$.*

Proof As in the scalar risk case, also here the conclusion follows directly from $\mathcal{G}^{(Kd)}(\mathfrak{r}, \mathfrak{u};\, B) \subset \mathcal{G}^{(Kd)}(\mathfrak{r}, \mathfrak{u};\, A)$. □

Assumption 3.11 (Standing Assumptions for Efficient Trade-Off; Vector Risk Case) *We assume the following properties:*

(i) $A \subset \mathbb{R}^{M+1}$ *is a set of admissible portfolios according to Definition 2.7.*
(ii) $\mathfrak{r}\colon A \to (\mathbb{R} \cup \{+\infty\})^K$ *is a lower semi-continuous extended-valued vector risk function satisfying* (r2) *in Assumption 3.1.*
(iii) $\mathfrak{u}\colon A \to \mathbb{R} \cup \{-\infty\}$ *is an upper semi-continuous extended-valued scalar utility function satisfying* (u2) *in Assumption 2.29.*
(iv) Assumption 3.4 concerning compact level sets holds for A, $\mathfrak{r}$, and $\mathfrak{u}$.

Similar as in Remark 2.66, by Proposition 3.6 (b), we can view the set $\mathcal{G}^{(Kd)}(\mathfrak{r}, \mathfrak{u};\ A)$ either as a *hypograph* on the space $(r, \mu) \in \mathbb{R}^{K+1}$ or as *epigraph* on the space of $\big((\widehat{r}^k, \mu), r_k\big)$, where $\widehat{r}^k$ is defined as

$$\widehat{r}^k := (r_1, \ldots, r_{k-1}, r_{k+1}, \ldots, r_K) \in \mathbb{R}^{K-1}. \tag{3.1.7}$$

By the characterization of the epigraph and hypograph (see Propositions A.5 and A.7), the set $\mathcal{G}^{(Kd)}(\mathfrak{r}, \mathfrak{u};\ A)$ naturally defines $K+1$ functions mapping $\mathbb{R}^K$ to $\mathbb{R} \cup \{\pm\infty\}$, namely,

$$\nu(r) := \sup\left\{\mu\ :\ (r, \mu) \in \mathcal{G}^{(Kd)}(\mathfrak{r}, \mathfrak{u};\ A) \subset \mathbb{R}^{K+1}\right\} \tag{3.1.8}$$

$$= \sup\{\mathfrak{u}(x) : \mathfrak{r}(x) \leq r, x \in A\}, \qquad r \in \mathbb{R}^K; \tag{3.1.9}$$

see also (2.3.12) and (2.3.14) for the scalar risk case, and for $k = 1, \ldots, K$

$$\gamma_k(\hat{r}^k, \mu) := \inf\left\{r_k\ :\ (r, \mu) \in \mathcal{G}^{(Kd)}(\mathfrak{r}, \mathfrak{u};\ A) \subset \mathbb{R}^{K+1}\right\} \tag{3.1.10}$$

$$= \inf\{\mathfrak{r}_k(x)\ :\ \mathfrak{u}(x) \geq \mu, \mathfrak{r}_i(x) \leq r_i, i \neq k, x \in A\}, \quad (\widehat{r}^k, \mu) \in \mathbb{R}^K, \tag{3.1.11}$$

cf. (2.3.13) and (2.3.15).

The next result is a similar version of Proposition 2.68 and Lemma 2.69.

Proposition 3.12 (Functions Related to the Efficient Frontier; Vector Risk Case) *Assume for a set $A \subset \mathbb{R}^{M+1}$ of admissible portfolios, as well as for extended-valued vector risk and scalar utility functions $\mathfrak{r}$ and $\mathfrak{u}$ that Assumption 3.11 holds. Then*

(a) $\mathcal{G}^{(Kd)}(\mathfrak{r}, \mathfrak{u};\ A) \neq \emptyset$.

(b) $\nu\colon \mathbb{R}^K \to \mathbb{R} \cup \{-\infty\}$ *in* (3.1.8) *and* $\gamma_k\colon \mathbb{R}^K \to \mathbb{R} \cup \{+\infty\}, k = 1, \ldots, K$, *in* (3.1.10) *are well-defined extended-valued functions. Moreover, the function* $(\widehat{r}^k, \mu) \mapsto \gamma_k(\widehat{r}^k, \mu)$ *is decreasing in* $\widehat{r}^k$ *(coordinate-wise) and increasing in* μ *and lower semi-continuous as well as proper convex, and the function* $r \mapsto \nu(r)$ *is increasing (coordinate-wise), upper semi-continuous, and proper concave.*

(c) $\mathcal{G}_{\text{eff}}^{(Kd)}(\mathfrak{r}, \mathfrak{u};\ A) \neq \emptyset$.

(d) Furthermore, for any $(r^*, \mu^*) \in \mathcal{G}_{\text{eff}}^{(Kd)}(\mathfrak{r}, \mathfrak{u};\ A)$, *we have*

$$[r_1^*, \infty) \times \ldots \times [r_K^*, \infty) \subset \operatorname{dom}(\nu) := \{r \in \mathbb{R}^K : \nu(r) > -\infty\}$$

and

$$[r_1^*, \infty) \times \ldots \times [r_{k-1}^*, \infty) \times [r_{k+1}^*, \infty) \times \ldots \times [r_K^*, \infty) \times (-\infty, \mu^*] \subset \operatorname{dom}(\gamma_k),$$

where $\operatorname{dom}(\gamma_k) := \{(\widehat{r}^k, \mu) \in \mathbb{R}^K : \gamma_k(\hat{r}^k, \mu) < \infty\}$.

Proof *ad (a).* Apparently $\mathcal{G}^{(Kd)}(\mathfrak{r}, \mathfrak{u};\, A) \neq \emptyset$, since $\operatorname{dom}(\mathfrak{r}) \cap \operatorname{dom}(\mathfrak{u}) \neq \emptyset$ by Assumption 3.11 (iv).

ad (b). A similar compactness argument as used in the proof of Proposition 2.68 (b) yields

$$\nu(r) < \infty \quad \text{for all} \quad r \in \mathbb{R}^K \text{ and } \gamma_k(\mu) > -\infty \quad \text{for all} \quad \mu \in \mathbb{R} \tag{3.1.12}$$

and $k = 1, \ldots, K$. Hence, $\nu\colon \mathbb{R}^K \to \mathbb{R} \cup \{-\infty\}$ and $\gamma_k\colon \mathbb{R}^K \to \mathbb{R} \cup \{+\infty\}$ are well-defined.

The remaining assertions follow immediately: Using the representations (3.1.9) and (3.1.11), it is obvious that ν is increasing (coordinate-wise) and γ_k is increasing in μ, but decreasing in $\widehat{r}^k$. Moreover, $\mathcal{G}^{(Kd)}(\mathfrak{r}, \mathfrak{u};\, A)$ is closed and convex according to Proposition 3.6. Thus, semi-continuity follows directly from the characterization of the epigraph and hypograph (see Propositions A.5 and A.7).

Furthermore, ν and γ_k are proper by (a) and since both are well-defined. Finally, concavity of ν and convexity of γ_k are consequences of Lemma A.13.

Alternatively, we could also directly apply Proposition A.19 to (3.1.9) and (3.1.11) in order to derive the concavity and convexity of ν and γ_k, respectively.

ad (c). This is an immediate consequence of the well-definedness of ν and γ_k (see (3.1.12)) and the proof of Proposition 2.68 (c).

ad (d). Follows from a similar argument as the one used in Lemma 2.69 and is therefore omitted. $\square$

In the following, we use

$$\widehat{\mathbb{R}}_k^K := \left\{\widehat{r}^k = (r_1, \ldots, r_{k-1}, r_{k+1}, \ldots, r_K) \,:\, r_i \in \mathbb{R}\right\} \cong \mathbb{R}^{K-1}.$$

Before we continue, we want to note that for $k = 1, \ldots, K$

$$\begin{aligned}\operatorname{dom}(\gamma_k) &= \left\{(\widehat{r}^k, \mu) \in \widehat{\mathbb{R}}_k^K \times \mathbb{R} \,:\, \gamma_k(\widehat{r}^k, \mu) < +\infty\right\}\\ &\subset \widehat{\mathbb{R}}_k^K \times \mathbb{R} \cong \mathbb{R}^{k-1} \times \{0\} \times \mathbb{R}^{K-k} \times \mathbb{R} =: \mathcal{X}_k^{K+1} \subset \mathbb{R}^{K+1}\end{aligned}$$

and

$$\begin{aligned}\operatorname{dom}(\nu) &= \left\{r \in \mathbb{R}^K \,:\, \nu(r) > -\infty\right\}\\ &\subset \mathbb{R}^K \cong \mathbb{R}^K \times \{0\} =: \mathcal{X}_{K+1}^{K+1} \subset \mathbb{R}^{K+1}\end{aligned}$$

can be embedded into K-dimensional subspaces $\mathcal{X}_k^{K+1}$, $k = 1, \ldots, K+1$ of $\mathbb{R}^{K+1}$. Therefore, the projection $\operatorname{Proj}_{\mathcal{X}_k^{K+1}}\colon \mathbb{R}^{K+1} \to \mathcal{X}_k^{K+1} \subset \mathbb{R}^{K+1}$ is well-defined and can be identified with the elimination of the kth variable, which we denote in the following as $\mathrm{Proj}_{\widehat{\mathbb{R}}_k^K \times \mathbb{R}}$ (elimination of r_k) and $\mathrm{Proj}_{\mathbb{R}^K}$ (elimination of μ).

Furthermore, for $k = 1, \ldots, K$ define the mapping $\widehat{P}_k : \mathbb{R}^{K+1} \to \mathbb{R}^{K+1}$ by

$$\widehat{P}_k\left(\widehat{r}^k, \mu, r_k\right) = (r, \mu).$$

With this notation introduced, we can now give a series of results all concerning the representation of the efficient frontier in the vector risk case. The vector-valued variant of Theorem 2.70 reads as follows.

Theorem 3.13 (Representation of Efficient Frontier 1; Vector Risk Case) *Assume for a set $A \subset \mathbb{R}^{M+1}$ of admissible portfolios that the vector risk function $\mathfrak{r}\colon A \to (\mathbb{R} \cup \{+\infty\})^K$ and the scalar utility function $\mathfrak{u}\colon A \to \mathbb{R} \cup \{-\infty\}$ satisfy Assumption 3.11. Then, risk-utility values of Pareto efficient portfolios, i.e., points in the set $\mathcal{G}_{\text{eff}}^{(Kd)}(\mathfrak{r}, \mathfrak{u}; A)$, have the following representation in terms of the graphs of functions ν and γ_k, for $k = 1, \ldots, K$, defined in (3.1.8) and (3.1.10):*

$$\mathcal{G}_{\text{eff}}^{(Kd)}(\mathfrak{r}, \mathfrak{u}; A) = \mathrm{graph}(\nu) \cap \bigcap_{k=1}^{K} \widehat{P}_k\left[\mathrm{graph}(\gamma_k)\right]. \tag{3.1.13}$$

In addition, for $K \geq 2$ we claim ($K = 1$; see Theorem 2.70):

$$\mathcal{G}_{\text{eff}}^{(Kd)}(\mathfrak{r}, \mathfrak{u}; A) = \mathrm{graph}\left(\nu|_{\text{dom}}\right) \cap \bigcap_{k=1}^{K} \widehat{P}_k\left[\mathrm{graph}\left(\gamma_k|_{\text{dom}}\right)\right], \tag{3.1.14}$$

where we set

$$\mathrm{graph}(\nu|_{\text{dom}}) := \left\{(r, \nu(r)) \,:\, r \in \mathrm{dom}(\nu) \subset \mathbb{R}^K\right\}$$

and

$$\widehat{P}_k\left[\mathrm{graph}(\gamma_k|_{\text{dom}})\right] := \left\{\widehat{P}_k\left[(\widehat{r}^k, \mu, \gamma_k(\widehat{r}^k, \mu))\right] \,:\, (\widehat{r}^k, \mu) \in \mathrm{dom}(\gamma_k)\right\}$$

with

$$r = (r_1, \ldots, r_K) \in \mathbb{R}^K,\ \widehat{r}^k = (r_1, \ldots, r_{k-1}, r_{k+1}, \ldots, r_K) \in \widehat{\mathbb{R}}_k^K$$

and

$$\begin{aligned}\widehat{P}_k\left[(\widehat{r}^k, \mu, \gamma_k(\widehat{r}^k, \mu))\right] = \\ \left(r_1, \ldots, r_{k-1}, \gamma_k\left(\widehat{r}^k, \mu\right), r_{k+1}, \ldots, r_K, \mu\right) \in \mathbb{R}^K \times \mathbb{R}\end{aligned} \tag{3.1.15}$$

for $\left(\widehat{r}^k, \mu\right) \in \mathrm{dom}\left(\gamma_k\right)$. Furthermore, we claim that

$$\mathrm{graph}(\nu|_{\mathrm{dom}}) \subset \mathcal{G}^{(Kd)}(\mathfrak{r}, \mathfrak{u};\ A) \tag{3.1.16}$$

and

$$\widehat{P}_k \left[\mathrm{graph}(\gamma_k|_{\mathrm{dom}})\right] \subset \mathcal{G}^{(Kd)}(\mathfrak{r}, \mathfrak{u};\ A) \tag{3.1.17}$$

holds for all $k \in \{1, \ldots, K\}$.

Proof The proof of (3.1.13) goes much along the lines of the proof of Theorem 2.70 (see also Theorem 4 in Part I [78]). In the case $K = 1$, (3.1.13) equals (2.3.18) and was already proved there. So let $K \geq 2$. We start to prove (3.1.13), but can already use partial results in the proof for further claims.

ad "⊃" for (3.1.13): Let

$$(r^*, \mu^*) \in \mathrm{graph}(\nu) \cap \bigcap_{k=1}^{K} \widehat{P}_k \left[\mathrm{graph}(\gamma_k)\right].$$

Since $\mathrm{graph}(\nu) = \left\{(r, \nu(r)) : r \in \mathbb{R}^K\right\}$, $\nu(r) \in [-\infty, \infty)$, $\gamma_k(\widehat{r}^k, \mu) \in (-\infty, \infty]$ and

$$\begin{aligned} &\widehat{P}_k \left[\mathrm{graph}(\gamma_k)\right] \\ &\quad = \left\{(r_1, \ldots, r_{k-1}, \gamma_k(\widehat{r}^k, \mu), r_{k+1}, \ldots, r_K, \mu) : (\widehat{r}^k, \mu) \in \widehat{\mathbb{R}}_k^K \times \mathbb{R}\right\}, \end{aligned}$$

necessarily $(r^*, \mu^*) \in \mathbb{R}^K \times \mathbb{R}$ holds. In particular (3.1.14) holds, once (3.1.13) is proved. On our way to prove "⊃" of (3.1.13), we next need to show that

$$(r^*, \mu^*) \overset{!}{\in} \mathcal{G}^{(Kd)}(\mathfrak{r}, \mathfrak{u};\ A). \tag{3.1.18}$$

Using $(r^*, \mu^*) \in \mathrm{graph}(\nu)$, we get from (3.1.9)

$$\mu^* = \nu\left(r^*\right) = \sup\left\{\mathfrak{u}(x) : \mathfrak{r}(x) \leq r^*,\ x \in A\right\} \tag{3.1.19}$$

and using $(r^*, \mu^*) \in \widehat{P}_k \left[\mathrm{graph}(\gamma_k)\right]$ and (3.1.11) for $k = 1, \ldots, K$ follows

$$r_k^* = \gamma_k\left(\widehat{r^*}^k,\ \mu^*\right) = \inf\left\{\mathfrak{r}_k(x) : \mathfrak{u}(x) \geq \mu^*,\ \mathfrak{r}_i(x) \leq r_i^*,\ i \neq k,\ x \in A\right\}. \tag{3.1.20}$$

With (3.1.19), we can select a sequence $x^{(n)} \in A$ such that $\mathfrak{r}(x^{(n)}) \leq r^*$ and $\mathfrak{u}(x^{(n)}) \nearrow \mu^*$. By Assumption 3.4 (b), $\mathcal{B}_A^{(Kd)}(\mathfrak{r} \leq r^*;\ \mathfrak{u} \geq \mu^* - 1)$ is compact (cf. (3.1.3)).

Hence, without loss of generality, we may assume $x^{(n)} \to x^* \in A$ as $n \to \infty$ with $\mathfrak{r}(x^*) \leq r^*$ and $\mathfrak{u}(x^*) \geq \mu^*$ since $\mathfrak{r}$ has lower semi-continuous components by Assumption 3.11 (ii) and $\mathfrak{u}$ is upper semi-continuous (see Assumption 3.11 (iii)). Hence, (3.1.18) follows. This argument already proves (3.1.16) (note that $(r^*, \mu^*) \in \widehat{P}_k[\operatorname{graph}(\gamma_k)]$ is not needed for this argument). Furthermore, a similar argument, which we leave to the reader, gives (3.1.17).

In addition, in the above situation, we get $\mathfrak{r}(x^*) = r^*$, since $\mathfrak{r}_k(x^*) < r_k^*$ for some $k \in \{1, \ldots, K\}$ would contradict (3.1.20).

We still have to show that (r^*, μ^*) is indeed on the efficient frontier. So let $(r^+, \mu^+) \in \mathcal{G}^{(Kd)}(\mathfrak{r}, \mathfrak{u}; A)$. If $\mu^+ > \mu^*$ and $r^+ \leq r^*$, then by (3.1.8)

$$\nu\left(r^+\right) = \sup\left\{\mu : \left(r^+, \mu\right) \in \mathcal{G}^{(Kd)}(\mathfrak{r}, \mathfrak{u}; A)\right\} \geq \mu^+ > \mu^* \overset{(3.1.19)}{=} \nu(r^*)$$

contradicting that ν is increasing according to Proposition 3.12 (b). On the other hand, if $r^+ \leq r^*$ with $r_k^+ < r_k^*$ for some $k \in \{1, \ldots, K\}$ and $\mu^+ \geq \mu^*$, then by (3.1.10)

$$\begin{aligned} \gamma_k\left(\widehat{r^+}^k, \mu^+\right) &= \inf\left\{r_k : \widehat{P}_k\left[\widehat{r^+}^k, \mu^+, r_k\right] \in \mathcal{G}^{(Kd)}(\mathfrak{r}, \mathfrak{u}; A)\right\} \\ &\leq r_k^+ < r_k^* \overset{(3.1.20)}{=} \gamma_k(\widehat{r^*}^k, \mu^*) \end{aligned}$$

contradicting that γ_k is decreasing in $\widehat{r}^k$ and increasing in μ. Thus, $(r^*, \mu^*) \in \mathcal{G}_{\text{eff}}^{(Kd)}(\mathfrak{r}, \mathfrak{u}; A)$ and "$\supset$" for (3.1.13) is proved.

ad "$\subset$" for (3.1.13): Let

$$(r^*, \mu^*) \in \mathcal{G}_{\text{eff}}^{(Kd)}(\mathfrak{r}, \mathfrak{u}; A) \subset \mathcal{G}^{(Kd)}(\mathfrak{r}, \mathfrak{u}; A) \subset \mathbb{R}^K \times \mathbb{R}.$$

Then, there exists some efficient portfolio $x^* \in A$ with $r^* = \mathfrak{r}(x^*)$ and $\mu^* = \mathfrak{u}(x^*)$. Moreover, the efficiency gives that the supremum in (3.1.19) is attained by x^*, i.e.,

$$\nu(r^*) = \mathfrak{u}(x^*) = \mu^* \quad \text{giving } (r^*, \mu^*) \in \operatorname{graph}(\nu).$$

Similarly, it follows that the infimum in (3.1.20) is attained by x^* as well, i.e.,

$$\gamma_k\left(\widehat{r^*}^k, \mu^*\right) = \mathfrak{r}_k(x^*) = r_k^* \quad \text{which yields} \quad (r^*, \mu^*) \in \widehat{P}_k\left[\operatorname{graph}(\gamma_k)\right].$$

Altogether (3.1.13) and therefore also (3.1.14) are by now proved. This concludes the proof since (3.1.16) and (3.1.17) had already been a byproduct of our arguments earlier. □

The next theorem is the vector risk analog of Corollary 2.71 and Theorem 2.76 (c). Here, we can identify sets on which the efficient frontier might be represented as a single graph.

Theorem 3.14 (Representation of Efficient Frontier 2; Vector Risk Case) *In the situation of Theorem 3.13, moreover, in the following holds true for $K \geq 2$:*

(a) Setting

$$N = N(\mathfrak{r}, \mathfrak{u};\, A) := \mathrm{Proj}_{\mathbb{R}^K}\left(\mathcal{G}_{\mathrm{eff}}^{(Kd)}(\mathfrak{r}, \mathfrak{u};\, A)\right) = \left\{r \in \mathbb{R}^K \,:\, \exists\, \mu \in \mathbb{R} \text{ with } (r, \mu) \in \mathcal{G}_{\mathrm{eff}}^{(Kd)}(\mathfrak{r}, \mathfrak{u};\, A)\right\}, \tag{3.1.21}$$

we have

$$N = \mathrm{Proj}_{\mathbb{R}^K}\left(\mathrm{graph}\left(\nu|_{\mathrm{dom}}\right)\right) \cap \bigcap_{k=1}^{K} \mathrm{Proj}_{\mathbb{R}^K}\left(\widehat{P}_k\left[\mathrm{graph}(\gamma_k|_{\mathrm{dom}})\right]\right) \subset \mathrm{dom}(\nu) \tag{3.1.22}$$

and

$$\mathcal{G}_{\mathrm{eff}}^{(Kd)}(\mathfrak{r}, \mathfrak{u};\, A) = \mathrm{graph}\left(\nu|_N\right). \tag{3.1.23}$$

(b) For arbitrary but fixed $k \in \{1, \ldots, K\}$, we set

$$M_k = M_k(\mathfrak{r}, \mathfrak{u};\, A) := \mathrm{Proj}_{\widehat{\mathbb{R}}_k^K \times \mathbb{R}}\left(\mathcal{G}_{\mathrm{eff}}^{(Kd)}(\mathfrak{r}, \mathfrak{u};\, A)\right) \tag{3.1.24}$$

$$= \left\{(\widehat{r}^k, \mu) \,:\, \exists r_k \in \mathbb{R} \text{ with } (r, \mu) \in \mathcal{G}_{\mathrm{eff}}^{(Kd)}(\mathfrak{r}, \mathfrak{u};\, A)\right\}. \tag{3.1.25}$$

Then, we have

$$M_k = \mathrm{Proj}_{\widehat{\mathbb{R}}_k^K \times \mathbb{R}}\left(\mathrm{graph}(\nu|_{\mathrm{dom}})\right) \cap \bigcap_{\ell=1}^{K} \mathrm{Proj}_{\widehat{\mathbb{R}}_k^K \times \mathbb{R}} \widehat{P}_\ell\left[\mathrm{graph}(\gamma_\ell|_{\mathrm{dom}})\right] \subset \mathrm{Proj}_{\widehat{\mathbb{R}}_k^K \times \mathbb{R}} \widehat{P}_k\left[\mathrm{graph}(\gamma_k|_{\mathrm{dom}})\right] \subset \mathrm{dom}\left(\gamma_k\right) \tag{3.1.26}$$

and

$$\mathcal{G}_{\mathrm{eff}}^{(Kd)}(\mathfrak{r}, \mathfrak{u};\, A) = \widehat{P}_k\left[\mathrm{graph}\left(\gamma_k|_{M_k}\right)\right]. \tag{3.1.27}$$

Proof *ad (a).* Note that (3.1.22) is a simple consequence of (3.1.14) and the definition of N in (3.1.21) because

$$\mathrm{Proj}_{\mathbb{R}^K}\left(\mathrm{graph}(\nu|_{\mathrm{dom}})\right) = \mathrm{dom}(\nu).$$

Thus, it remains to show (3.1.23).

Claim 1:

$$\mathcal{G}_{\text{eff}}^{(Kd)}(\mathfrak{r}, \mathfrak{u};\, A) \overset{!}{\subset} \text{graph}\,(\nu|_N) = \{(r, \nu(r)) \,:\, r \in N\} \subset \mathbb{R}^K \times \mathbb{R} \quad (3.1.28)$$

So let $(r^*, \mu^*) \in \mathcal{G}_{\text{eff}}^{(Kd)}(\mathfrak{r}, \mathfrak{u};\, A)$. Then by (3.1.14), $(r^*, \mu^*) \in \text{graph}(\nu|_{\text{dom}})$, in particular $r^* \in \text{dom}(\nu)$ and $\nu\,(r^*) = \mu^* > -\infty$. So clearly

$$r^* \in \text{Proj}_{\mathbb{R}^K}\,(\text{graph}(\nu|_{\text{dom}}))\,. \quad (3.1.29)$$

In order to see that $r^* \in N$, by (3.1.22) we still have to show that

$$r^* \overset{!}{\in} \text{Proj}_{\mathbb{R}^K}\left(\widehat{P}_k\,[\text{graph}(\gamma_k|_{\text{dom}})]\right), \quad \text{for all } k = 1, \ldots, K. \quad (3.1.30)$$

Again from (3.1.14), for any $k \in \{1, \ldots, K\}$, it follows that there exists some $\left(\widehat{r^+}^k, \mu^+\right) \in \text{dom}\,(\gamma_k)$ with

$$\begin{aligned} (r^*, \mu^*) &\overset{(3.1.14)}{=} \widehat{P}_k\left[\left(\widehat{r^+}^k, \mu^+, \gamma_k\left(\widehat{r^+}^k, \mu^+\right)\right)\right] \\ &= \left(r_1^+, \ldots, r_{k-1}^+, \gamma_k\left(\widehat{r^+}^k, \mu^+\right), r_{k+1}^+, \ldots, r_K^+, \mu^+\right). \end{aligned}$$

Therefore, $\widehat{r^*}^k = \widehat{r^+}^k$, $\mu^* = \mu^+$, and $r_k^* = \gamma_k\left(\widehat{r^+}^k, \mu^+\right) = \gamma_k\left(\widehat{r^*}^k, \mu^*\right)$, i.e., (3.1.30) holds. Thus, together with (3.1.29) and (3.1.22) follows $r^* \in N$. Therefore, $(r^*, \mu^*) \in \text{graph}\,(\nu|_N)$ as claimed in (3.1.28).

Claim 2:

$$\text{graph}\,(\nu|_N) \overset{!}{\subset} \mathcal{G}_{\text{eff}}^{(Kd)}(\mathfrak{r}, \mathfrak{u};\, A) \quad (3.1.31)$$

Let $(r^*, \mu^*) \in \text{graph}\,(\nu|_N)$. Then by (3.1.22), $r^* \in N \subset \text{dom}(\nu)$ and $\mu^* = \nu\,(\mu^*) > -\infty$. Therefore, $(r^*, \mu^*) \in \text{graph}\,(\nu|_{\text{dom}})$ and by (3.1.16) we also get $(r^*, \mu^*) \in \mathcal{G}^{(Kd)}(\mathfrak{r}, \mathfrak{u};\, A)$.
In order to see (3.1.31), according to (3.1.14), we still need to show that

$$(r^*, \mu^*) \overset{!}{\in} \widehat{P}_k\,[\text{graph}(\gamma_k|_{\text{dom}})]\,, \qquad \text{for all} \quad k = 1, \ldots, K,$$

which means we have to show that

$$\left(\widehat{r^*}^k, \mu^*\right) \in \text{dom}\,(\gamma_k) \quad \text{and} \quad r_k^* = \gamma_k\left(\widehat{r^*}^k, \mu^*\right). \quad (3.1.32)$$

Clearly, $(r^*, \mu^*) \in \mathcal{G}^{(Kd)}(\mathfrak{r}, \mathfrak{u};\, A)$ implies

$$\gamma_k(\widehat{r^*}^k, \mu^*) = \inf\{r_k \in \mathbb{R} \,:\, \widehat{P}_k[\widehat{r^*}^k, \mu^*, r_k] \in \mathcal{G}^{(Kd)}(\mathfrak{r}, \mathfrak{u};\, A)\} \leq r_k^* < \infty \quad (3.1.33)$$

and therefore $(\widehat{r^*}^k, \mu^*) \in \text{dom}(\gamma_k)$.

Now, to show (3.1.32) for some fixed $k \in \{1, \ldots, K\}$, assume for the contrary $r_k^* > \gamma_k \left(\widehat{r^*}^k, \mu^*\right)$. From (3.1.22) we know that

$$r^* \in N \subset \mathrm{Proj}_{\mathbb{R}^K} \left(\widehat{P}_k \left[\mathrm{graph}(\gamma_k|_{\mathrm{dom}})\right]\right). \tag{3.1.34}$$

Thus, there must exist some $\mu^+ = \mu_k^+ \in \mathbb{R}$ with

$$(r^*, \mu^+) \in \widehat{P}_k \left[\mathrm{graph}(\gamma_k|_{\mathrm{dom}})\right] \overset{(3.1.17)}{\subset} \mathcal{G}^{(Kd)}(\mathfrak{r}, \mathfrak{u};\, A),$$

i.e., $\left(\widehat{r^*}^k, \mu^+\right) \in \mathrm{dom}\,(\gamma_k)$ and $r_k^* = \gamma_k \left(\widehat{r^*}^k, \mu^+\right) < \infty$. Therefore, $\gamma_k \left(\widehat{r^*}^k, \mu^*\right) < r_k^* = \gamma_k \left(\widehat{r^*}^k, \mu^+\right)$. But since

$$\mu^* = \nu\,(r^*) = \sup \left\{\mu \,:\, (r^*,\, \mu) \in \mathcal{G}^{(Kd)}(\mathfrak{r}, \mathfrak{u};\, A)\right\} \geq \mu^+,$$

this contradicts the fact that γ_k is increasing in μ (see Proposition 3.12).

Altogether, we proved

$$(r^*, \mu^*) \in \mathrm{graph}(\nu|_{\mathrm{dom}}) \cap \bigcap_{k=1}^{K} \widehat{P}_k \left[\mathrm{graph}(\gamma_k|_{\mathrm{dom}})\right] \overset{(3.1.14)}{=} \mathcal{G}_{\mathrm{eff}}^{(Kd)}(\mathfrak{r}, \mathfrak{u};\, A),$$

which finishes the proof of (3.1.31) and thus (3.1.23) is proved.

ad (b). This can be shown very similarly to the proof of (a) and is therefore left to the reader. □

Here and in the following, we use the convention that $\widehat{r}^k$ (a vector in $\mathbb{R}^K$, where the kth component is deleted) followed by r_k (a scalar for the kth component) is to be interpreted as a completed vector in $\mathbb{R}^K$, i.e.,

$$\left(\widehat{r}^k, r_k\right) := (r_1, \ldots, r_{k-1}, r_k,\, r_{k+1}, \ldots, r_K) = r \in \mathbb{R}^K \tag{3.1.35}$$

and similarly $\left(\widehat{r}^k, \gamma_k(./.)\right) := (r_1, \ldots, r_{k-1}, \gamma_k(./.),\, r_{k+1}, \ldots, r_K) \in \mathbb{R}^K$.

Let in the following $\widehat{r}^{k,\ell} \in \mathbb{R}^{K-2}$ be the vector $r \in \mathbb{R}^K$, where the kth and the ℓth component is deleted. We set

$$\widehat{\mathbb{R}}_{k,\ell}^K := \left\{\widehat{r}^{k,\ell} = (r_1, \ldots, r_{k-1}, r_{k+1}, \ldots, r_{\ell-1}, r_{\ell+1}, \ldots, r_K) \,:\, r_i \in \mathbb{R}\right\} \cong \mathbb{R}^{K-2}.$$

With this further notation, we can give our third result on the representation on the efficient frontier in the vector risk case which is a generalization of Theorem 2.76 (b).

Corollary 3.15 (Representation of Efficient Frontier 3; Vector Risk Case) *Under the same assumptions as in Theorem 3.13, the following holds true for $K \geq 2$:*

(a) $\nu\colon N \to \mathbb{R}$ *is strictly increasing in each component;* $\gamma_k\colon M_k \to \mathbb{R}$ *is strictly increasing in* μ *and strictly decreasing in* $\widehat{r}_k$ *(coordinate-wise).*

(b) $\gamma_k\colon M_k \to \mathbb{R}$, $k = 1,\dots,K$, *and* $\nu\colon N \to \mathbb{R}$ *are (for fixed* $\widehat{r}^k \in \widehat{\mathbb{R}}^K_k$*) inverse to each other in the following sense:*

$$\gamma_k\left(\widehat{r}^k,\ \nu\left(\widehat{r}^k, r_k\right)\right) = r_k \quad \textit{for all} \quad r = \left(\widehat{r}^k, r_k\right) \in N \subset \operatorname{dom}(\nu) \tag{3.1.36}$$

$$\nu\left(\widehat{r}^k, \gamma_k\left(\widehat{r}^k, \mu\right)\right) = \mu \quad \textit{for all} \quad \left(\widehat{r}^k, \mu\right) \in M_k \subset \operatorname{dom}(\gamma_k). \tag{3.1.37}$$

(c) $\gamma_k\colon M_k \to \mathbb{R}$ *and* $\gamma_\ell\colon M_\ell \to \mathbb{R}$ $(\ell \neq k)$ *are* $\left(\textit{for fixed } \left(\widehat{r}^{k,\ell}, \mu\right) \in \widehat{\mathbb{R}}^K_{k,\ell} \times \mathbb{R}\right)$ *inverse to each other in the following sense:*

$$\gamma_\ell\left(\widehat{r}^{k,\ell}, \gamma_k\left(\widehat{r}^{k,\ell}, r_\ell, \mu\right), \mu\right) = r_\ell \quad \textit{for all} \quad \left(\widehat{r}^k, \mu\right) \in M_k \subset \operatorname{dom}(\gamma_k) \tag{3.1.38}$$

with a similar convention as above, i.e., $\left(\widehat{r}^{k,\ell},\ r_\ell\right) := \widehat{r}^k$ *and* $\left(\widehat{r}^{k,\ell},\ r_k\right) := \widehat{r}^\ell$. *Note that* (3.1.38) *also holds true, when* k *and* ℓ *are interchanged.*

Proof *ad (a).* We only show the assertion for ν since the one for γ_k is shown similarly. From Proposition 3.12 we already know that ν is increasing coordinate-wise. Assume $\nu\colon N \to \mathbb{R}$ were not strictly increasing in some component. Then there exist $r^* \in N$ and $r^* + \varepsilon e_k \in N$ (for some $k \in \{1,\dots,K\}$ and $\varepsilon > 0; e_k \in \mathbb{R}^K$ k–th unit vector) such that $\nu(r^*) = \nu\left(r^* + \varepsilon e_k\right) =: \mu^*$. Thus, by (3.1.23), $(r^*, \mu^*) \in \mathcal{G}^{(Kd)}_{\text{eff}}(\mathfrak{r}, \mathfrak{u};\, A)$ as well as $(r^* + \varepsilon e_k, \mu^*) \in \mathcal{G}^{(Kd)}_{\text{eff}}(\mathfrak{r}, \mathfrak{u};\, A)$, which is a contradiction since then no portfolio $x \in A$ with $\mathfrak{r}(x) = r^* + \varepsilon e_k$, and $\mathfrak{u}(x) = \mu^*$ can be efficient.

ad (b). Let $r = \left(\widehat{r}^k, r_k\right) \in N \subset \operatorname{dom}(\nu)$ be given for some $k \in \{1,\dots,K\}$ fixed. Then $\mu := \nu\left(\widehat{r}^k, r_k\right) \in \mathbb{R}$ and by (3.1.23), $(r, \mu) \in \mathcal{G}^{(Kd)}_{\text{eff}}(\mathfrak{r}, \mathfrak{u};\, A)$. Hence, by (3.1.27), $(\widehat{r}_k, \mu) \in M_k \subset \operatorname{dom}(\gamma_k)$ and $\gamma_k\left(\widehat{r}_k, \mu\right) = r_k$. This implies (3.1.36). The proof of (3.1.37) is similar.

ad (c). To show (3.1.38), let $\left(\widehat{r}^k, \mu\right) \in M_k \subset \operatorname{dom}(\gamma_k)$ be given for some $k \in \{1,\dots,K\}$. Set

$$r_k := \gamma_k\left(\widehat{r}^k, \mu\right) = \gamma_k\left(\widehat{r}^{k,\ell}, r_\ell, \mu\right) < \infty \tag{3.1.39}$$

for some $\ell \in \{1,\dots,K\}$, $\ell \neq k$.

Then by (3.1.27), for this k, we have $(r, \mu) = (\widehat{r}_k, r_k, \mu) \in \mathcal{G}^{(Kd)}(\mathfrak{r}, \mathfrak{u};\, A)$. Hence, again by (3.1.27), now for ℓ, we get $(\widehat{r}_\ell, \mu) \in M_\ell \subset \operatorname{dom}(\gamma_\ell)$ and $\gamma_\ell\left(\widehat{r}_\ell, \mu\right) = r_\ell$. This together with (3.1.39) implies (3.1.38) since $\widehat{r}_\ell = \left(\widehat{r}^{k,\ell}, r_k\right)$. □

3.1.3 Partial Continuity of the Representing Functions

In the case of a scalar risk function, we obtained continuity for γ and ν wherever their graphs represent the efficient frontier(see Lemma 2.72 and also Theorem 2.76). However, this is different in the situation with a vector risk function.

As shown in Proposition 3.12, $r \mapsto \nu(r)$ is upper semi-continuous concave, and $\left(\widehat{r}^k, \mu\right) \mapsto \gamma_k\left(\widehat{r}^k, \mu\right)$ is lower semi-continuous convex. According to [90, Theorem 10.1] both ν and γ_k are continuous in the (relative) interior of their domains (cf. also Theorem A.14). Nevertheless, discontinuities on the boundary are possible.

Following the idea of the example in [90] after Theorem 10.1, a counterexample for continuity of $\nu\colon N \to \mathbb{R}$ and $\gamma_k\colon M_k \to \mathbb{R}$ can be constructed.

Example 3.16 (Counterexample for Continuity of $\nu\colon N \to \mathbb{R}$ and $\gamma_k\colon M_k \to \mathbb{R}$) *Let $K = 2$ and*

$$\mathcal{G} = \mathcal{G}^{(2d)}(\mathfrak{r}, \mathfrak{u};\, A) = \{(r_1, r_2, \mu)\,:\, r_1, r_2 \geq 0, \mu \leq \sqrt{r_1 r_2}\} \tag{3.1.40}$$

(see Fig. 3.1). Note that $\mathcal{G}^{(2d)}(\mathfrak{r}, \mathfrak{u};\, A)$ is closed and convex and has all necessary properties given in Proposition 3.6. Then we have

$$\mathcal{G}^{(2d)}_{\text{eff}}(\mathfrak{r}, \mathfrak{u};\, A) = \{(r_1, r_2, \mu)\,:\, r_1, r_2 > 0,\ \mu = \sqrt{r_1 r_2}\} \cup \{(0,0,0)\}, \tag{3.1.41}$$

$$N = \{(r_1, r_2)\,:\, r_1, r_2 > 0\} \cup \{(0,0)\}, \tag{3.1.42}$$

$$M_1 = \{(r_2, \mu)\,:\, r_2, \mu > 0\} \cup \{(0,0)\}, \tag{3.1.43}$$

$$M_2 = \{(r_1, \mu)\,:\, r_1, \mu > 0\} \cup \{(0,0)\}. \tag{3.1.44}$$

Function $\nu\colon N \to \mathbb{R}$ with $\nu(r_1, r_2) = \sqrt{r_1 r_2}$ obviously is continuous. On the other hand, function $\gamma_1\colon M_1 \to \mathbb{R}$ defined by

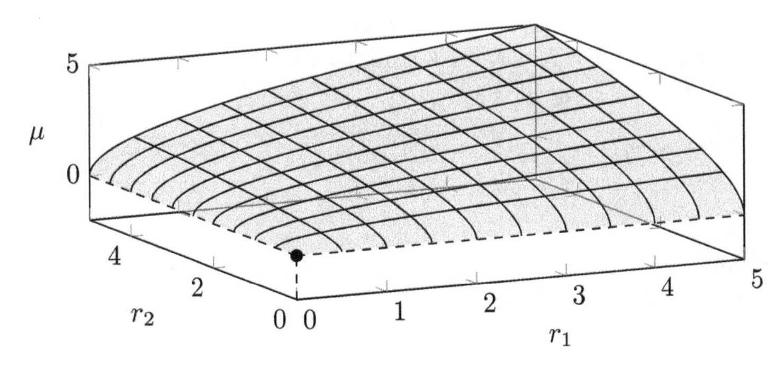

Fig. 3.1 Plot of $\mathcal{G}^{(2d)}(\mathfrak{r}, \mathfrak{u};\, A)$ in Example 3.16

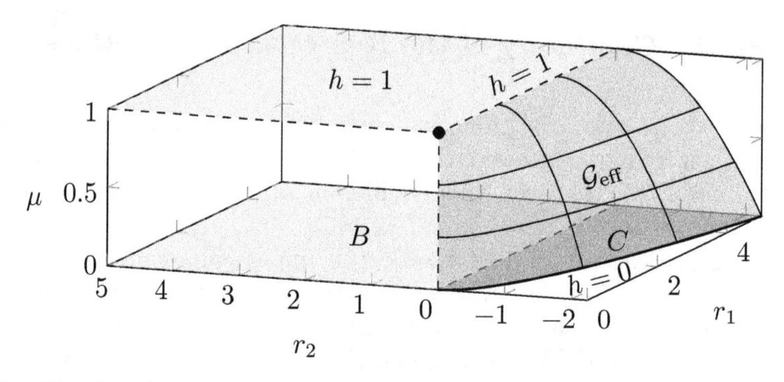

Fig. 3.2 Graph of h

$$\gamma_1(r_2, \mu) = \begin{cases} \mu^2/r_2, & \text{if } r_2 > 0, \\ 0, & \text{if } r_2 = \mu = 0 \end{cases}$$

is not continuous in $(0,0)$, *because for each* $\alpha > 0$, *we have* $\gamma_1(\mu^2/\alpha, \mu) \to \alpha \neq \gamma_1(0,0)$ *as* $\mu \to 0$. *The same can be shown for* γ_2. *In a similar way, an example can be constructed which leads to a discontinuous function* ν.

Example 3.17 (Second Counterexample for Continuity of $\nu \colon N \to \mathbb{R}$ **and** $\gamma_k \colon M_k \to \mathbb{R}$**)** *Let* $B := \{(r_1, r_2) : r_1 \geq 0, r_2 \geq 0\}$ *be the positive quadrant in risk space and* $C := \{(r_1, r_2) : r_1 \geq r_2^2, r_2 < 0\}$. *We define* $h : \mathbb{R}^2 \to \mathbb{R} \cup \{-\infty\}$ *as*

$$h(r_1, r_2) := \begin{cases} 1, & \text{if } (r_1, r_2) \in B, \\ 1 - \frac{r_2^2}{r_1}, & \text{if } (r_1, r_2) \in C, \\ -\infty, & \text{otherwise.} \end{cases}$$

Then $h|_C \in [0,1)$ *with* $h|_C(r_1, r_2) = 0 \Leftrightarrow r_1 = r_2^2, r_2 < 0$. *It is easy to see that* h *is concave and upper semi-continuous. For* $(r_1, r_2) \in (B \cup C) \setminus \{(0,0)\}$, h *is even continuous but it is not continuous in* $(r_1, r_2) = (0,0)$ *(see Fig. 3.2). Setting*

$$\mathcal{G} = \mathcal{G}^{(2d)}(\mathfrak{r}, \mathfrak{u}; A) := \{(r_1, r_2, \mu) \in \mathbb{R}^3 : \mu \leq h(r_1, r_2), (r_1, r_2) \in B \cup C\}$$
$$= \text{hypo}(h) \subset \mathbb{R}^3,$$

we have that $\mathcal{G}^{(2d)}(\mathfrak{r}, \mathfrak{u}; A)$ *is closed and convex and satisfies all properties of Proposition 3.6. Here we have*

$$\begin{aligned}\mathcal{G}_{\text{eff}}^{(2d)}(\mathfrak{r}, \mathfrak{u};\, A) &= \{(r_1, r_2, \mu) \,:\, \mu = h(r_1, r_2),\ (r_1, r_2) \in C \cup \{(0,0)\}\}\,, \\ N &= C \cup \{(0,0)\}\,, \\ M_1 &= \{(r_2, \mu) \,:\, 0 \le \mu \le 1,\ r_2 < 0\} \cup \{(0,1)\}\,, \\ M_2 &= \{(r_1, \mu) \,:\, 0 \le \mu \le 1,\ r_1 > 0\} \cup \{(0,1)\}\,.\end{aligned}$$

ν defined in (3.9) coincides with h, and thus $\nu\colon N \to \mathbb{R}$ is not continuous, but $\mathcal{G}_{\text{eff}}^{(2d)}(\mathfrak{r}, \mathfrak{u};\, A) = \operatorname{graph}(\nu|_N)$ is a smooth manifold. On the other hand, the function $\gamma_1 : M_1 \to \mathbb{R}$ defined by

$$\gamma_1(r_2, \mu) = \begin{cases} \frac{r_2^2}{1-\mu}, & \text{if } 0 \le \mu < 1,\ r_2 < 0, \\ 0, & \text{if } (r_2, \mu) = (0,1) \end{cases}$$

is continuous in $(r_2, \mu) = (0,1)$. And similarly, the function $\gamma_2\colon M_2 \to \mathbb{R}$ defined by

$$\gamma_2(r_1, \mu) = \begin{cases} -\sqrt{(1-\mu) r_1}, & \text{if } 0 \le \mu < 1,\ r_1 > 0, \\ 0, & \text{if } (r_1, \mu) = (0,1) \end{cases}$$

is also continuous in $(r_1, \mu) = (0,1)$.

Note that in the above example, $\mathcal{G}_{\text{eff}}^{(2d)}(\mathfrak{r}, \mathfrak{u};\, A)$ was not a closed set in $\mathbb{R}^3$. This is in contrast to the scalar risk case where the efficient frontier is always a closed set in $\mathbb{R}^2$ (see Remark 2.84). Thus, we conclude:

Remark 3.18 *The set $\mathcal{G}_{\text{eff}}^{(Kd)}(\mathfrak{r}, \mathfrak{u};\, A)$ for $K \ge 2$ is in general not a closed subset of $\mathbb{R}^{K+1}$.*

Our goal in the following therefore is a weaker continuity of ν and γ_k on the boundary of N and M_k, respectively, which holds only on one-dimensional lines parallel to the coordinate axes. In order to see that, let us assume our standing assumptions, i.e., Assumption 3.11. Recall the definition of ν from (3.1.9):

$$\nu(r) = \nu^{(Kd)}(r) = \sup\{\mathfrak{u}(x) \,:\, \mathfrak{r}(x) \le r,\ x \in A\}\,, \tag{3.1.45}$$

for $r = (\widehat{r}_k, r_k) \in \mathbb{R}^K$. Fixing $k \in \{1, \ldots, K\}$ and $\widehat{r}_k \in \mathbb{R}^{K-1}$ for the moment, we define

$$A_{\widehat{r}_k} := \{x \in A \,:\, \widehat{\mathfrak{r}}_k(x) \le \widehat{r}_k\} \subset A, \tag{3.1.46}$$

where $\mathfrak{r} = (\widehat{\mathfrak{r}}_k, \mathfrak{r}_k)\,, A_{\widehat{\mathfrak{r}}_k}$ is closed and convex. Assuming $A_{\widehat{r}_k} \ne \emptyset$, we obtain from (3.1.45)

$$\nu(r) = \sup\left\{\mathfrak{u}(x) \,:\, \mathfrak{r}_k(x) \le r_k,\ x \in A_{\widehat{r}_k}\right\} =: \nu_k^{(1d)}(r_k) = \nu_k^{(1d)}\left(r_k; A_{\widehat{r}_k}\right) \tag{3.1.47}$$

with $\nu_k^{(1d)}: \mathbb{R} \to \mathbb{R} \cup \{-\infty\}$ stemming from a risk-utility optimization problem with the scalar-valued convex risk function $\mathfrak{r}_k: A \to \mathbb{R} \cup \{+\infty\}$ according to our theory in Chap. 2, but restricted to $A_{\widehat{r}_k}$. In order to apply the scalar risk theory here, we need to provide the standing assumptions of Chap. 2, i.e., Assumption 2.64 for $\mathfrak{r}_k$, $\mathfrak{u}$, and $A_{\widehat{r}_k}$. It is easy to see that, given Assumption 3.11, this indeed holds, in case we additionally require the following assumption:

Assumption 3.19 (Further Compact Level Set 2; $k \in \{1, \ldots, K\}$ and $\widehat{r}_k \in \mathbb{R}^{K-1}$ fixed) *We assume*

(a) $\operatorname{dom}(\mathfrak{u}_{|A_{\widehat{r}_k}}) \cap \operatorname{dom}(\mathfrak{r}_{k|A_{\widehat{r}_k}}) \neq \emptyset$ *and*
(b) for all $(s, \mu) \in \mathbb{R}^2$, *the sets* $\mathcal{B}_{A_{\widehat{r}_k}}(\mathfrak{r}_k \leq s;\, \mathfrak{u} \geq \mu)$ *from* (2.3.4) *are compact.*

Remark 3.20 *Assumption 3.19 seems a bit unhandy since it depends strongly on* $A_{\widehat{r}_k}$. *But fortunately it can be replaced by the stronger Assumption 3.5.*

Lemma 3.21 *Let Assumption 3.11 be given and Assumption 3.5 hold for some* $k \in \{1, \ldots, K\}$ *for* $\mathfrak{r}$, $\mathfrak{u}$, *and* A. *Let furthermore* $x \in A$ *be any efficient portfolio for* $\mathfrak{r}$, $\mathfrak{u}$ *and* A *with* $r := \mathfrak{r}(x) = (\widehat{r}_k, r_k) \in \mathbb{R}^K$. *Then Assumption 3.19 holds for this* k *and* $\widehat{r}_k$.

Proof Apparently $x \in A_{\widehat{r}_k}$ and since the efficient portfolio x has finite risk-utility, $x \in \operatorname{dom}\left(\mathfrak{u}_{|A_{\widehat{r}_k}}\right) \cap \operatorname{dom}\left(\mathfrak{r}_{|A_{\widehat{r}_k}}\right)$ follows, which gives Assumption 3.19 (a). On the other hand $A_{\widehat{r}_k} \subset A$ and therefore Assumption 3.19 (b) is weaker than Assumption 3.5 (b). □

Thus, in case Assumptions 3.5 and 3.11 are given, $\nu_k^{(1d)}$ in (3.1.47) is upper semi-continuous concave, according to Proposition 2.68 (b). Furthermore, following Theorem 2.76 (c), there exists an interval $I = I_{\widehat{r}_k} \subset \mathbb{R}$ such that the efficient frontier $\mathcal{G}_{\text{eff}}^{(1d)} = \mathcal{G}_{\text{eff}}^{(1d)}\left(\mathfrak{r}_k, \mathfrak{u}; A_{\widehat{r}_k}\right)$ corresponding to the scalar-valued risk function $\mathfrak{r}_k$ is given by the graph of $\nu_k^{(1d)}$, i.e.,

$$\mathcal{G}_{\text{eff}}^{(1d)}\left(\mathfrak{r}_k, \mathfrak{u}; A_{\widehat{r}_k}\right) = \operatorname{graph}\left(\nu_k^{(1d)}\big|_{I_{\widehat{r}_k}}\right) \tag{3.1.48}$$

and, moreover by Theorem 2.76 (a), $\nu_k^{(1d)}$ is continuous on $I_{\widehat{r}_k}$. Using (3.1.47), this already gives the continuity of $\nu: N \to \mathbb{R}$ along one-dimensional lines parallel to the coordinate axes, provided we can show that for all arbitrary but fixed $\widehat{r}_k^* = (r_1^*, \ldots, r_{k-1}^*, r_{k+1}^*, \ldots, r_K^*) \in \mathbb{R}^{K-1}$, the following inclusion holds true:

$$\begin{aligned}
&\{r \in N : \widehat{r}_k = (r_1, \ldots, r_{k-1}, r_{k+1}, \ldots, r_K) = \widehat{r}_k^*\} \\
&\quad \overset{!}{\subset} \left(\widehat{r}_k^*, I_{\widehat{r}_k^*}\right) := \left\{\left(r_1^*, \ldots, r_{k-1}^*, s, r_{k+1}^*, \ldots, r_K^*\right) : s \in I_{\widehat{r}_k^*}\right\}.
\end{aligned} \tag{3.1.49}$$

In order to see (3.1.49), we start to discuss the relation between (K–dim) efficient portfolios for $\mathfrak{r}, \mathfrak{u}$, and A, and (1–dim) efficient portfolios for $\mathfrak{r}_k$, $\mathfrak{u}$, and $A_{\widehat{r}_k} \subset A$ as in (3.1.48).

Lemma 3.22 *Under Assumption 3.11, let $x \in \mathrm{dom}(\mathfrak{r}) \cap \mathrm{dom}(\mathfrak{u}) \subset A$ be given and set $r := \mathfrak{r}(x) = (\widehat{r}_k, r_k) \in \mathbb{R}^K$. Then the following properties are equivalent:*

(a) $x \in A$ is an efficient portfolio for $\mathfrak{r}, \mathfrak{u}$, and A with K-dimensional risk vector $\mathfrak{r}$.
(b) $x \in A_{\widehat{r}_k} \neq \emptyset$ is an efficient portfolio for $\mathfrak{r}_k, \mathfrak{u}$, and $A_{\widehat{r}_k}$ with scalar-valued risk function $\mathfrak{r}_k$ for all $k = 1, \ldots, K$.

Proof "(a) $\Rightarrow$ (b)." If $x \in A$ with (a), then for $\widehat{r}_k := \widehat{\mathfrak{r}}_k(x)$, we get $x \in A_{\widehat{r}_k} \neq \emptyset$. Since x is efficient for the K-dimensional risk vector $\mathfrak{r}$, by Definition 3.7 it follows for all $k \in \{1, \ldots, K\}$ that there cannot exist a portfolio $x' \in A_{\widehat{r}_k} \subset A$, such that either

$$\mathfrak{r}_k(x') \leq \mathfrak{r}_k(x) \quad \text{and} \quad \mathfrak{u}(x') > \mathfrak{u}(x)$$

or,

$$\mathfrak{r}_k(x') < \mathfrak{r}_k(x) \quad \text{and} \quad \mathfrak{u}(x') \geq \mathfrak{u}(x),$$

holds true. Hence, $x \in A_{\widehat{r}_k}$ is efficient for $\mathfrak{r}_k$, $\mathfrak{u}$, and $A_{\widehat{r}_k}$ (see Definition 2.59) and thus (b) follows.

"(b) $\Leftarrow$ (a)." Let $x \in A_{\widehat{r}_k}$ with $\widehat{r}_k := \widehat{\mathfrak{r}}_k(x)$ be given such that x is efficient for $\mathfrak{r}_k$, $\mathfrak{u}$, and $A_{\widehat{r}_k}$ for all $k = 1, \ldots, K$. We have to show that there does not exist some $x' \in A$ with

(α) $\quad \mathfrak{r}(x') \leq \mathfrak{r}(x) \quad \text{and} \quad \mathfrak{u}(x') > \mathfrak{u}(x)$

or

(β) $\quad \mathfrak{r}(x') \leq \mathfrak{r}(x), \ \mathfrak{r}(x') \neq \mathfrak{r}(x) \quad \text{and} \quad \mathfrak{u}(x') \geq \mathfrak{u}(x).$

Assuming the contrary, i.e., there exists some $x' \in A$ with (α) or (β), immediately gives a contradiction. For example, in case (β), there must exist some $k^* \in \{1, \ldots, K\}$ such that $x' \in A_{\widehat{r}_{k^*}}$ for $r := \mathfrak{r}(x) = (\widehat{r}_{k^*}, r_{k^*})$ with

$$\mathfrak{r}_{k^*}(x') < \mathfrak{r}_{k^*}(x) \quad \text{and} \quad \mathfrak{u}(x') \geq \mathfrak{u}(x).$$

But that means that $x \in A_{\widehat{r}_{k^*}}$ is not efficient for $\mathfrak{r}_{k^*}$, $\mathfrak{u}$, and $A_{\widehat{r}_{k^*}}$, a contradiction. The proof of case (a) is similar. □

The missing assertion (3.1.49) is now immediate with the following lemma:

Lemma 3.23 *Under Assumption 3.11 and Assumption 3.5 for all $k = 1, \ldots, K$, we get*

$$N = \bigcap_{k=1}^{K} \left[\bigcup_{\widehat{r}_k \in \mathbb{R}^{K-1}} \left(\widehat{r}_k, I_{\widehat{r}_k} \right) \right] \subset \mathbb{R}^K.$$

Proof By Lemma 3.21, Assumption 3.19 is satisfied in all relevant situations and thus (3.1.48) holds, i.e., $\mathcal{G}_{\text{eff}}^{(1d)}\left(\mathfrak{r}_k, \mathfrak{u}; A_{\widehat{r}_k}\right) = \text{graph}\left(\nu_k^{(1d)}|_{I_{\widehat{r}_k}}\right)$. Moreover, by (3.1.23), we have $\mathcal{G}_{\text{eff}}^{(Kd)}(\mathfrak{r}, \mathfrak{u}; A) = \text{graph}(\nu|_N)$. Therefore, all efficient points for $\mathfrak{r}$, $\mathfrak{u}$, and A in risk-utility space lie on the graph of $\nu|_N$. On the other hand, by Lemma 3.22, these efficient points are also efficient points for $\mathfrak{r}_k$, $\mathfrak{u}$, and $A_{\widehat{r}_k}$, $k = 1, \ldots, K$, and vice versa! Hence, in risk-utility space, they lie on all graphs of $\nu_k^{(1d)}|_{I_{\widehat{r}_k}}$, for $k = 1, \ldots, K$ (cf. (3.1.47) and (3.1.48)). This concludes the proof. □

Lemma 3.23 provides the missing link to the (partial) continuity of ν.

Corollary 3.24 (Partial Continuity of ν and γ_k) *Under the assumptions of Lemma 3.23, $\nu \colon N \to \mathbb{R}$ and $\gamma_k \colon M_k \to \mathbb{R}$ are continuous along one-dimensional lines parallel to the coordinate axes and in the relative interior of N and M_k, respectively.*

Proof The arguments for $\nu \colon N \to \mathbb{R}$ are given above. Also, the continuity in the (relative) interior follows from Theorem A.14 or [90, Theorem 10.1], as already remarked earlier. The arguments for $\gamma_k \colon M_k \to \mathbb{R}$,

$$\gamma_k(\widehat{r}_k, \mu) = \inf\left\{\mathfrak{r}_k(x) : \mathfrak{u}(x) \geq \mu,\ x \in A_{\widehat{r}_k}\right\} =: \gamma_k^{(1d)}(\mu) = \gamma_k^{(1d)}\left(\mu;\ A_{\widehat{r}_k}\right) \tag{3.1.50}$$

need again Theorem 2.76 (c), which provides an interval $J = J_{\widehat{r}_k} \subset \mathbb{R}$ such that the efficient frontier $\mathcal{G}_{\text{eff}}^{(1d)}\left(\mathfrak{r}_k, \mathfrak{u}; A_{\widehat{r}_k}\right)$ corresponding to the scalar-valued risk function $\mathfrak{r}_k$ is given by the graph of $\gamma_k^{(1d)}$, i.e., with $\widehat{P}(x_1, x_2) = (x_2, x_1)$,

$$\mathcal{G}_{\text{eff}}^{(1d)}\left(\mathfrak{r}_k, \mathfrak{u}; A_{\widehat{r}_k}\right) = \widehat{P}\left[\text{graph}\left(\gamma_k^{(1d)}|_{J_{\widehat{r}_k}}\right)\right],$$

with $\gamma_k^{(1d)}$ continuous on $J_{\widehat{r}_k}$. The remaining arguments, in particular an analog to Lemma 3.23 stating

$$M_k = \bigcap_{k=1}^{K} \left[\bigcup_{\widehat{r}_k \in \mathbb{R}^{K-1}} \left(\widehat{r}_k,\ J_{\widehat{r}_k} \right) \right] \subset \widehat{\mathbb{R}}_k^K \times \mathbb{R},$$

where $\left(\widehat{r}_k,\ J_{\widehat{r}_k}\right) := \left\{(r_1, \ldots, r_{k-1}, r_{k+1}, \ldots, r_K, \mu) : \mu \in J_{\widehat{r}_k}\right\}$, are similar as the ones before for ν. Thus, we leave them to the reader. □

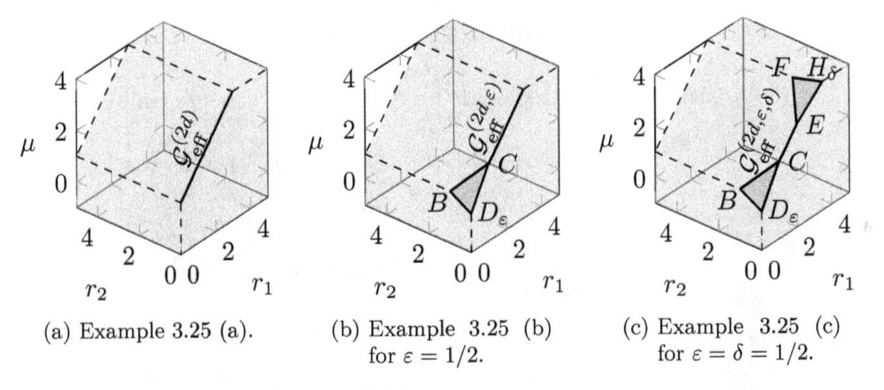

(a) Example 3.25 (a).

(b) Example 3.25 (b) for $\varepsilon = 1/2$.

(c) Example 3.25 (c) for $\varepsilon = \delta = 1/2$.

Fig. 3.3 Illustration for Example 3.25

We want to close this discussion on the parameterizations $\nu\colon N \to \mathbb{R}$ and $\gamma_k\colon M_k \to \mathbb{R}$ of the efficient frontier by some examples. Although the one-dimensional restrictions $\nu_k^{(1d)}\colon I_{\widehat{r_k}} \to \mathbb{R}$ and $\gamma_k^{(1d)}\colon J_{\widehat{r_k}} \to \mathbb{R}$ are defined on intervals and thus on convex sets, N and M_k in general need not be convex. Even worse, the sets $N \cap \left(\widehat{r}_k, I_{\widehat{r_k}}\right)$ and $M_k \cap \left(\widehat{r}_k, J_{\widehat{r_k}}\right)$ may in some cases even not be connected, as we see below.

Example 3.25 (Example for Nonconvex N and M_k)

(a) Let $K = 2$ and

$$\begin{aligned}\mathcal{G}^{(2d)} &= \mathcal{G}^{(2d)}(\mathfrak{r}, \mathfrak{u};\ A)\\ &:= \{(r_1, r_2, \mu)\ :\ r_1, r_2 \geq 0,\ \mu \leq \min\{4,\ 1 + r_1\}\} \subset \mathbb{R}^3 \quad (3.1.51)\end{aligned}$$

is a closed, convex, and "eligible" (in the sense of Proposition 3.6) set of the risk-utility space. We obtain

$$\begin{aligned}\mathcal{G}^{(2d)}_{\text{eff}}(\mathfrak{r}, \mathfrak{u};\ A) &= \{(r_1, 0, \mu)\ :\ \mu = 1 + r_1,\ r_1 \in [0,3]\},\\ N &= \{(r_1, 0)\ :\ r_1 \in [0,3]\},\ \nu(r_1, 0) = 1 + r_1,\\ M_1 &= \{(0, \mu)\ :\ \mu \in [1,4]\},\ \gamma_1(0, \mu) = \mu - 1,\\ M_2 &= \{(r_1, \mu)\ :\ \mu = 1 + r_1,\ r_1 \in [0,3]\},\ \gamma_2(r_1, \mu) = 0\end{aligned}$$

(see Fig. 3.3a).

(b) Next we add one more constraint to $\mathcal{G}^{(2d)}$. Note that the points $B := (0,1,1)$, $C := (1,0,2)$, and $D := (0,0,1)$ build a triangle on the facet $\{\mu = 1 + r_1\} \cap \mathcal{G}^{(2d)}$ of $\mathcal{G}^{(2d)}$. Letting

$$\begin{aligned}\mathcal{G}^{(2d,\varepsilon)} &:= \mathcal{G}^{(2d,\varepsilon)}(\mathfrak{r}, \mathfrak{u}; A)\\ &:= \mathcal{G}^{(2d)} \cap \{(r_1, r_2, \mu)\ :\ (1+\varepsilon) r_1 + \varepsilon r_2 - \mu \geq \varepsilon - 1\}\end{aligned}$$

for $\varepsilon > 0$ *small,* $\mathcal{G}^{(2d,\varepsilon)}$ *is still closed, convex, and eligible, but* $\mathcal{G}^{(2d,\varepsilon)}$ *has a new facet built by the triangle* B, C, *and* $D_\varepsilon := (0,0,1-\varepsilon)$, *which we denote by* $\Delta(B,C,D_\varepsilon)$. *All points on this triangle are efficient. We get*

$$\begin{aligned}\mathcal{G}_{\text{eff}}^{(2d,\varepsilon)}(\mathfrak{r},\mathfrak{u};A) &= \{(r_1,0,\mu)\,:\,\mu = 1+r_1,\ r_1\in[1,3]\}\cup\Delta(B,C,D_\varepsilon)\,,\\ \Delta(B,C,D_\varepsilon) &= \Big\{(r_1,r_2,\mu)\,:\,r_1,r_2\geq 0,\ r_1+r_2\leq 1,\\ &\qquad\qquad \mu = (1+\varepsilon)r_1+\varepsilon r_2-\varepsilon+1\Big\}\end{aligned}$$

(see Fig. 3.3b).
For instance, the parameterization of $\mathcal{G}_{\text{eff}}^{(2d,\varepsilon)}$ *on* N *is given by*

$$\begin{aligned}N = N^{(2d,\epsilon)} &:= N_a\cup N_b,\\ N_a &:= \{(r_1,0)\,:\,r_1\in[1,3]\}\,,\\ N_b &:= \{(r_1,r_2)\,:\,r_1,r_2\geq 0,\ r_1+r_2\leq 1\}\end{aligned}$$

and $\nu\colon N\to\mathbb{R}$

$$(r_1,r_2)\mapsto\begin{cases}\mu = 1+r_1, & \textit{for } r_2 = 0 \textit{ and } (r_1,0)\in N_a,\\ \mu = (1+\varepsilon)r_1+\varepsilon r_2-\varepsilon+1, & \textit{for } (r_1,r_2)\in N_b.\end{cases}$$

Clearly, $N\subset\mathbb{R}^2$ *is not convex, as are* M_1 *and* M_2 *(which is left to the reader).*

(c) *At last, we construct an example such that* $N\cap\left(\widehat{r}_k, J_{\widehat{r}_k}\right)$ *is not connected. Note again that the points* $E := (2,0,3)$, $F = (3,1,4)$, *and* $H = (3,0,4)$ *built another triangle on the facet* $\{\mu = 1+r_1\}\cap\mathcal{G}^{(2d,\varepsilon)}$ *of* $\mathcal{G}^{(2d,\varepsilon)}$. *Setting*

$$\begin{aligned}\mathcal{G}^{(2d,\varepsilon,\delta)} &:= \mathcal{G}^{(2d,\varepsilon,\delta)}(\mathfrak{r},\mathfrak{u};A)\\ &:= \mathcal{G}^{(2d,\varepsilon)}\cap\{(r_1,r_2,\mu)\,:\,r_1+\delta r_2-(1+\delta)\mu\geq -1-3\delta\}\end{aligned}$$

for $\delta > 0$ *small, again* $\mathcal{G}^{(2d,\varepsilon,\delta)}$ *is still closed, convex, and eligible with one more facet built by the triangle* E, F, *and* $H_\delta := (3+\delta,0,4)$. *Also all points on the triangle* $\Delta(E,F,H_\delta)$ *are efficient. Therefore*

$$\begin{aligned}\mathcal{G}_{\text{eff}}^{(2d,\varepsilon,\delta)} &= \mathcal{G}_{\text{eff}}^{(2d,\varepsilon,\delta)}(\mathfrak{r},\mathfrak{u};A)\\ &= \{(r_1,0,\mu)\,:\,\mu = 1+r_1,\ r_1\in[1,2]\}\cup\Delta(B,C,D_\varepsilon)\\ &\qquad\cup\Delta(E,F,H_\delta)\,,\end{aligned}$$

with

$$\begin{aligned}&\Delta(E,F,H_\delta)\\ &\quad= \left\{(r_1,r_2,\mu)\,:\,\mu = \frac{1}{1+\delta}\left(r_1+\delta r_2+1+3\delta\right),(r_1,r_2)\in N_\delta\right\},\end{aligned}$$

where

$$N_\delta := \left\{(r_1, r_2) \,:\, 0 \le r_2 \le r_1 - 2,\ \frac{1}{\delta} r_1 + r_2 \le 1 + \frac{3}{\delta}\right\}$$

(see Fig. 3.3c).
Finally, we have with N_b *from (b)*

$$N = N^{(2d,\varepsilon,\delta)} := N_b \cup \underbrace{\{(r_1, 0) \,:\, r_1 \in [1,2]\}}_{=: N_c} \cup N_\delta$$

and $\nu \colon N \to \mathbb{R}$

$$(r_1, r_2) \mapsto \begin{cases} \mu = (1+\varepsilon) r_1 + \varepsilon r_2 - \varepsilon + 1, & \text{for } (r_1, r_2) \in N_b, \\ \mu = 1 + r_1, & \text{for } r_2 = 0 \text{ and } (r_1, 0) \in N_c, \\ \mu = \frac{1}{1+\delta}(r_1 + \delta r_2 + 1 + 3\delta), & \text{for } (r_1, r_2) \in N_\delta \end{cases}$$

parameterizes the efficient frontier $\mathcal{G}_{\text{eff}}^{(2d,\varepsilon,\delta)}$.
Note that $N = N^{(2d,\varepsilon,\delta)}$ *when restricted, e.g., to* $\widehat{r}_1 = r_2 = \frac{1}{2}$ *contains two disjoint intervals and thus* $N \cap (\widehat{r}_1, I_{\widehat{r}_1})$ *in this case is not connected.*

Remark 3.26 *In all observed examples,* N *and* M_k *but also* $\mathcal{G}_{\text{eff}}^{(Kd)}(\mathfrak{r}, \mathfrak{u}; A)$ *are connected. It is, however, not yet clear whether or not this holds true in general. We discuss that question in the next section.*

3.2 Connectedness of the Efficient Frontier

Here we aim to show path-connectedness of $\mathcal{G}_{\text{eff}}^{(Kd)} = \mathcal{G}_{\text{eff}}^{(Kd)}(\mathfrak{r}, \mathfrak{u}; A)$ with multidimensional risk vector $\mathfrak{r}$ (see Lemma 2.72 and Corollary 2.73 for the one-dimensional case). The standing assumptions for this section are collected below. In contrast to the standing assumptions of Sect. 3.1 (Assumption 3.11), we here need additionally non-negativity of the risk components (or just boundedness from below) and some strictness condition to ensure uniqueness of some efficient portfolios stemming from an auxiliary scalar risk efficient frontier.

Assumption 3.27 (Standing Assumptions for Connectedness; Vector Risk Case) *Assume for a set* $A \subset \mathbb{R}^{M+1}$ *of admissible portfolios (cf. Definition 2.7) that the vector risk function* $\mathfrak{r} = (\mathfrak{r}_1, \ldots, \mathfrak{r}_K) \colon A \to (\mathbb{R} \cup \{+\infty\})^K$, $K \ge 2$, *and the scalar utility function* $\mathfrak{u} \colon A \to \mathbb{R} \cup \{-\infty\}$ *satisfy Assumption 3.11, in particular Assumption 3.4 concerning compact level sets holds as well. We assume furthermore:*

(i) The components of the risk vector are all non-negative, i.e., $\mathfrak{r}_i \colon A \to \mathbb{R}_{\ge 0} \cup \{+\infty\}$, $i = 1, \ldots, K$.

(ii) Either (r2s) *holds for* all $\mathfrak{r}_i$, $i = 1, \ldots, K$ *(see Assumption 2.20), or* $\mathfrak{u}$ *satisfies* (u2s) *in Assumption 2.29.*

Instead of pursuing the connectedness directly for arbitrary $K \geq 2$, we start with a case study for $K = 2$.

3.2.1 Case Study for Two-Dimensional Risk Vectors

In the following, we consider a two-dimensional risk vector

$$\mathfrak{r} = \mathfrak{r}(x) = (\mathfrak{r}_1, \mathfrak{r}_2)(x), \quad x \in A, \tag{3.2.1}$$

for a set $A \subset \mathbb{R}^{M+1}$ of admissible portfolios.

Lemma 3.28 *Let* $\mathfrak{r}$ *in* (3.2.1) *satisfy (i) of Assumption 3.27 for* $K = 2$. *For fixed* $\alpha, \beta > 0$ *set*

$$\mathfrak{r}^{(1d)} = \mathfrak{r}^{(1d)}_{\alpha,\beta}(x) := \max\{\alpha\mathfrak{r}_1(x), \beta\mathfrak{r}_2(x)\} \geq 0, \quad x \in A. \tag{3.2.2}$$

Then $\mathfrak{r}^{(1d)}: A \to \mathbb{R}_{\geq 0} \cup \{+\infty\}$ *is an extended-valued scalar risk function which is lower semi-continuous and satisfies* (r2) *of Assumption 2.20.*

Addition: If all the components of $\mathfrak{r}$ additionally satisfy (r2s) of Assumption 2.20, then $\mathfrak{r}^{(1d)}$ also satisfies (r2s).

Proof Obvious. In particular, the proper convexity of $\mathfrak{r}^{(1d)}$ follows immediately from the corresponding properties of $\mathfrak{r}_1$ and $\mathfrak{r}_2$ (see Assumption 3.11 (ii) and Assumption 3.4 (i)). □

Remark 3.29

(a) In the following, it is sometimes necessary to assume besides Assumption 3.27 that $\mathfrak{r}_i$, $i \in [1, 2]$ *are even strictly positive, say*

$$\mathfrak{r}_i(x) > \kappa \quad \textit{for some} \quad \kappa \geq 0 \quad \textit{and all} \quad x \in A. \tag{3.2.3}$$

If this is not the case, in order to satisfy (3.2.3), *we may w.l.o.g. replace* $\mathfrak{r}_i$ *with* $\mathfrak{r}_i + 1$, *since this* does not *change connectedness properties of* $\mathcal{G}^{(2d)}_{\text{eff}} = \mathcal{G}^{(2d)}_{\text{eff}}(\mathfrak{r}, \mathfrak{u}; A)$.

(b) Similarly, in case all risk components are just bounded from below, by adding suitable constants to the risk components, the connectedness question of $\mathcal{G}^{(2d)}_{\text{eff}}$ *may be reduced to the non-negative risk vector case as in Assumption 3.27. We nevertheless stick to the non-negativity assumption, because it is needed to construct some auxiliary scalar risk-efficient frontiers (see Lemma 3.31 and Lemma 3.33).*

(c) In case a component $\mathfrak{r}_i$ *of* $\mathfrak{r}$ *is not strictly convex (for instance, the standard deviation* (2.1.24)*) but* $\mathfrak{r}_i^2$ *satisfies* (r2s)*, the connectedness result below is still applicable with* $\mathfrak{r}_i^2$ *as risk component (cf. Remark 2.23 and Lemma 2.26). Note that the so transformed efficient frontier belongs to the same efficient portfolios as the original one for* $\mathfrak{r}_i$*.*

Using the upper semi-continuous extended-valued scalar utility function $\mathfrak{u}\colon A \to \mathbb{R} \cup \{-\infty\}$ satisfying (u2) of Assumption 2.29 (i.e., (iii) of Assumption 3.11), we state the corresponding *portfolio optimization problems* $(P_{\min}^{\alpha,\beta})$ and $(P_{\max}^{\alpha,\beta})$ for the admissible set $A \subset \mathbb{R}^{M+1}$ as follows: We define for fixed $\mu \in \mathbb{R}$,

$$\min_{x \in A} \mathfrak{r}^{(1d)}(x) \quad \text{subject to} \quad \mathfrak{u}(x) \geq \mu, \tag{$P_{\min}^{\alpha,\beta}$}$$

and for fixed $c \in \mathbb{R}_{\geq 0}$,

$$\max_{x \in A} \mathfrak{u}(x) \quad \text{subject to} \quad \mathfrak{r}^{(1d)}(x) \leq c. \tag{$P_{\max}^{\alpha,\beta}$}$$

$(P_{\min}^{\alpha,\beta})$ and $(P_{\max}^{\alpha,\beta})$ are equivalent in the sense that an efficient solution of $(P_{\min}^{\alpha,\beta})$ for $\mu \in \mathbb{R}$ is also an efficient solution for $(P_{\max}^{\alpha,\beta})$ for some appropriate $c = c(\mu)$, and vice versa (see Sect. 2.3).

Remark 3.30 *Note that under the condition* (3.2.3)*, there cannot be any solutions of* $(P_{\max}^{\alpha,\beta})$ *for* $c = 0$*, since* $\mathfrak{r}^{(1d)}(x) > \kappa \max\{\alpha, \beta\} \geq 0$ *for all* $x \in A$*.*

Using Corollary 2.73, Theorem 2.76, and Theorem 2.79, we obtain several helpful properties of the efficient frontier $\mathcal{G}_{\text{eff}}\left(\mathfrak{r}^{(1d)}, \mathfrak{u};\, A\right)$.

Lemma 3.31 (Efficient Frontier for $\mathfrak{r}^{(1d)}$, $\mathfrak{u}$, and A) *Let Assumption 3.27 be satisfied for* $K = 2$ *with* $\mathfrak{r} = (\mathfrak{r}_1, \mathfrak{r}_2)$ *from* (3.2.1) *and let* $\mathfrak{r}^{(1d)} = \mathfrak{r}_{\alpha,\beta}^{(1d)}$ *be as in* (3.2.2) *for fixed* $\alpha, \beta > 0$*. Then there exists a non-empty efficient frontier* $\mathcal{G}_{\text{eff}}^{(1d)} := \mathcal{G}_{\text{eff}}(\mathfrak{r}^{(1d)}, \mathfrak{u};\, A) \subset \mathbb{R}^2$ *for* $(P_{\min}^{\alpha,\beta})$ *and* $(P_{\max}^{\alpha,\beta})$*, which is either a single point or path-connected (not necessarily bounded) and can be parameterized by a graph of continuous functions*

$$\nu^{(1d)}\colon I^{(1d)} \to J^{(1d)} \quad \text{or} \quad \gamma^{(1d)}\colon J^{(1d)} \to I^{(1d)}, \tag{3.2.4}$$

for intervals $I^{(1d)} \subset \mathbb{R}$ *and* $J^{(1d)} \subset \mathbb{R}$*, respectively (see Fig. 3.4).*

Furthermore, each point $(c, \mu) \in \mathcal{G}_{\text{eff}}^{(1d)}$ *corresponds to a unique efficient portfolio* $X^{(1d)}(c, \mu) \in A$ *and the mapping*

$$(c, \mu) \mapsto X^{(1d)}(c, \mu) \tag{3.2.5}$$

is continuous on $\mathcal{G}_{\text{eff}}^{(1d)}$*. Thus, there exist continuous paths of unique efficient frontier portfolios in A, namely,*

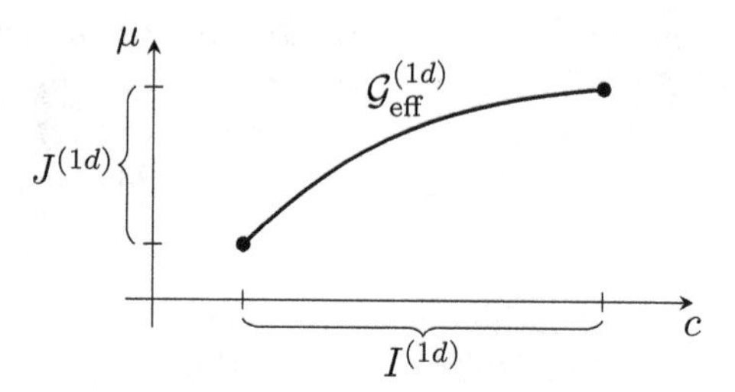

Fig. 3.4 Sketch of $\mathcal{G}_{\text{eff}}^{(1d)} := \mathcal{G}_{\text{eff}}(\mathfrak{r}^{(1d)}, \mathfrak{u}; A)$

$$x^* = x^*(c) := X^{(1d)}\left(c, \nu^{(1d)}(c)\right) \in A, \quad c \in I^{(1d)}, \tag{3.2.6}$$

and

$$y^* = y^*(\mu) := X^{(1d)}\left(\gamma^{(1d)}(\mu), \mu\right) \in A, \quad \mu \in J^{(1d)}, \tag{3.2.7}$$

where x^* is the unique solution of $(P_{\max}^{\alpha,\beta})$ realizing the risk-utility values $\left(c, \nu^{(1d)}(c)\right) \in \mathcal{G}_{\text{eff}}^{(1d)}$ and y^* is the unique solution of $(P_{\min}^{\alpha,\beta})$ realizing the risk-utility values $\left(\gamma^{(1d)}(\mu), \mu\right) \in \mathcal{G}_{\text{eff}}^{(1d)}$, i.e., we have

$$\begin{aligned} \left(\mathfrak{r}^{(1d)}\left(x^*(c)\right), \mathfrak{u}\left(x^*(c)\right)\right) &= \left(c, \nu^{(1d)}(c)\right) \quad \textit{and} \\ \left(\mathfrak{r}^{(1d)}\left(y^*(\mu)\right), \mathfrak{u}\left(y^*(\mu)\right)\right) &= \left(\gamma^{(1d)}(\mu), \mu\right). \end{aligned} \tag{3.2.8}$$

Proof Follows immediately from Corollary 2.73, Theorem 2.76, and Theorem 2.79 since given Assumption 3.27 implies that $\mathfrak{r}^{(1d)}, \mathfrak{u}$, and A satisfy Assumption 2.64. Moreover, according to Assumption 3.27 (ii), $\mathfrak{r}^{(1d)}$ defined in (3.2.2) satisfies (r2s) of Assumption 2.20 (see Addition of Lemma 3.28) or $\mathfrak{u}$ satisfies, (u2s) of Assumption 2.29.

Note that in particular Assumption 2.64 (iv), i.e., Assumption 2.56 concerning compact level sets of $\mathfrak{r}^{(1d)} = \mathfrak{r}_{\alpha,\beta}^{(1d)}, \mathfrak{u}$, and A, holds true. To be more precise, $\text{dom}\left(\mathfrak{r}^{(1d)}\right) \cap \text{dom}(\mathfrak{u}) \neq \emptyset$ is an immediate consequence of $\text{dom}(\mathfrak{r}) \cap \text{dom}(\mathfrak{u}) \neq \emptyset$ (see Assumption 3.4 (a)). And concerning the level sets, we have

$$\begin{aligned} \mathcal{B}_A\left(\mathfrak{r}^{(1d)} \leq c; \mathfrak{u} \geq \mu\right) &\overset{(2.3.4)}{=} \left\{x \in A : \mathfrak{r}^{(1d)}(x) \leq c,\ \mathfrak{u}(x) \geq \mu\right\} \\ &\subset \left\{x \in A : \mathfrak{r}_1(x) \leq \frac{c}{\alpha},\ \mathfrak{r}_2(x) \leq \frac{c}{\beta},\ \mathfrak{u}(x) \geq \mu\right\} \end{aligned}$$

which is compact as a consequence of Assumption 3.4 (b).

Thus, the topological structure of $\mathcal{G}_{\text{eff}}^{(1d)}$, in particular path-connectedness, follows from Corollary 2.73, the uniqueness and continuity of the efficient

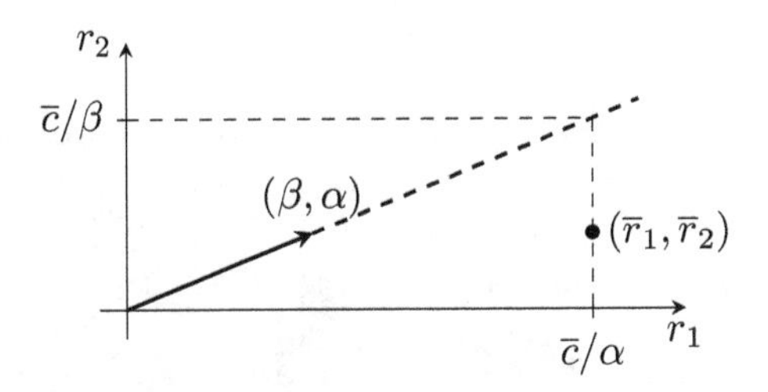

Fig. 3.5 $(\overline{r}_1, \overline{r}_2)$ with (3.2.9)

portfolio map (3.2.5) results from Theorem 2.79, and all the other representations result from Theorem 2.76. □

Remark 3.32 *Note that since* $\mathfrak{r}^{(1d)} = \mathfrak{r}^{(1d)}_{\alpha,\beta}$ *depends on* (α, β), *all of the above depends on* $\alpha, \beta > 0$ *as well, i.e., for instance,* $I^{(1d)} = I^{(1d)}_{\alpha,\beta}$, $X^{(1d)} = X^{(1d)}_{\alpha,\beta}$ *and* $\mathcal{G}^{(1d)}_{\text{eff}} = \mathcal{G}^{(1d)}_{\text{eff}}(\alpha, \beta)$, *but also* $\nu^{(1d)} = \nu^{(1d)}_{\alpha,\beta}$ *and* $\gamma^{(1d)} = \gamma^{(1d)}_{\alpha,\beta}$.

While having reconsidered the properties of the efficient frontier $\mathcal{G}^{(1d)}_{\text{eff}} = \mathcal{G}_{\text{eff}}(\mathfrak{r}^{(1d)}, \mathfrak{u};\ A) \subset \mathbb{R}^2$ in Lemma 3.31, we next look for a linkage to the efficient frontier $\mathcal{G}^{(2d)}_{\text{eff}} = \mathcal{G}^{(2d)}_{\text{eff}}(\mathfrak{r}, \mathfrak{u};\ A) \subset \mathbb{R}^3$.

Lemma 3.33 (Relation of $\mathcal{G}_{\text{eff}}(\mathfrak{r}^{(1d)}, \mathfrak{u};\ A)$ and $\mathcal{G}^{(2d)}_{\text{eff}}(\mathfrak{r}, \mathfrak{u};\ A)$) *Let Assumption 3.27 be satisfied with* $\mathfrak{r} = (\mathfrak{r}_1, \mathfrak{r}_2)$ *from* (3.2.1), *such that additionally* (3.2.3) *is true. For* $\mathfrak{r}^{(1d)} = \mathfrak{r}^{(1d)}_{\alpha,\beta}$ *defined in* (3.2.2) *for fixed* $\alpha, \beta > 0$, *let furthermore* $(\overline{c}, \overline{\mu}) \in \mathcal{G}^{(1d)}_{\text{eff}} = \mathcal{G}_{\text{eff}}\left(\mathfrak{r}^{(1d)}, \mathfrak{u};\ A\right)$ *be given, i.e.,* $\overline{\mu} = \nu^{(1d)}(\overline{c})$ *for* $\overline{c} \in I^{(1d)} = I^{(1d)}_{\alpha,\beta} \subset \mathbb{R}_{>0}$ *(see Remark 3.30). We then claim that there exist* $\overline{r}_1, \overline{r}_2 > 0$ *with*

$$\max\{\alpha\overline{r}_1,\ \beta\overline{r}_2\} = \overline{c} > 0 \tag{3.2.9}$$

and $(\overline{r}_1, \overline{r}_2, \overline{\mu}) \in \mathcal{G}^{(2d)}_{\text{eff}} := \mathcal{G}^{(2d)}_{\text{eff}}(\mathfrak{r}, \mathfrak{u};\ A) \subset \mathbb{R}^3$ *(see Fig. 3.5).*

Proof Let $\overline{x} := x^*(\overline{c}) \in A$ be the unique efficient portfolio with $c = \overline{c}$ for $(P^{\alpha,\beta}_{\max})$ from (3.2.6). In particular due to (3.2.8),

$$\mathfrak{r}^{(1d)}(\overline{x}) = \overline{c} > 0 \quad \text{and} \quad \mathfrak{u}(\overline{x}) = \overline{\mu}. \tag{3.2.10}$$

Claim: Setting $\overline{r}_1 := \mathfrak{r}_1(\overline{x})$, $\overline{r}_2 := \mathfrak{r}_2(\overline{x})$, then

$$(\overline{r}_1, \overline{r}_2, \overline{\mu}) \overset{!}{\in} \mathcal{G}^{(2d)}_{\text{eff}} = \mathcal{G}^{(2d)}_{\text{eff}}(\mathfrak{r}, \mathfrak{u};\ A) \quad \text{holds.} \tag{3.2.11}$$

Note that with (3.2.10), in particular (3.2.9) holds true because $\mathfrak{r}^{(1d)} = \max\{\alpha\mathfrak{r}_1, \beta\mathfrak{r}_2\}$. Together with (3.2.3), we obtain that

$$0 < \alpha \overline{r}_1 \leq \overline{c} \quad \text{and} \quad 0 < \beta \overline{r}_2 \leq \overline{c}, \tag{3.2.12}$$

in particular $\overline{r}_1, \overline{r}_2 > 0$. Note that at least one of the inequalities in (3.2.12) must be equal, i.e., either $\alpha \overline{r}_1 = \overline{c}$ or $\beta \overline{r}_2 = \overline{c}$ holds true. Moreover, again since $\overline{x} \in A$ is the unique efficient portfolio for $(P_{\max}^{\alpha,\beta})$ with $c = \overline{c}$, we infer that besides $\overline{x}$, there does not exist some $x' \in A$ with

$$\alpha \mathfrak{r}_1(x') \leq \overline{c}, \ \beta \mathfrak{r}_2(x') \leq \overline{c} \quad \text{and} \quad \mathfrak{u}(x') \geq \overline{\mu}. \tag{3.2.13}$$

Using (3.2.12), we see that besides $\overline{x}$, there also does not exist some $x' \in A$ with

$$\mathfrak{r}_1(x') \leq \overline{r}_1, \ \mathfrak{r}_2(x') \leq \overline{r}_2 \quad \text{and} \quad \mathfrak{u}(x') \geq \overline{\mu}. \tag{3.2.14}$$

Therefore, since $\overline{x} \in A$ satisfies all inequalities in (3.2.14) with equality, there does not exist some $x' \in A$ with either

$$\mathfrak{r}_1(x') \leq \overline{r}_1, \ \mathfrak{r}_2(x') \leq \overline{r}_2 \quad \text{and} \quad \mathfrak{u}(x') > \overline{\mu}. \tag{I}$$

or

$$\mathfrak{r}_1(x') \leq \overline{r}_1, \ \mathfrak{r}_2(x') \leq \overline{r}_2, \ (\mathfrak{r}_1(x'), \mathfrak{r}_2(x')) \neq (\overline{r}_1, \overline{r}_2) \quad \text{and} \quad \mathfrak{u}(x') \geq \overline{\mu}. \tag{II}$$

Exchanging $\overline{r}_1, \overline{r}_2$, and $\overline{\mu}$ with $\mathfrak{r}_1(\overline{x}), \mathfrak{r}_2(\overline{x})$, and $\mathfrak{u}(\overline{x})$ in (I) and (II), we can apply Definition 3.7 and obtain that $\overline{x} \in A$ is an efficient portfolio for the two-dimensional problem $\mathcal{G}^{(2d)} := \mathcal{G}^{(2d)}(\mathfrak{r}, \mathfrak{u}; A) \subset \mathbb{R}^3$ and $(\overline{r}_1, \overline{r}_2, \overline{\mu}) \in \mathcal{G}_{\text{eff}}^{(2d)} = \mathcal{G}_{\text{eff}}^{(2d)}(\mathfrak{r}, \mathfrak{u}; A)$. □

In the proof of Lemma 3.33, we used $x^* = x^*(c) = X^{(1d)}(c, \nu^{(1d)}(c)) \in A$, $c \in I^{(1d)}$, from (3.2.6) which is a continuous connected curve in A with no self-intersections. Thus, we obtain:

Corollary 3.34 (Continuous Connected Curve in $\mathcal{G}_{\text{eff}}^{(2d)}(\mathfrak{r}, \mathfrak{u}; A)$) *Let Assumption 3.27 hold true for $K = 2$ with additionally* (3.2.3) *and, moreover, components $\mathfrak{r}_i \colon A \to \mathbb{R}_{\geq 0} \cup \{+\infty\}$ $(i = 1, 2)$ which are continuous on* $\text{dom}(\mathfrak{r}) = \text{dom}(\mathfrak{r}_1) \cap \text{dom}(\mathfrak{r}_2)$. *Then*

$$(\mathfrak{r}(x^*(c)), \mathfrak{u}(x^*(c))) \in \mathcal{G}_{\text{eff}}^{(2d)}, \quad c \in I^{(1d)} = I_{\alpha,\beta}^{(1d)}, \tag{3.2.15}$$

is a continuous connected curve in $\mathcal{G}_{\text{eff}}^{(2d)} = \mathcal{G}_{\text{eff}}^{(2d)}(\mathfrak{r}, \mathfrak{u}; A)$ with no self-intersections.

Proof Follows immediately from the arguments above and the fact that under our conditions efficient portfolios for a given risk-utility value in $\mathcal{G}_{\text{eff}}^{(1d)}$ are unique (see Lemma 3.31). So continuity of x^* in (3.2.6) implies trivially that $\mathfrak{r} \circ x^*$ is continuous since $x^*(c) \in \text{dom}(\mathfrak{r})$ for all $c \in I^{(1d)}$. Moreover, $\mathfrak{u} \circ x^* = \nu^{(1d)}$ is already known to be continuous; see again Lemma 3.31, in particular (3.2.8). □

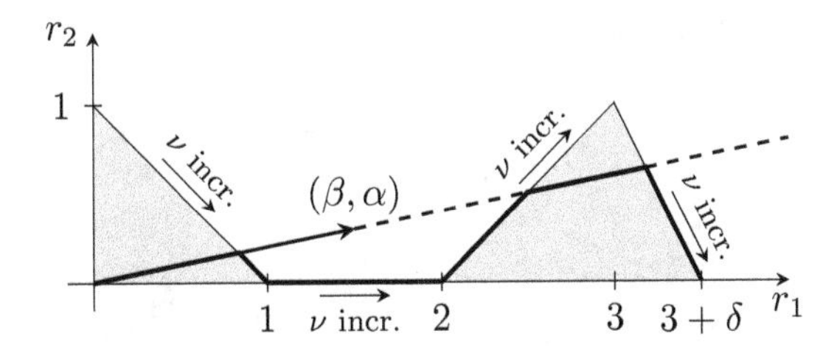

Fig. 3.6 Risk projection of the path in (3.2.15) for a fixed vector (α, β)

Remark 3.35

(a) *In the above argument, we only used continuity of* $\mathfrak{r}$ *on its domain; for the utility function* $\mathfrak{u}\colon A \to \mathbb{R} \cup \{-\infty\}$, *no further assumptions—besides the ones from Assumption 3.27—are necessary.*

(b) *In case condition* (3.2.3) *fails, risk values zero are possible and it might happen that* $(\overline{c} = 0, \overline{\mu}) \in \mathcal{G}_{\text{eff}}^{(1d)}$. *Using arguments similar to the proof of Lemma 3.33, one finds that then* $(\overline{r}_1 = 0, \overline{r}_2 = 0, \overline{\mu}) \in \mathcal{G}_{\text{eff}}^{(2d)}$.

(c) *We further note that even if condition* (3.2.3) *fails, the arguments given in Lemma 3.33 may still be used for* $\overline{c} \in I^{(1d)} \setminus \{0\}$ *to construct points* $(\overline{r}_1, \overline{r}_2, \overline{\mu})$ *in* $\mathcal{G}_{\text{eff}}^{(2d)}$ *(only* $\overline{r}_1$, $\overline{r}_2$ *positive might fail). Therefore, together with (b), even without* (3.2.3), *the continuous map into* $\mathcal{G}_{\text{eff}}^{(2d)}$ *in* (3.2.15) *can be constructed for all* $c \in I^{(1d)}$.

Since the path $x^* = x^*(c) = X^{(1d)}\left(c, \nu^{(1d)}(c)\right)$ (see (3.2.6)) is a continuous image of $I^{(1d)}$ as well as of $\mathcal{G}_{\text{eff}}^{(1d)}$, Corollary 3.34 helps us to distinguish certain curves in $\mathcal{G}_{\text{eff}}^{(2d)}$ as continuous image of $I^{(1d)}$ or $\mathcal{G}_{\text{eff}}^{(1d)}$.

Example 3.36 *Remember the picture of* $\mathcal{G}_{\text{eff}}^{(2d,\varepsilon,\delta)}$ *from Example 3.25 (c) (see Fig. 3.3c) with two-dimensional risk vector. Although here condition* (3.2.3) *does not hold, according to Remark 3.35 (c), the map in* (3.2.15) *can still be constructed for all* $c \in I^{(1d)}$. *We visualize the risk part of this map in Fig. 3.6.*

Besides the shaded set $N = N^{(2d,\varepsilon,\delta)}$ *from Example 3.25 (c), we here depict the points* $\mathfrak{r}(x^*(c)) = (\overline{r}_1, \overline{r}_2)(c)$, $c \in I^{(1d)} = I_{\alpha,\beta}^{(1d)}$ *from Lemma 3.33 or Corollary 3.34 for fixed* $\alpha, \beta > 0$ *in bold face.*

The following lemma is some kind of "inverse" to Lemma 3.33: It, however, works only on the positive quadrant, i.e., for $\overline{r}_1, \overline{r}_2 > 0$, and thus now condition (3.2.3) is essential.

Lemma 3.37 (Inverse Relation Between the Two Efficient Frontiers $\mathcal{G}_{\text{eff}}(\mathfrak{r}^{(1d)}, \mathfrak{u}; A)$ **and** $\mathcal{G}_{\text{eff}}^{(2d)}(\mathfrak{r}, \mathfrak{u}; A)$**)** *Let Assumption 3.27 be satisfied with* $\mathfrak{r} = (\mathfrak{r}_1, \mathfrak{r}_2)$ *from* (3.2.1) *such that additionally* (3.2.3) *holds. Let, furthermore,* $(\overline{r}_1, \overline{r}_2, \overline{\mu}) \in \mathcal{G}_{\text{eff}}^{(2d)} = \mathcal{G}_{\text{eff}}^{(2d)}(\mathfrak{r}, \mathfrak{u}; A)$ *be given, in particular* $\overline{r}_1, \overline{r}_2 > 0$. *We*

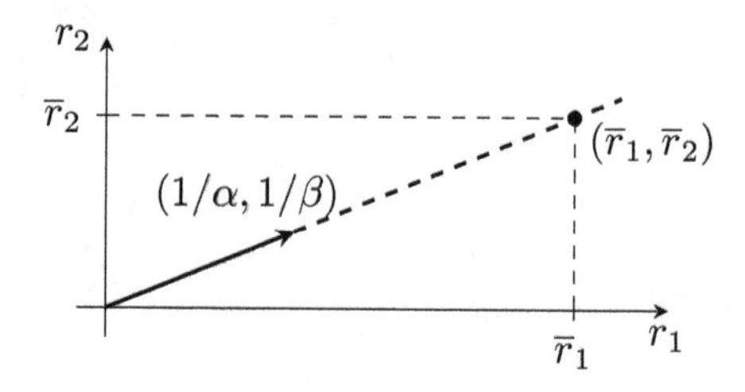

Fig. 3.7 Situation of Lemma 3.37

choose $(\alpha, \beta) \in \mathbb{R}^2_{>0}$, *w.l.o.g.* $(\alpha, \beta) \in \mathcal{S}^1$ *(unit sphere in* $\mathbb{R}^2$*), i.e.,* $\alpha^2 + \beta^2 = 1$*, such that* $\left(\alpha^{-1}, \beta^{-1}\right)$ *is on the same ray through* $(0, 0)$*, as* $(\overline{r}_1, \overline{r}_2)$ *(see Fig. 3.7). That is, there exists some* $\overline{c} > 0$ *with*

$$\frac{\overline{c}}{\alpha} = \overline{r}_1 \quad \textit{and} \quad \frac{\overline{c}}{\beta} = \overline{r}_2, \tag{3.2.16}$$

such that in particular

$$\overline{c} = \max \left\{ \alpha \overline{r}_1, \beta \overline{r}_2 \right\}. \tag{3.2.17}$$

Then the point $(\overline{c}, \overline{\mu})$ *is efficient for* $(P^{\alpha,\beta}_{\max})$ *with* $c = \overline{c}$*, i.e.,* $(\overline{c}, \overline{\mu}) \in \mathcal{G}^{(1d)}_{\text{eff}}(\alpha, \beta) = \mathcal{G}_{\text{eff}}\left(\mathfrak{r}^{(1d)}_{\alpha,\beta}, \mathfrak{u};\, A\right)$.

Proof Let $(\overline{r}_1, \overline{r}_2, \overline{\mu}) \in \mathcal{G}^{(2d)}_{\text{eff}} = \mathcal{G}^{(2d)}_{\text{eff}}(\mathfrak{r}, \mathfrak{u};\, A)$ be given and denote $\overline{x} \in A$ the unique efficient portfolio for the two-dimensional risk vector $\mathfrak{r} = (\mathfrak{r}_1, \mathfrak{r}_2)$ with $\mathfrak{r}_1(\overline{x}) = \overline{r}_1$, $\mathfrak{r}_2(\overline{x}) = \overline{r}_2$, and $\mathfrak{u}(\overline{x}) = \overline{\mu}$. By definition of the efficient frontier, there does not exist some $x' \in A$ with (I) or (II) (see proof of Lemma 3.33).

Claim: Then for $\mathfrak{r}^{(1d)} = \mathfrak{r}^{(1d)}_{\alpha,\beta}$, there does not exist some $x' \in A$ with

$$\mathfrak{r}^{(1d)}\left(x'\right) \leq \overline{c} \quad \text{and} \quad \mathfrak{u}\left(x'\right) > \overline{\mu} \tag{I'}$$

or

$$\mathfrak{r}^{(1d)}\left(x'\right) < \overline{c} \quad \text{and} \quad \mathfrak{u}\left(x'\right) \geq \overline{\mu}. \tag{II'}$$

To see that with an indirect argument, assume, for instance, (I') holds for some $x' \in A$. Then, by (3.2.2), we have

$$\mathfrak{r}^{(1d)}\left(x'\right) = \max \left\{ \alpha \mathfrak{r}_1\left(x'\right), \beta \mathfrak{r}_2\left(x'\right) \right\} \leq \overline{c},$$

in particular

$$\alpha \mathfrak{r}_1\left(x'\right) \leq \overline{c} \quad \text{and} \quad \beta \mathfrak{r}_2\left(x'\right) \leq \overline{c}.$$

Using (3.2.16) and (3.2.17) follows $\mathfrak{r}_1(x') \leq \overline{r}_1$ and $\mathfrak{r}_2(x') \leq \overline{r}_2$, contradicting (I). Similarly, assuming (II') holds for some $x' \in A$, contradicts (II) and thus the "Claim" is proved.

Finally, since $\overline{x} \in A$ satisfies $\mathfrak{r}^{(1d)}(\overline{x}) = \overline{c}$ and $\mathfrak{u}(\overline{x}) = \overline{\mu}$, we conclude from (I') and (II') that $(\overline{c}, \overline{\mu}) \in \mathcal{G}_{\text{eff}}^{(1d)} = \mathcal{G}_{\text{eff}}^{(1d)}(\alpha, \beta)$. □

With Lemma 3.37, we know that arbitrary but fixed points in $\mathcal{G}_{\text{eff}}^{(2d)} = \mathcal{G}_{\text{eff}}^{(2d)}(\mathfrak{r}, \mathfrak{u}; A)$ can be found in the image of the map given in (3.2.15) on $I^{(1d)} = I_{\alpha,\beta}^{(1d)}$ or, via (3.2.6), as image of a map on $\mathcal{G}_{\text{eff}}^{(1d)} = \mathcal{G}_{\text{eff}}^{(1d)}(\alpha, \beta)$ for suitable (α, β).

To show *path-connectedness* of $\mathcal{G}_{\text{eff}}^{(2d)}$, we have to construct a path in $\mathcal{G}_{\text{eff}}^{(2d)}$ between two (possibly) different but arbitrary points:

$$g^{(i)} := \left(r_1^{(i)}, r_2^{(i)}, \mu^{(i)}\right) \in \mathcal{G}_{\text{eff}}^{(2d)}, \quad i = 1, 2. \tag{3.2.18}$$

So let $\left(1/\alpha^{(i)}, 1/\beta^{(i)}\right) \in \mathcal{S}^1$ be on the same ray through $(0, 0)$ as $\left(r_1^{(i)}, r_2^{(i)}\right)$, $i = 1, 2$. Then $g^{(i)}$ is in the image of the map given in (3.2.15) on $I_{\alpha^{(i)},\beta^{(i)}}^{(1d)}$. In case $\left(\alpha^{(1)}, \beta^{(1)}\right) = \left(\alpha^{(2)}, \beta^{(2)}\right)$, we can use the continuous map (3.2.15) and the fact that $I^{(1d)}$ and thus $\mathcal{G}_{\text{eff}}^{(1d)}$ is path-connected to construct a continuous path in $\mathcal{G}_{\text{eff}}^{(2d)}$ between $g^{(1)}$ and $g^{(2)}$.

Otherwise, i.e., in case $\left(\alpha^{(1)}, \beta^{(1)}\right) \neq \left(\alpha^{(2)}, \beta^{(2)}\right)$, we intend to make a homotopy between the images of $I_{\alpha^{(i)},\beta^{(i)}}^{(1d)}$ or $\mathcal{G}_{\text{eff}}^{(1d)}\left(\alpha^{(i)}, \beta^{(i)}\right)$, $i = 1, 2$, inside $\mathcal{G}_{\text{eff}}^{(2d)}$.

To see that, note again, see Remark 3.32, that the parametrizations in (3.2.4), (3.2.6), and (3.2.7) all depend on α and β, i.e.,

$$\nu_{\alpha,\beta}^{(1d)} : I_{\alpha,\beta}^{(1d)} \to J_{\alpha,\beta}^{(1d)} \quad \text{and} \quad \gamma_{\alpha,\beta}^{(1d)} : J_{\alpha,\beta}^{(1d)} \to I_{\alpha,\beta}^{(1d)} \tag{3.2.19}$$

as well as

$$x^* = x_{\alpha,\beta}^*(c) = X_{\alpha,\beta}^{(1d)}\left(c, \nu_{\alpha,\beta}^{(1d)}(c)\right), \quad c \in I_{\alpha,\beta}^{(1d)} \tag{3.2.20}$$

and

$$y^* = y_{\alpha,\beta}^*(\mu) = X_{\alpha,\beta}^{(1d)}\left(\gamma_{\alpha,\beta}^{(1d)}(\mu), \mu\right), \quad \mu \in J_{\alpha,\beta}^{(1d)}. \tag{3.2.21}$$

If all the above dependencies on α and β were continuous, for example, $x^* = x_{\alpha,\beta}^*$ continuous for varying (α, β) from $\left(\alpha^{(1)}, \beta^{(1)}\right)$ to $\left(\alpha^{(2)}, \beta^{(2)}\right)$, then, together with the map in (3.2.15), this would provide the missing link between the images of $I_{\alpha^{(1)},\beta^{(1)}}^{(1d)}$ and $I_{\alpha^{(2)},\beta^{(2)}}^{(1d)}$ as well as between the images of $\mathcal{G}_{\text{eff}}^{(1d)}\left(\alpha^{(1)}, \beta^{(1)}\right)$ and $\mathcal{G}_{\text{eff}}^{(1d)}\left(\alpha^{(2)}, \beta^{(2)}\right)$. However, in general continuity

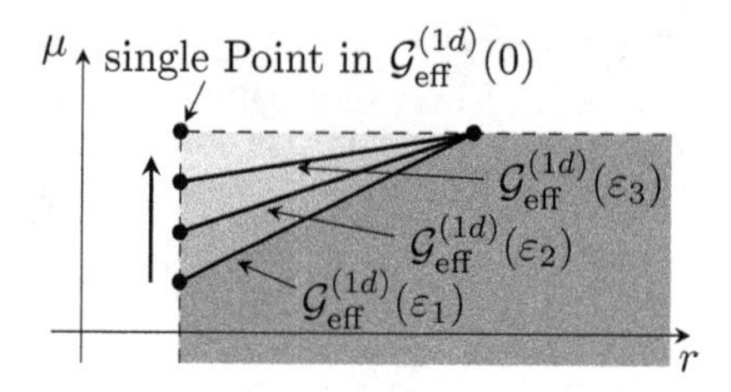

Fig. 3.8 Discontinuous transformation of $\mathcal{G}_{\text{eff}}^{(1d)}(\varepsilon)$ to $\mathcal{G}_{\text{eff}}^{(1d)}(0)$

in (3.2.19), (3.2.20), or (3.2.21) with respect to (α, β) may fail, even when risk and utility are changed continuously, as illustrated in Fig. 3.8, where we use $\varepsilon \geq 0$ as parameter instead of (α, β) and argue in the risk-utility space.

Fortunately, in our situation, the same risk function used in (3.2.2), namely,

$$\mathfrak{r}_{\alpha,\beta}^{(1d)}(x) = \max\{\alpha\mathfrak{r}_1(x), \beta\mathfrak{r}_2(x)\}, \quad x \in A,$$

with $\mathfrak{r} = (\mathfrak{r}_1, \mathfrak{r}_2)$ from (3.2.1) is of a very special type in that it is continuous and monotonically increasing in both α and β. Moreover, the used utility function is not at all depending on those two parameters. So using Definition 2.81 and Proposition 2.82 with $\mathfrak{r}_{\alpha,\beta}^{(1d)} : A \to \mathbb{R}_{\geq 0} \cup \{+\infty\}$, we get under Assumption 3.27 for $K = 2$, the following endpoints of $I_{\alpha,\beta}^{(1d)}$ and $J_{\alpha,\beta}^{(1d)}$:

$$c_{\min}^{\alpha,\beta} := \inf I_{\alpha,\beta}^{(1d)} = \inf\left\{\mathfrak{r}_{\alpha,\beta}^{(1d)}(x) \,:\, \mathfrak{u}(x) > -\infty,\ x \in A\right\} \geq 0, \tag{3.2.22}$$

$$\begin{aligned}\mu_{\max} := \sup J_{\alpha,\beta}^{(1d)} &= \sup\left\{\mathfrak{u}(x),\ \mathfrak{r}_{\alpha,\beta}^{(1d)}(x) < \infty,\quad x \in A\right\} \\ &= \sup\{\mathfrak{u}(x),\ x \in \operatorname{dom}(\mathfrak{r})\} > -\infty,\end{aligned} \tag{3.2.23}$$

$$\mu_{\min}^{\alpha,\beta} := \inf J_{\alpha,\beta}^{(1d)} = \lim_{c \searrow c_{\min}^{\alpha,\beta}} \sup\left\{\mathfrak{u}(x) \,:\, \mathfrak{r}_{\alpha,\beta}^{(1d)}(x) \leq c,\ x \in A\right\} \leq \mu_{\max}, \tag{3.2.24}$$

$$c_{\max}^{\alpha,\beta} := \sup I_{\alpha,\beta}^{(1d)} = \lim_{\mu \nearrow \mu_{\max}} \inf\left\{\mathfrak{r}_{\alpha,\beta}^{(1d)}(x) \,:\, \mathfrak{u}(x) \geq \mu,\ x \in A\right\} \geq c_{\min}^{\alpha,\beta}, \tag{3.2.25}$$

Note that $\mu_{\max}$ does not depend on α and β, since $\operatorname{dom}\left(\mathfrak{r}_{\alpha,\beta}^{(1d)}\right) = \operatorname{dom}(\mathfrak{r}_1) \cap \operatorname{dom}(\mathfrak{r}_2) = \operatorname{dom}(\mathfrak{r}) \subset A$. Furthermore, $c_{\min}^{\alpha,\beta}$ and $c_{\max}^{\alpha,\beta}$ are continuous and monotone increasing in α and β since the side conditions in (3.2.22) and (3.2.25) do not depend on the parameters.

Similarly,

$$\gamma_{\alpha,\beta}^{(1d)} = \gamma_{\alpha,\beta}^{(1d)}(\mu) = \inf\left\{\mathfrak{r}_{\alpha,\beta}^{(1d)}(x) \,:\, \mathfrak{u}(x) \geq \mu,\ x \in A\right\}, \quad \mu \in J_{\alpha,\beta}^{(1d)}, \tag{3.2.26}$$

is continuous in α, β and monotone increasing in α as well as in β. However, $\mu_{\min}^{\alpha,\beta}$ is in general neither monotone nor continuous in α and β. The discontinuity of $\mu_{\min}^{\alpha,\beta}$ can be seen, for instance, by an analog of Example 3.17, with N transformed (and the set C squished) to fit into the positive quadrant.

Before we move on to show the path-connectedness of $\mathcal{G}_{\text{eff}}^{(2d)}$, we need several lemmata, in particular more information on $\mu_{\min}^{\alpha,\beta}$.

Lemma 3.38 *Let Assumption 3.27 be satisfied for* $\mathfrak{r} = (\mathfrak{r}_1, \mathfrak{r}_2)$ *from* (3.2.1). *Consider* $\mathfrak{r}_{\alpha,\beta}^{(1d)}$ *from* (3.2.2), *for any point* (α, β) *in the unit sphere* $\mathcal{S}^1 \subset \mathbb{R}^2$ *with* $\alpha, \beta > 0$. *Let another point* $(\widetilde{\alpha}, \widetilde{\beta}) \neq (0,0)$ *be on the ray through* (α, β) *and* $(0,0)$, *i.e.,*

$$(\widetilde{\alpha}, \widetilde{\beta}) = \widetilde{c} \cdot (\alpha, \beta) \quad \textit{for some} \quad \widetilde{c} > 0. \tag{3.2.27}$$

Then the following holds:

(i) $\mathfrak{r}_{\widetilde{\alpha},\widetilde{\beta}}^{(1d)}(x) = \widetilde{c} \cdot \mathfrak{r}_{\alpha,\beta}^{(1d)}(x)$ *for all* $x \in A$,
(ii) $c_{\min}^{\widetilde{\alpha},\widetilde{\beta}} = \widetilde{c} \cdot c_{\min}^{\alpha,\beta}$,
(iii) $\mu_{\min}^{\widetilde{\alpha},\widetilde{\beta}} = \mu_{\min}^{\alpha,\beta}$.

Proof Follows immediately from the definitions. □

Much harder is the following lemma. It will be crucial for the connectedness proof to come (see Corollary 3.41).

Lemma 3.39 *Let Assumption 3.27 be satisfied with* $\mathfrak{r} = (\mathfrak{r}_1, \mathfrak{r}_2)$ *from* (3.2.1) *such that additionally* (3.2.3) *holds. Assume, moreover, that* $\mu_{\max} \in \mathbb{R} \cup \{+\infty\}$ *from* (3.2.23) *is not assumed by any* $x \in A$. *For any two points* $(\alpha^{(i)}, \beta^{(i)}) \in \mathcal{S}^1 \cap \mathbb{R}^2_{>0}$, $i = 1, 2$, *with* $\alpha^{(1)} < \alpha^{(2)}$, *set*

$$\mathcal{B} := \left\{ (\alpha, \beta) \in \mathcal{S}^1 \, : \, \alpha^{(1)} \leq \alpha \leq \alpha^{(2)}, \, \beta^{(2)} \leq \beta \leq \beta^{(1)} \right\}. \tag{3.2.28}$$

Then there exists a constant $\mu^* = \mu^*_{\mathcal{B}}(\mathfrak{r}, \mathfrak{u}; A) \in \mathbb{R}$ *such that*

$$\sup_{(\alpha,\beta) \in \mathcal{B}} \mu_{\min}^{\alpha,\beta} < \mu^* < \mu_{\max}. \tag{3.2.29}$$

In particular, we obtain $\mu^* \in J_{\alpha,\beta}^{(1d)}$ *for all* $(\alpha, \beta) \in \mathcal{B}$ *(see Fig. 3.9).*

Proof Assume on the contrary that (3.2.29) does not hold, i.e., there exists a maximizing sequence $(\alpha_n, \beta_n) \in \mathcal{B} \subset \mathcal{S}^1$ with

$$\mu_{\min}^{\alpha_n,\beta_n} \to \mu_{\max} \quad \text{as} \quad n \to \infty. \tag{3.2.30}$$

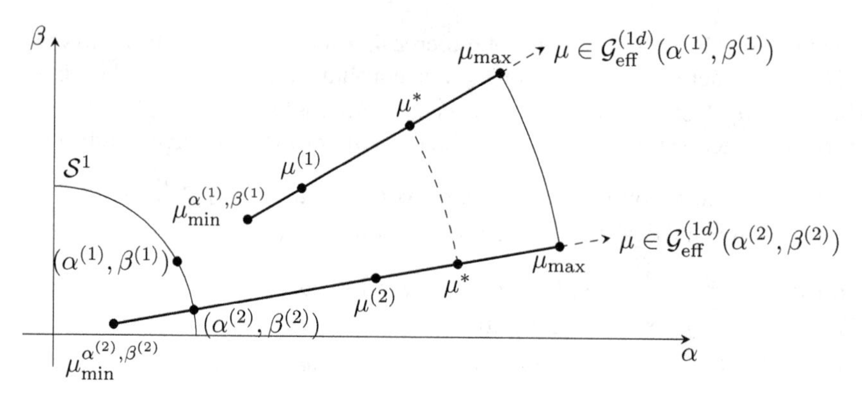

Fig. 3.9 α/μ space

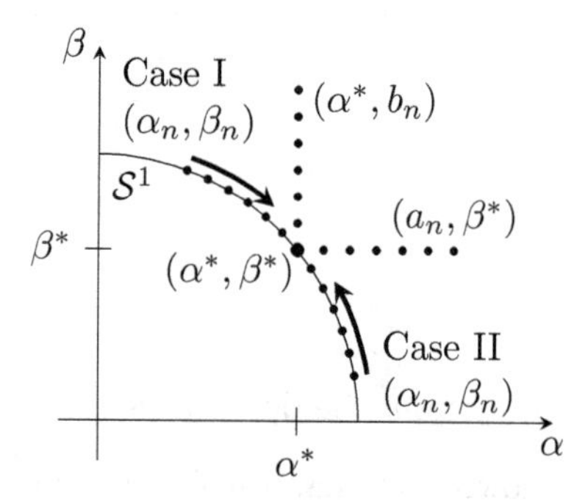

Fig. 3.10 Case I (left) and Case II (right) of (α^*, β^*)

Using compactness of $\mathcal{B} \subset \mathcal{S}^1$, we may w.l.o.g. assume that $(\alpha_n, \beta_n) \to (\alpha^*, \beta^*) \in \mathcal{B}$ as $n \to \infty$ and, since $\mu_{\max}$ is not assumed, necessarily

$$\mu_{\min}^{\alpha^*,\beta^*} < \mu_{\max} \tag{3.2.31}$$

must hold. To see this, notice that by Lemma 3.31, we have $\mathcal{G}_{\text{eff}}^{(1d)}(\alpha^*, \beta^*) \neq \emptyset$ and therefore $J_{\alpha^*,\beta^*}^{(1d)} \neq \emptyset$ as well. If (3.2.31) would not hold, then $\mu_{\min}^{\alpha^*,\beta^*} = \mu_{\max}$, which forces $\mathcal{G}_{\text{eff}}^{(1d)}(\alpha^*, \beta^*)$ to be a single point. But this is impossible since $\mu_{\max} \in \mathbb{R} \cup \{+\infty\}$ from (3.2.23) is not assumed by any $x \in A$ by assumption. Next, we can choose a subsequence of (α_n, β_n), again denoted (α_n, β_n), such that the whole subsequence either lies to the "left" (Case I) or to the "right" (Case II) of (α^*, β^*) seen from the origin, cf. Fig. 3.10.

Using (3.2.30), (3.2.31), and the scaling from Lemma 3.38 (iii), we obtain with $b_n := \frac{\beta_n}{\alpha_n}\alpha^*$ and $a_n := \frac{\alpha_n}{\beta_n}\beta^*$ either

Case I: $b_n \searrow \beta^* \quad (n \to \infty)$ and $\lim_{n\to\infty} \mu_{\min}^{\alpha_n,\beta_n} = \lim_{n\to\infty} \mu_{\min}^{\alpha^*,b_n} > \mu_{\min}^{\alpha^*,\beta^*}$,

or

Case II: $a_n \searrow \alpha^* \quad (n \to \infty)$ and $\lim_{n\to\infty} \mu_{\min}^{\alpha_n,\beta_n} = \lim_{n\to\infty} \mu_{\min}^{a_n,\beta^*} > \mu_{\min}^{\alpha^*,\beta^*}$.

Note that (α^*, b_n) and (a_n, β^*) no longer lie in $\mathcal{S}^1$.

ad Case I: In this case, there exists an $\varepsilon > 0$ such that, w.l.o.g.,

$$\mu_{\min}^{\alpha^*,b_n} \geq \mu_{\min}^{a^*,b^*} + \varepsilon \qquad \text{for all } n \in \mathbb{N}. \tag{3.2.32}$$

By definition (3.2.24),

$$\mu_{\min}^{\alpha^*,b_n} = \lim_{c \searrow c_{\min}^{\alpha^*,b_n}} \sup\left\{\mathfrak{u}(x) \,:\, \mathfrak{r}_{\alpha^*,b_n}^{(1d)}(x) \leq c,\ x \in A\right\}.$$

For fixed $n \in \mathbb{N}$, we choose $c_n \in \mathbb{R}$ such that both

$$c_{\min}^{\alpha^*,b_n} + \frac{1}{n} \geq c_n \geq c_{\min}^{\alpha^*,b_n} \to c_{\min}^{\alpha^*,b^*} \qquad (n \to \infty) \tag{3.2.33}$$

and

$$\mu_{\min}^{\alpha^*,b_n} - \frac{\varepsilon}{2} \leq \sup\left\{\mathfrak{u}(x) \,:\, \mathfrak{r}_{\alpha^*,b_n}^{(1d)}(x) \leq c_n,\ x \in A\right\} \leq \mu_{\min}^{\alpha^*,b_n} + \frac{\varepsilon}{2} \tag{3.2.34}$$

hold true. Using that $\mathfrak{r}_{\alpha,\beta}^{(1d)}$ is monotone increasing in α and β, we obtain from $b_n \geq \beta^*$ for all $n \in \mathbb{N}$ that

$$\begin{aligned}
\sup&\left\{\mathfrak{u}(x) \,:\, \mathfrak{r}_{\alpha^*,\beta^*}^{(1d)}(x) \leq c_n,\ x \in A\right\} \\
&\geq \sup\left\{\mathfrak{u}(x) \,:\, \mathfrak{r}_{\alpha^*,b_n}^{(1d)}(x) \leq c_n,\ x \in A\right\} \\
&\overset{(3.2.34)}{\geq} \mu_{\min}^{\alpha^*,b_n} - \frac{\varepsilon}{2} \overset{(3.2.32)}{\geq} \mu_{\min}^{\alpha^*,\beta^*} + \frac{\varepsilon}{2}.
\end{aligned}$$

This, however, is a contradiction, because with (3.2.24) and (3.2.33),

$$\mu_{\min}^{\alpha^*,\beta^*} = \lim_{n \to \infty} \sup\left\{\mathfrak{u}(x) \,:\, \mathfrak{r}_{\alpha^*,\beta^*}^{(1d)}(x) \leq c_n,\ x \in A\right\}$$

follows. Thus, (3.2.29) is proved in this case.

The proof of *Case II* is similar and therefore omitted. □

Last, but not least, the efficient portfolio map is continuous in α and β in the following sense.

Lemma 3.40 (Efficient Portfolio Map; Continuity with Respect to Parameters) *Let Assumption 3.27 for $K = 2$ be given. Then, using $\mathcal{B} \subset \mathcal{S}^1$ from (3.2.28), and using the efficient portfolio map $X^{(1d)} = X_{\alpha,\beta}^{(1d)}$ from (3.2.5) and $\gamma_{\alpha,\beta}^{(1d)}$ from (3.2.26), the map*

$$z_{\alpha,\beta}^* \colon \mathcal{B} \to A, \quad (\alpha, \beta) \mapsto z_{\alpha,\beta}^*(\overline{\mu}) := X_{\alpha,\beta}^{(1d)}\left(\gamma_{\alpha,\beta}^{(1d)}(\overline{\mu}), \overline{\mu}\right)$$

is continuous in α *(with* $\beta = \sqrt{1-\alpha^2}$*), provided* $\overline{\mu}$ *is fixed and satisfies* $\overline{\mu} \in J^{(1d)}_{\alpha,\beta}$ *for all* $(\alpha, \beta) \in \mathcal{B}$.

Proof The proof is along the lines of the proof of Theorem 2.79 (b) and therefore omitted. □

Now we have collected all ingredients to obtain the connectedness of $\mathcal{G}^{(2d)}_{\text{eff}}$.

Corollary 3.41 (Path-Connected Efficient Frontier; Two-Dim. Vector Risk) *Let Assumption 3.27 hold true for* $K = 2$ *with additionally* (3.2.3) *and, moreover, components* $\mathfrak{r}_i \colon A \to \mathbb{R}_{>0} \cup \{+\infty\}$ $(i = 1, 2)$ *which are continuous on* $\text{dom}(\mathfrak{r})$. *Then the set* $\mathcal{G}^{(2d)}_{\text{eff}} = \mathcal{G}^{(2d)}_{\text{eff}}(\mathfrak{r}, \mathfrak{u};\, A) \subset \mathbb{R}^3$ *is path-connected.*

Proof We discuss again two cases.

Case I: $\mu_{\max}$ *is assumed* by some $\overline{x} \in A$, i.e., $\mu_{\max} = \mathfrak{u}(\overline{x}) = \max J^{(1d)}_{\alpha,\beta} \in \mathbb{R}$.

Let as in (3.2.18) $g^{(i)} = \left(r_1^{(i)}, r_2^{(i)}, \mu^{(i)}\right) \in \mathcal{G}^{(2d)}_{\text{eff}}$, $i = 1, 2$ be two arbitrary but fixed points in $\mathcal{G}^{(2d)}_{\text{eff}}$. As seen before, there are parameters $\left(\alpha^{(i)}, \beta^{(i)}\right) \in \mathcal{S}^1$, $i = 1, 2$ such that $g^{(i)}$ is in the image of the map (3.2.15) on $I^{(1d)}_{\alpha^{(i)},\beta^{(i)}}$ or $\mathcal{G}^{(1d)}_{\text{eff}}\left(\alpha^{(i)}, \beta^{(i)}\right)$ (see Fig. 3.9). Since the latter are connected, $g^{(i)}$ is connected within $\mathcal{G}^{(2d)}_{\text{eff}}$ to

$$\Big(\mathfrak{r}\left(\widetilde{z}^{(i)}\right),\, \underbrace{\mathfrak{u}\left(\widetilde{z}^{(i)}\right)}_{=\mu_{\max}}\Big) \in \mathcal{G}^{(2d)}_{\text{eff}},$$

where $\widetilde{z}^{(i)} = X^{(1d)}_{\alpha^{(i)},\beta^{(i)}}\left(\gamma^{(1d)}_{\alpha^{(i)},\beta^{(i)}}(\mu_{\max}),\, \mu_{\max}\right)$, $i = 1, 2$ (with $X^{(1d)} = X^{(1d)}_{\alpha,\beta}$ from (3.2.5)) are two efficient portfolios.

Finally, by the continuity of $\gamma^{(1d)}_{\alpha,\beta}$ in α and β (see (3.2.26)) and by Lemma 3.40 applied with $\overline{\mu} := \mu_{\max}$, $\left(\mathfrak{r}(\widetilde{z}^{(i)}), \mathfrak{u}(\widetilde{z}^{(i)})\right) \in \mathcal{G}^{(2d)}_{\text{eff}}$, $i = 1, 2$ are also connected in $\mathcal{G}^{(2d)}_{\text{eff}}$ by the path

$$(\mathfrak{r}(z_\lambda), \mathfrak{u}(z_\lambda)) \in \mathcal{G}^{(2d)}_{\text{eff}}, \quad \lambda \in [0, 1], \tag{3.2.35}$$

where $z_\lambda = X^{(1d)}_{\alpha_\lambda,\beta_\lambda}\left(\gamma^{(1d)}_{\alpha_\lambda,\beta_\lambda}(\mu_{\max}),\, \mu_{\max}\right)$ and $(\alpha_\lambda, \beta_\lambda) \in \mathcal{S}^1$, $\lambda \in [0, 1]$, connect $\left(\alpha^{(1)}, \beta^{(1)}\right)$ with $\left(\alpha^{(2)}, \beta^{(2)}\right)$ in $\mathcal{S}^1$. The path in (3.2.35) is continuous in λ since by assumption $\mathfrak{r}$ is continuous on $\text{dom}(\mathfrak{r})$ and continuity of z_λ follows from Lemma 3.40. On the other hand, $\mathfrak{u}(z_\lambda) = \mu_{\max}$ is constant anyway.

Altogether we obtain a continuous path connecting $g^{(1)}$ with $g^{(2)}$ within $\mathcal{G}^{(2d)}_{\text{eff}}$.

Case II: $\mu_{\max}$ *is not assumed.*

In this case, we have $\mu_{\max} = \sup J^{(1d)}_{\alpha,\beta}$ and $J^{(1d)}_{\alpha,\beta}$ is a right-open nondegenerate interval in which the left-hand side $\mu^{\alpha,\beta}_{\min}$ varies with α and β. In this case, a similar path as in *Case I* may be constructed via

$$\tilde{z}^{(i)} = X^{(1d)}_{\alpha^{(i)},\,\beta^{(i)}}\left(\gamma^{(1d)}_{\alpha^{(i)},\beta^{(i)}}\left(\mu^*\right), \mu^*\right)$$

with μ^* from Lemma 3.39 and applying Lemma 3.40 with $\overline{\mu} = \mu^*$ (see again Fig. 3.9). □

Remark 3.42 *As stated before, for the path-connectedness of $\mathcal{G}^{(2d)}_{\text{eff}}$, it suffices to assume Assumption 3.27 for $\mathfrak{r} = (\mathfrak{r}_1, \mathfrak{r}_2)$ whose components $\mathfrak{r}_i \colon A \to \mathbb{R}_{\geq 0} \cup \{+\infty\}$, $i = 1, 2$, are continuous on* $\operatorname{dom}(\mathfrak{r})$. *In fact, even only bounded from below risk components would work as well (see Remark 3.29 (b)).*

3.2.2 Connectivity for Higher-Dimensional Risk Vectors

The connectedness proof in the last subsection was only worked out for a $K = 2$ dimensional risk vector. However, for a general K-dimensional risk vector $\mathfrak{r} = (\mathfrak{r}_1, \dots, \mathfrak{r}_K)$ with $K \geq 2$, we again can construct a (K-parameter) family of scalar risk functions:

$$\mathfrak{r}^{(1d)}_{(\alpha_1,\dots,\alpha_K)}(x) := \max\left\{\alpha_1 \mathfrak{r}_1(x), \dots, \alpha_K \mathfrak{r}_K(x)\right), \quad x \in A$$

with parameters $\alpha_1, \dots, \alpha_K > 0$ and reasoning similarly as above we obtain:

Theorem 3.43 (Path-Connected Efficient Frontier; Vector Risk) *Let Assumption 3.27 be satisfied for $K \geq 2$. Assume additionally that the K-dimensional risk vector function $\mathfrak{r} = (\mathfrak{r}_1, \dots, \mathfrak{r}_K)$ has components $\mathfrak{r}_i \colon A \to \mathbb{R}_{\geq 0} \cup \{+\infty\}$, $i = 1, \dots, K$, which are continuous on* $\operatorname{dom}(\mathfrak{r})$. *Then the efficient frontier $\mathcal{G}^{(K-\text{dim})}_{\text{eff}} := \mathcal{G}^{(Kd)}_{\text{eff}}(\mathfrak{r}, \mathfrak{u};\, A) \subset \mathbb{R}^{K+1}$ is path-connected.*

Proof Works similarly as the $K = 2$ case, where once again we may assume w.l.o.g. that $\mathfrak{r}_i(x) > \kappa \geq 0$ holds for all $x \in A$, $i = 1, \dots, K$ (see (3.2.3)).

For instance, in an analog of Lemma 3.37, for $(\overline{r}_1, \dots, \overline{r}_K, \overline{\mu}) \in \mathcal{G}^{(K-\text{dim})}_{\text{eff}} = \mathcal{G}^{(Kd)}_{\text{eff}}(\mathfrak{r}, \mathfrak{u};\, A)$, one has to choose $\left(\frac{1}{\alpha_1}, \dots, \frac{1}{\alpha_K}\right)$ on the same ray through $(0, \dots, 0) \in \mathbb{R}^K$ as $(\overline{r}_1, \dots, \overline{r}_K)$. We leave the details to the reader. □

3.3 Efficient Portfolios in the Vector Risk Case

We now turn to the efficient portfolios in the vector risk case (see Sect. 2.4 for the one-dimensional risk case).

3.3.1 Existence and Uniqueness of Efficient Portfolios

In general, corresponding to each efficient point in the (vector risk, utility) space, there exist (multiple) efficient portfolios. Uniqueness needs additional assumptions. Thus, efficient portfolios can be viewed as a multifunction of the efficient points in the (vector risk, utility) space. Below we show that this multifunction is upper semi-continuous. For the convenience of the reader, we recall the definition of upper semi-continuity for multifunctions (see [19, Section 5.1]).

Definition 3.44 (Upper Semi-continuity of Multifunctions) *Let a multifunction $F\colon \mathbb{R}^N \to 2^{\mathbb{R}^M}$ be given. The domain of F is defined by* $\mathrm{dom}(F) := \{y \in \mathbb{R}^N : F(y) \neq \emptyset\}$. *We say that F is* upper semi-continuous *at $\bar{y} \in \mathrm{dom}(F)$ if $(y_n, z_n) \in \mathrm{graph}(F)$ converges to $(\bar{y}, \bar{z})$ implies that $\bar{z} \in F(\bar{y})$.*

Remark 3.45 *It is not hard to check that a function $f\colon \mathbb{R} \to \mathbb{R} \cup \{-\infty\}$ is upper semi-continuous at $\bar{t} \in \mathrm{dom}(f) := \{t : f(t) > -\infty\}$ if and only if the multifunction $F(t) := (-\infty, f(t)]$ is upper semi-continuous.*

Now we can state precisely what has been alluded to above (see Theorem 2.79 for the one-dimensional risk case).

Theorem 3.46 (Uniqueness of Efficient Portfolios for the Vector Risk-Utility Trade-Off) *Let Assumption 3.11 be satisfied for $K \geq 2$. Denote $X^{(Kd)}(r, \mu) \subset A \subset \mathbb{R}^{M+1}$ the set of efficient portfolios corresponding to $(r, \mu) \in \mathcal{G}^{(Kd)}_{\text{eff}}(\mathfrak{r}, \mathfrak{u}; A)$. Then $X^{(Kd)}$ is an upper semi-continuous multifunction on its domain $\mathcal{G}^{(Kd)}_{\text{eff}}(\mathfrak{r}, \mathfrak{u}; A)$.*

In addition, suppose that either the vector risk function $\mathfrak{r}$ satisfies (r2s) *in Assumption 3.1 or the utility function satisfies condition* (u2s) *in Assumption 2.29. Then, $X^{(Kd)}$ is single-valued and continuous. In particular, $X^{(Kd)}\colon \mathcal{G}^{(Kd)}_{\text{eff}}(\mathfrak{r}, \mathfrak{u}; A) \to \mathbb{R}^{M+1}$ is also injective.*

Proof Note that Assumption 3.11 ensures that $\mathcal{G}^{(Kd)}_{\text{eff}}(\mathfrak{r}, \mathfrak{u}; A)$ is non-empty (cf. Proposition 3.12). Next, we show that the multifunction $X := X^{(Kd)}$ is upper semi-continuous on $\mathcal{G}^{(Kd)}_{\text{eff}}(\mathfrak{r}, \mathfrak{u}; A)$. Consider $(\bar{r}, \bar{\mu}) \in \mathcal{G}^{(Kd)}_{\text{eff}}(\mathfrak{r}, \mathfrak{u}; A)$ and let $(r^{(n)}, \mu^{(n)}) \in \mathcal{G}^{(Kd)}_{\text{eff}}(\mathfrak{r}, \mathfrak{u}; A)$ be a sequence that converges to $(\bar{r}, \bar{\mu}) \in \mathbb{R}^{K+1}$ and let $x^{(n)} \in X(r^{(n)}, \mu^{(n)})$ converge to x^*. Since A is closed, $x^* \in A$. Moreover, since $\mathfrak{r}$ is lower semi-continuous and $\mathfrak{u}$ is upper semi-continuous, we have $\mathfrak{r}(x^*) \leq \liminf_{n\to\infty} \mathfrak{r}(x^{(n)}) \leq \liminf_{n\to\infty} r^{(n)} = \bar{r}$ and $\mathfrak{u}(x^*) \geq \limsup_{n\to\infty} \mathfrak{u}(x^{(n)}) \geq \limsup_{n\to\infty} \mu^{(n)} = \bar{\mu}$. Since $(\bar{r}, \bar{\mu}) \in \mathcal{G}^{(Kd)}_{\text{eff}}(\mathfrak{r}, \mathfrak{u}; A)$, we get $\mathfrak{r}(x^*) = \bar{r}$ and $\mathfrak{u}(x^*) = \bar{\mu}$. Thus, x^* is an efficient portfolio so that $x^* \in X(\bar{r}, \bar{\mu})$. This verifies the upper semi-continuity of the multifunction X.

We now turn to the uniqueness. Let us focus on the case when the vector risk function $\mathfrak{r}$ satisfies condition (r2s), i.e., at least one component $\mathfrak{r}_k$, $k \in \{1, \ldots, K\}$ of $\mathfrak{r}$ satisfies (r2s) from Assumption 2.20. Suppose that portfolios

$y^* \neq z^*$ both belong to $X(r,\mu)$ for some fixed $(r,\mu) \in \mathcal{G}_{\text{eff}}^{(Kd)}(\mathfrak{r},\mathfrak{u};\, A)$. Then we must have $\mathfrak{r}(y^*) = \mathfrak{r}(z^*) = r$, $\mathfrak{u}(y^*) = \mathfrak{u}(z^*) = \mu$, and $y^*, z^* \in A$. Since A is convex, $x^* := (y^* + z^*)/2 \in A$. Condition (u2) from Assumption 3.11 (iii) implies that $\mathfrak{u}(x^*) \geq \mu$ and due to (r2) for $\mathfrak{r}$ from Assumption 3.11 (ii), we obtain $\mathfrak{r}(x^*) \leq r$ where the inequality is strict for at least one component due to (r2s). Hence, (r,μ) cannot lie on the efficient frontier, a contradiction. Thus, X is single-valued. Since we already know that X as a multifunction is upper semi-continuous, X must be continuous.

The proof of uniqueness for the case when the utility function satisfies condition (u2s) is similar. □

Corollary 3.47 (Topological Properties of the Efficient Portfolio Set) *In the situation of Theorem 3.14 and Theorem 3.43, the sets* $N = N(\mathfrak{r},\mathfrak{u};\, A)$ *from* (3.1.21) *and* $M_k = M_k(\mathfrak{r},\mathfrak{u};\, A)$, $k = 1,\ldots,K$, *from* (3.1.24) *are path-connected. Furthermore, the efficient portfolio map* $X = X^{(Kd)}(r,\mu)$, $(r,\mu) \in \mathcal{G}_{\text{eff}}^{(Kd)}(\mathfrak{r},\mathfrak{u};\, A)$ *from Theorem 3.46 is continuous in this situation, and thus, the set of efficient portfolios is path-connected as well. Moreover, the efficient portfolios can be parameterized as graph over* N *or* M_k*, respectively, i.e., both*

$$\overline{x}(r) := X^{(Kd)}\left(r, \nu(r)\right), \quad r \in N, \tag{3.3.1}$$

and

$$\overline{y}_k\left(\widehat{r}_k, \mu\right) := X^{(Kd)}\left(\widehat{r}_k, \gamma_k\left(\widehat{r}_k, \mu\right), \mu\right), \quad \left(\widehat{r}_k, \mu\right) \in M_k, \tag{3.3.2}$$

yield all efficient portfolios for a given $\mathfrak{r}$, $\mathfrak{u}$, *and* A.

Also, $\overline{x}$ *and* $\overline{y}_k$ *are continuous in the relative interior of* N *and* M_k *and also at boundary points of* N *and* M_k *approached from the interior along one-dimensional lines parallel to coordinate axes, respectively.*

Proof N as well as M_k, $k = 1,\ldots,K$ are path-connected as projections of the path–connected set $\mathcal{G}_{\text{eff}}^{(Kd)}(\mathfrak{r},\mathfrak{u};\, A)$ to a K-dimensional hyperplane of $\mathbb{R}^{K+1}$ (see (3.1.21) and (3.1.24)). Similarly, the set of efficient portfolios is path-connected as image of the continuous efficient portfolio map $X^{(Kd)}$. The remaining claims follow immediately from Corollary 3.24 and Theorem 3.46. □

3.3.2 Connections of Scalar Risk and Vector Risk Theory

In the following, we derive some helpful results connecting the scalar risk theory and the vector risk theory.

For a column vector $v \in \mathbb{R}^K_{\geq 0}$, we define $\mathfrak{r}^v := \langle v, \mathfrak{r}^\top \rangle = v^\top \cdot \mathfrak{r}^\top$ (see Remark 3.3). Then $\mathfrak{r}^v$ is a one-dimensional risk function that shares the properties of the components of $\mathfrak{r}$. It is easy to show that

$$\mathcal{G}^{(1d)}(\mathfrak{r}^v, \mathfrak{u}; A) = \{(v^\top \cdot r^\top, \mu) : (r, \mu) \in \mathcal{G}^{(Kd)}(\mathfrak{r}, \mathfrak{u}; A)\}. \tag{3.3.3}$$

However, the relationship between the Pareto efficient frontier of a portfolio problem involving a vector risk measure and its one-dimensional projection is not so evident. Let us define

$$\begin{aligned} \mu \mapsto \gamma^v(\mu) &:= \inf\{r : (r, \mu) \in \mathcal{G}^{(1d)}(\mathfrak{r}^v, \mathfrak{u}; A)\} \\ &= \inf\{\mathfrak{r}^v(x) : \mathfrak{u}(x) \geq \mu, \ x \in A\}, \end{aligned} \tag{3.3.4}$$

and

$$\begin{aligned} r \mapsto \nu^v(r) &:= \sup\{\mu : (r, \mu) \in \mathcal{G}^{(1d)}(\mathfrak{r}^v, \mathfrak{u}; A)\} \\ &= \sup\{\mathfrak{u}(x) : \mathfrak{r}^v(x) \leq r, \ x \in A\}. \end{aligned} \tag{3.3.5}$$

Using Proposition 2.82, we get

Proposition 3.48 (Endpoint Properties 1) *Let Assumption 3.11 be satisfied for $K \geq 2$ for a set of admissible portfolios A and extended-valued risk and utility functions $\mathfrak{r}$ and $\mathfrak{u}$, respectively. Then $\gamma^v : \mathbb{R} \to \mathbb{R} \cup \{+\infty\}$ and $\nu^v : \mathbb{R} \to \mathbb{R} \cup \{-\infty\}$ are well-defined. Set*

$$I^v := \mathrm{dom}(\nu^v) \cap \mathrm{range}(\gamma^v) \subset \mathbb{R} \quad \text{and} \quad J^v := \mathrm{dom}(\gamma^v) \cap \mathrm{range}(\nu^v) \subset \mathbb{R},$$

Then, I^v and J^v are intervals and according to Theorem 2.76 (c), both

$$\mathrm{graph}\left(\nu^v_{|I^v}\right) \quad \text{and} \quad \widehat{P}\left[\mathrm{graph}\left(\gamma^v_{|J^v}\right)\right] \tag{3.3.6}$$

represent the efficient frontier $\mathcal{G}^{(1d)}(\mathfrak{r}^v, \mathfrak{u}; A)$. We further define

$$\begin{aligned} r_{\min}(v) &:= \inf I^v, \\ r_{\max}(v) &:= \sup I^v, \\ \mu_{\min}(v) &:= \inf J^v, \end{aligned}$$

and

$$\mu_{\max}(v) := \sup J^v.$$

Then, the following holds:

$$r_{\min}(v) = \inf\{\mathfrak{r}^v(x) : \mathfrak{u}(x) > -\infty, \ x \in A\} < +\infty, \tag{3.3.7}$$

$$\mu_{\max}(v) = \sup\{\mathfrak{u}(x) : \mathfrak{r}^v(x) < +\infty, \ x \in A\} > -\infty, \tag{3.3.8}$$

$$\mu_{\min}(v) = \lim_{r \searrow r_{\min}(v)} \sup\{\mathfrak{u}(x) : \mathfrak{r}^v(x) \leq r, \ x \in A\} \leq \mu_{\max}(v), \tag{3.3.9}$$

and

$$r_{\max}(v) = \lim_{\mu \nearrow \mu_{\max}(v)} \inf\{\mathfrak{r}^v(x) : \mathfrak{u}(x) \geq \mu,\ x \in A\} \geq r_{\min}(v). \tag{3.3.10}$$

Proof We only have to guarantee that $\mathfrak{r}^v$, $\mathfrak{u}$, and A satisfy Assumption 2.64. But this is evident with Assumption 3.11. Then, γ^v and ν^v are well-defined according to Proposition 2.68 and I^v and J^v are intervals with Corollary 2.74. Finally, Proposition 2.82 yields Eqs. (3.3.7) to (3.3.10). □

Using the properties of sup and inf, we can derive the following properties for the quantities defined in Proposition 3.48.

Proposition 3.49 (Endpoint Properties 2) *We set $\mu^*_{\max} := \mu_{\max}(\vec{1})$, where $\vec{1} = (1, \ldots, 1)^\top \in \mathbb{R}^K_{\geq 0}$ and $\mu^{**}_{\max} := \sup\{\mathfrak{u}(x) : x \in A\} \in \mathbb{R} \cup \{+\infty\}$. The quantities defined in Proposition 3.48 have the following properties:*

(a) $\mu_{\max}(v) = \mu^*_{\max}$ is independent of v for all $v \in \operatorname{int} \mathbb{R}^K_{\geq 0}$ since then $\operatorname{dom}(\mathfrak{r}^v) = \operatorname{dom}(\mathfrak{r})$.
(b) $\mu^{**}_{\max} \geq \mu_{\max}(v) \geq \mu^*_{\max}$ for $v \in \partial \mathbb{R}^K_{\geq 0}$ because here $\operatorname{dom}(\mathfrak{r}) \subset \operatorname{dom}(\mathfrak{r}^v)$.
(c) Both $r_{\min}(v)$ and $r_{\max}(v)$ are positive homogeneous and super-additive on $\mathbb{R}^K_{\geq 0}$, i.e., for instance,

$$r_{\min}(u + v) \geq r_{\min}(u) + r_{\min}(v), \quad \text{for all} \quad u, v \in \mathbb{R}^K_{\geq 0}.$$

(d) $\mu_{\min}(v)$ is invariant with respect to the variable multiplying by a positive constant, i.e.,

$$\mu_{\min}(tv) = \mu_{\min}(v), \ \text{for all}\ t > 0.$$

Moreover, for all $u, v \in \mathbb{R}^K_{\geq 0}$, $\mu_{\min}(u + v) \geq \min(\mu_{\min}(u), \mu_{\min}(v))$.

Proof Items (a) and (b) are straightforward. We verify (c). Let us focus on $r_{\min}(v)$ since the arguments for $r_{\max}(v)$ are similar.

$$\begin{aligned} r_{\min}(u + v) &= \inf\{\mathfrak{r}^{u+v}(x) : \mathfrak{u}(x) > -\infty, x \in A\} \\ &= \inf\{\mathfrak{r}^u(x) + \mathfrak{r}^v(x) : \mathfrak{u}(x) > -\infty, x \in A\} \\ &\geq \inf\{\mathfrak{r}^u(x) : \mathfrak{u}(x) > -\infty, x \in A\} \\ &\quad + \inf\{\mathfrak{r}^v(x) : \mathfrak{u}(x) > -\infty, x \in A\} \\ &= r_{\min}(u) + r_{\min}(v). \end{aligned} \tag{3.3.11}$$

Now, we turn to (d). For any $t > 0$,

$$\begin{aligned} \mu_{\min}(tv) &= \lim_{r \searrow r_{\min}(tv)} \sup\{\mathfrak{u}(x)] : \mathfrak{r}^{tv}(x) \leq r, x \in A\} \\ &= \lim_{r \searrow t r_{\min}(v)} \sup\{\mathfrak{u}(x) : t\mathfrak{r}^v(x) \leq r, x \in A\} \\ &= \lim_{r/t \searrow r_{\min}(v)} \sup\{\mathfrak{u}(x) : \mathfrak{r}^v(x) \leq r/t, x \in A\} = \mu_{\min}(v). \end{aligned} \tag{3.3.12}$$

This verifies that $\mu_{\min}(v)$ is invariant with respect to the variable multiplying by a positive constant. Finally, since $r_{\min}(u+v) \geq r_{\min}(u) + r_{\min}(v)$, we have

$$\begin{aligned}
\mu_{\min}(u+v) = & \lim_{r\searrow r_{\min}(u+v)} \sup\{\mathfrak{u}(x) : \mathfrak{r}^{u+v}(x) \leq r, x \in A\} \qquad (3.3.13)\\
\geq & \lim_{r\searrow r_{\min}(u)+r_{\min}(v)} \sup\{\mathfrak{u}(x) : \mathfrak{r}^{u}(x) + \mathfrak{r}^{v}(x) \leq r, x \in A\}\\
\geq & \lim_{\substack{r_u\searrow r_{\min}(u)\\ r_v\searrow r_{\min}(v)}} \sup\{\mathfrak{u}(x) : \mathfrak{r}^{u}(x) + \mathfrak{r}^{v}(x) \leq r_u + r_v, x \in A\}\\
\geq & \lim_{\substack{r_u\searrow r_{\min}(u)\\ r_v\searrow r_{\min}(v)}} \sup\{\mathfrak{u}(x) : \mathfrak{r}^{u}(x) \leq r_u, \mathfrak{r}^{v}(x) \leq r_v, x \in A\}\\
= & \lim_{\substack{r_u\searrow r_{\min}(u)\\ r_v\searrow r_{\min}(v)}} \sup \Big[\{\mathfrak{u}(x) : \mathfrak{r}^{u}(x) \leq r_u, x \in A\} \\
& \qquad \cap \{\mathfrak{u}(x) : \mathfrak{r}^{v}(x) \leq r_v, x \in A\}\Big]\\
= & \lim_{\substack{r_u\searrow r_{\min}(u)\\ r_v\searrow r_{\min}(v)}} \min \Big[\sup\{\mathfrak{u}(x) : \mathfrak{r}^{u}(x) \leq r_u, x \in A\},\\
& \qquad \sup\{\mathfrak{u}(x) : \mathfrak{r}^{v}(x) \leq r_v, x \in A\}\Big]\\
= & \min \Big[\lim_{r_u\searrow r_{\min}(u)} \sup\{\mathfrak{u}(x) : \mathfrak{r}^{u}(x) \leq r_u, x \in A\},\\
& \qquad \lim_{r_v\searrow r_{\min}(v)} \sup\{\mathfrak{u}(x) : \mathfrak{r}^{v}(x) \leq r_v, x \in A\}\Big]\\
= & \min(\mu_{\min}(u), \mu_{\min}(v)),
\end{aligned}$$

which completes the proof. □

3.3.3 Markowitz Portfolios with Tracking Error

We want to close this section by continuing the discussion of the Markowitz portfolios with the close-to-benchmark constraint from Example 2.85, but now with a vector risk point of view.

Example 3.50 (Continuation: Markowitz Portfolios with Close-to-Benchmark Constraint; Tracking Error) *As already mentioned in Remark 2.87, we can view* (2.4.33)–(2.4.36) *from a vector risk point of view, i.e., with*

$$A := \left\{x = \left(0, \widehat{x}^\top\right)^\top \in \mathbb{R}^{M+1} : \widehat{S}_0^\top \widehat{x} = 1\right\}, \qquad (3.3.14)$$

the expected utility $\mathfrak{u}(x) = \mathrm{E}\left[\widehat{S}_1^\top \widehat{x}\right]$, *half the variance*

$$\mathfrak{r}_1(x) = \mathfrak{r}_{\mathrm{Var}}(x) = \frac{1}{2}\sigma^2\left(\widehat{S}_1^\top \widehat{x}\right) = \frac{1}{2}\widehat{x}^\top \Sigma \widehat{x} \qquad (3.3.15)$$

as first risk function and additionally the benchmark related tracking error risk function

$$\mathfrak{r}_2(x) = \mathfrak{r}_{\text{Track}}(x) = \frac{1}{2}\left(\widehat{x} - \widehat{x}^*\right)^\top \Sigma \left(\widehat{x} - \widehat{x}^*\right), \tag{3.3.16}$$

where $\Sigma \in \mathbb{R}^{M \times M}$ *is the positive definite covariant matrix for a financial market* $S_\bullet$ *as in Definition 2.3 (see* (2.4.57)*). Thus, we study a two-parameter problem:*

$$\frac{1}{2}\sigma^2 = \min_{\left(0,\widehat{x}^\top\right)^\top \in A} \frac{1}{2}\widehat{x}^\top \Sigma \widehat{x} \tag{3.3.17}$$

subject to

$$\mathrm{E}\left[\widehat{S}_1^\top \widehat{x}\right] \geq \mu \tag{3.3.18}$$

$$\frac{1}{2}\left(\widehat{x} - \widehat{x}^*\right)^\top \Sigma \left(\widehat{x} - \widehat{x}^*\right) \leq \varrho, \tag{3.3.19}$$

which, in the notation of (3.1.11), *reads as*

$$\gamma_1(\varrho, \mu) := \inf\left\{\mathfrak{r}_1(x) \,:\, \mathfrak{u}(x) \geq \mu,\ \mathfrak{r}_2(x) \leq \varrho,\ x = \left(0, \widehat{x}^\top\right)^\top \in A\right\} \tag{3.3.20}$$

for $\varrho \geq 0$ *and* $\mu \in \mathbb{R}$. *We note that* $\gamma_1(\varrho, \mu) = \gamma_\varrho(\mu)$ *(see* (2.4.56)*).* .

The question is, which points on the graph of γ_1 *represent* (2d)*-efficient portfolios for* (3.3.17)–(3.3.19) *in terms of Definition 3.7; i.e., which points are contained in* $\mathcal{G}_{\text{eff}}^{(2d)}(\mathfrak{r}_1, \mathfrak{r}_2, \mathfrak{u};\ A)$*?*

A partial answer has already been arranged in Example 2.85 under the condition

$$\widehat{x}_{\text{minVar}} \in D_\varrho = \left\{\widehat{x} \in \mathbb{R}^M \,:\, \frac{1}{2}\left(\widehat{x} - \widehat{x}^*\right)^\top \Sigma \left(\widehat{x} - \widehat{x}^*\right) \leq \varrho\right\}, \tag{3.3.21}$$

cf. (2.4.52), *for the minimum variance portfolio*

$$x_{\text{minVar}} = \left(0, \widehat{x}_{\text{minVar}}^\top\right)^\top \in \left\{x = (0, \widehat{x}^\top)^\top : \widehat{S}_1^\top \widehat{x} = 1\right\} \subset \mathbb{R}^{M+1}. \tag{3.3.22}$$

For further reference, we denote the (vector risk, utility) values of x_{minVar} *by*

$$\varrho_{\text{minVar}} := \mathfrak{r}_{\text{Track}}(x_{\text{minVar}}),\ \mu_{\text{minVar}} = \mathrm{E}\left[\widehat{S}_1^\top \widehat{x}_{\text{minVar}}\right] \tag{3.3.23}$$

and

$$\frac{1}{2}\sigma_{\text{minVar}}^2 := \mathfrak{r}_{\text{Var}}\left(x_{\text{minVar}}\right). \tag{3.3.24}$$

Then, clearly $\widehat{x}_{\text{minVar}} \in D_{\varrho_{\text{minVar}}}$ *and* ϱ_{minVar} *is the minimal* ϱ *with that property, i.e.,*

$$\varrho_{\text{minVar}} = \min\left\{\varrho \colon \widehat{x}_{\text{minVar}} \in D_\varrho\right\}. \tag{3.3.25}$$

Thus, the condition $\widehat{x}_{\text{minVar}} \in D_\varrho$ *of Example 2.85 is satisfied if and only if* $\varrho \geq \varrho_{\text{minVar}}$. *In the following discussion of* $\mathcal{G}^{(2d)}_{\text{eff}}(\mathfrak{r}_1, \mathfrak{r}_2, \mathfrak{u};\, A)$, *we distingush two cases.*

Case *(A):* $\varrho \geq \varrho_{\text{minVar}}$. *Note that for* $\varrho \geq \varrho_{\text{minVar}}$ *fixed, we have given a sketch of* $\mathcal{G}^{(1d)}_{\text{eff}}(\varrho) := \mathcal{G}^{(1d)}_{\text{eff}}(\mathfrak{r} = \mathfrak{r}_1, \mathfrak{u};\, A_\varrho)$ *in Fig. 2.6b, which coincides locally with the graph of* $\gamma_1(\varrho, \cdot)$ *since by* (2.4.53) *it is*

$$A_\varrho = \left\{ x = \left(0, \widehat{x}^\top\right)^\top \in \mathbb{R}^{M+1} \,:\, \widehat{S}_0^\top \widehat{x} = 1 \text{ and } \widehat{x} \in D_\varrho \right\}. \tag{3.3.26}$$

However, at least for $\varrho > \varrho_{\text{minVar}}$, *all but one of the points of* $\mathcal{G}^{(1d)}_{\text{eff}}(\varrho)$ *which lie on the Markowitz bullet cannot be efficient for the two-parameter problem in* (3.3.20), *because the graph of* γ_1 *is here locally constant in* ϱ *since the Markowitz efficient portfolios cannot be improved any more. The only point on the Markowitz bullet which is efficient for the two-parameter problem* (3.3.20) *as well for fixed* $\varrho > \varrho_{\text{minVar}}$ *is*

$$(\gamma_1(\varrho, \mu^*), \varrho, \mu^*) \in \mathcal{G}^{(2d)}_{\text{eff}}(\mathfrak{r}_1, \mathfrak{r}_2, \mathfrak{u};\, A), \tag{3.3.27}$$

where $\mu^* = \mu^*(\varrho)$ *from Remark 2.86 (see again Fig. 2.6).*

Similarly, all remaining (1d)*-efficient points* $(\gamma_\varrho(\mu), \mu) \in \mathcal{G}^{(1d)}_{\text{eff}}(\varrho)$ *of Fig. 2.6b for* $\mu \in (\mu^*, \mu_{\max}]$ *with*

$$\mu_{\max} = \mu_{\max}(\varrho) := \sup\left\{ \mathrm{E}\left[\widehat{S}_1^\top \widehat{x}\right] \,:\, (0, \widehat{x}^\top)^\top \in A_\varrho \right\} \in \mathbb{R}, \tag{3.3.28}$$

from (2.4.55) *are also* (2d)*-efficient for the vector risk problem* (3.3.20). *This follows since* γ_1, *for such* μ *fixed, is locally strictly decreasing in* ϱ *until a point representing a Markowitz efficient portfolio is reached (cf. Corollary 3.15), i.e., together with* (3.3.27), *we obtain*

$$(\gamma_1(\varrho, \mu), \varrho, \mu) \in \mathcal{G}^{(2d)}_{\text{eff}}(\mathfrak{r}_1, \mathfrak{r}_2, \mathfrak{u};\, A), \quad \textit{for } \varrho > \varrho_{\text{minVar}} \textit{ and } \mu \in [\mu^*(\varrho), \mu_{\max}(\varrho)]. \tag{3.3.29}$$

On the other hand, for $\varrho = \varrho_{\text{minVar}}$, *the mininum variance portfolio* $\widehat{x}_{\text{minVar}}$ *is contained in* $D_{\varrho_{\text{minVar}}}$. *But as unique global minimizer, it is certainly also efficient for the two-parameter problem in* (3.3.20). *Therefore, for* $\varrho = \varrho_{\text{minVar}}$, *we obtain that the (vector risk, utility) values of* x_{minVar} *are contained in* $\mathcal{G}^{(2d)}_{\text{eff}}(\mathfrak{r}_1, \mathfrak{r}_2, \mathfrak{u};\, A)$ *as well, i.e., we get* $\gamma_1\left(\varrho_{\text{minVar}}, \mu_{\text{minVar}}\right) = \mathfrak{r}_{\text{Var}}\left(x_{\text{minVar}}\right) = \frac{1}{2}\sigma^2_{\text{minVar}}$ *and*

$$\left(\frac{1}{2}\sigma^2_{\text{minVar}},\, \varrho_{\text{minVar}}, \mu_{\text{minVar}}\right) \in \mathcal{G}^{(2d)}_{\text{eff}}\left(\mathfrak{r}_1, \mathfrak{r}_2, \mathfrak{u};\, A\right). \tag{3.3.30}$$

Similar to (3.3.29), *we also have*

$$\left(\gamma_1(\varrho_{\mathrm{minVar}},\mu), \varrho_{\mathrm{minVar}}, \mu\right) \in \mathcal{G}_{\mathrm{eff}}^{(2d)}(\mathfrak{r}_1, \mathfrak{r}_2, \mathfrak{u};\ A)$$
$$\text{for all } \mu \in \left[\mu^*(\varrho_{\mathrm{minVar}}), \mu_{\max}(\varrho_{\mathrm{minVar}})\right]. \quad (3.3.31)$$

Therefore, for $\varrho = \varrho_{\mathrm{minVar}}$*, there are actually two (2d)-efficient points on the efficient frontier for* (3.3.20) *which stem from Markowitz efficient portfolios,* (3.3.30) *and the left endpoint of* (3.3.31).

Case (B): $\varrho < \varrho_{\mathrm{minVar}}$. *In this case, we have* $\widehat{x}_{\mathrm{minVar}} \notin D_\varrho$ *(see* (3.3.25)*). In order to consider efficient points of* (3.3.20) *for* $\varrho < \varrho_{\mathrm{minVar}}$*, we have to divide further. We set*

$$\varrho_{\mathrm{minTrack}} := \inf\left\{\varrho \geq 0 \,:\, D_\varrho \cap \left\{\widehat{S}_0^\top \widehat{x} = 1\right\} \neq \emptyset\right\} \qquad (3.3.32)$$

and

$$\varrho^{\triangleleft} := \inf\left\{\varrho \geq 0 \,:\, D_\varrho \cap Y \neq \emptyset\right\} \geq \varrho_{\mathrm{minTrack}}, \qquad (3.3.33)$$

where $Y \subset \left\{\widehat{S}_1^\top \widehat{x} = 1\right\} \subset \mathbb{R}^M$ *is the ray of Markowitz efficient portfolios (see Remark 2.86).*

We only want to consider the case when

$$\varrho_{\mathrm{minTrack}} < \varrho^{\triangleleft} < \varrho_{\mathrm{minVar}}, \qquad (3.3.34)$$

because the remaining cases are even simpler. For $\varrho = \varrho_{\mathrm{minTrack}}$*, there is a unique portfolio* $\widehat{x}_{\mathrm{minTrack}} \in D_\varrho \cap \{\widehat{S}_0^\top \widehat{x} = 1\}$ *realizing the minimal tracking error, and for* $\varrho < \varrho_{\mathrm{minTrack}}$*, there is none. Thus,* $x_{\mathrm{minTrack}} = (0, \widehat{x}_{\mathrm{minTrack}}^\top)^\top \in \mathbb{R}^{M+1}$ *is a (2d)-efficient portfolio for* (3.3.20) *which satisfies* $\mathfrak{r}_{\mathrm{Track}}(x_{\mathrm{minTrack}}) = \varrho_{\mathrm{minTrack}}$. *We further set*

$$\mu_{\mathrm{minTrack}} := \mathrm{E}\left[\widehat{S}^\top \widehat{x}_{\mathrm{minTrack}}\right] \quad \text{and} \quad \frac{1}{2}\sigma^2_{\mathrm{minTrack}} := \mathfrak{r}_{\mathrm{Var}}\left(x_{\mathrm{minTrack}}\right), \qquad (3.3.35)$$

and obtain $\gamma_1(\varrho_{\mathrm{minTrack}}, \mu_{\mathrm{minTrack}}) = \frac{1}{2}\sigma^2_{\mathrm{minTrack}}$. *Therefore,*

$$\left(\frac{1}{2}\sigma^2_{\mathrm{minTrack}}, \varrho_{\mathrm{minTrack}}, \mu_{\mathrm{minTrack}}\right) \in \mathcal{G}_{\mathrm{eff}}^{(2d)}(\mathfrak{r}_1, \mathfrak{r}_2, \mathfrak{u};\ A). \qquad (3.3.36)$$

The remaining situations are sketched in Fig. 3.11, where $\mu_{\max} = \mu_{\max}(\varrho)$ *was given in* (3.3.28) *and*

$$\mu_{\min} = \mu_{\min}(\varrho) := \inf\left\{\mathrm{E}\left[\widehat{S}_1^\top \widehat{x}\right] \,:\, \left(0, \widehat{x}^\top\right)^\top \in A_\varrho\right\} \in \mathbb{R}, \qquad (3.3.37)$$

with A_ϱ *given in* (3.3.26). *It should be noted that for* $\varrho = \varrho^{\triangleleft}$ *there is a unique portfolio* $x^{\triangleleft} = \left(0, \widehat{x}^{\triangleleft\top}\right)^\top \in \mathbb{R}^{M+1}$ *which corresponds to a point on the*

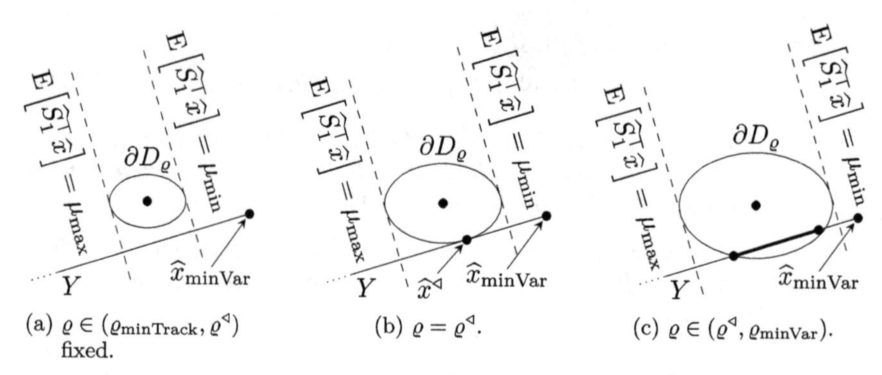

(a) $\varrho \in (\varrho_{\mathrm{minTrack}}, \varrho^\lhd)$ fixed. (b) $\varrho = \varrho^\lhd$. (c) $\varrho \in (\varrho^\lhd, \varrho_{\mathrm{minVar}})$.

Fig. 3.11 Illustration for Example 3.50, all projections to $\left\{\widehat{x} : \widehat{S}_0^\top \widehat{x} = 1\right\}$

Markowitz bullet (cf. Fig. 3.11b). Using $\mu^\lhd := \mathrm{E}\left[\widehat{S}_1^\top \widehat{x}^\lhd\right]$, *we get*

$$\frac{1}{2}\left(\sigma^\lhd\right)^2 = \mathfrak{r}_{\mathrm{Var}}\left(x^\lhd\right) = \gamma_1\left(\varrho^\lhd, \mu^\lhd\right) \ \textit{and}\ \left(\frac{1}{2}\left(\sigma^\lhd\right)^2, \varrho^\lhd, \mu^\lhd\right) \in \mathcal{G}_{\mathrm{eff}}^{(2d)}(\mathfrak{r}_1, \mathfrak{r}_2, \mathfrak{u};\ A). \tag{3.3.38}$$

Finally, we can draw the graph of $\gamma = \gamma_\varrho$,

$$\gamma_\varrho(\mu) = \inf\left\{\mathfrak{r}_1\left(\widehat{x}\right) = \frac{1}{2}\widehat{x}^\top \Sigma \widehat{x} : \mathfrak{u}(x) = \mathrm{E}\left[\widehat{S}_1^\top \widehat{x}\right] \geq \mu,\ x = \left(0, \widehat{x}^\top\right)^\top \in A_\varrho\right\} \tag{3.3.39}$$

for $\varrho < \varrho_{\mathrm{minVar}}$ *fixed (see Fig. 3.12 for the (1d)-efficient points on* $\mathrm{graph}(\gamma_\varrho)$*).*

Similarly as for $\varrho \geq \varrho_{\mathrm{minVar}}$, *also for* $\varrho < \varrho_{\mathrm{minVar}}$, *not all points in* $\mathcal{G}_{\mathrm{eff}}^{(1d)}(\mathfrak{r}_1, \mathfrak{u};\ A_\varrho)$ *remain efficient for the two-parameter problem* (3.3.20) *(see Fig. 3.12c), where all but two points on the Markowitz bullet are no longer (2d)-efficient.*

We conclude with a sketch of the set $M_1 = M_1(\mathfrak{r}_1, \mathfrak{r}_2, \mathfrak{u};\ A)$ *defined in* (3.1.24) *which is the projection of* $\mathcal{G}_{\mathrm{eff}}^{(2d)}(\mathfrak{r}_1, \mathfrak{r}_2, \mathfrak{u};\ A)$ *to the* (ϱ, μ)*-space (see Fig. 3.13a). Together with the graphs of* $\gamma_1(\varrho, \cdot) = \gamma_\varrho$ *in Fig. 3.12 restricted to* M_1, *this should give a good impression of the form of the efficient frontier* $\mathcal{G}_{\mathrm{eff}}^{(2d)}(\mathfrak{r}_1, \mathfrak{r}_2, \mathfrak{u};\ A)$.

Note that it may be possible for some ϱ *that* $\mu_{\min}(\varrho) < \mu_{\mathrm{minVar}}$ *(cf. again Fig. 3.13a), although* $\gamma_1(\varrho, \mu) = \gamma_\varrho(\mu) \geq \frac{1}{2}\sigma_{\mathrm{minVar}}$ *necessarily holds true for all* ϱ, *cf. Remark 2.86, in particular* (2.4.54).

Finally, in Fig. 3.13b, we get a similar sketch of $N = N(\mathfrak{r}_1, \mathfrak{r}_2, \mathfrak{u};\ A)$ *defined in* (3.1.21) *which is the projection of* $\mathcal{G}_{\mathrm{eff}}^{(2d)}(\mathfrak{r}_1, \mathfrak{r}_2, \mathfrak{u};\ A)$ *to the* $(\frac{1}{2}\sigma^2, \varrho)$*-space.*

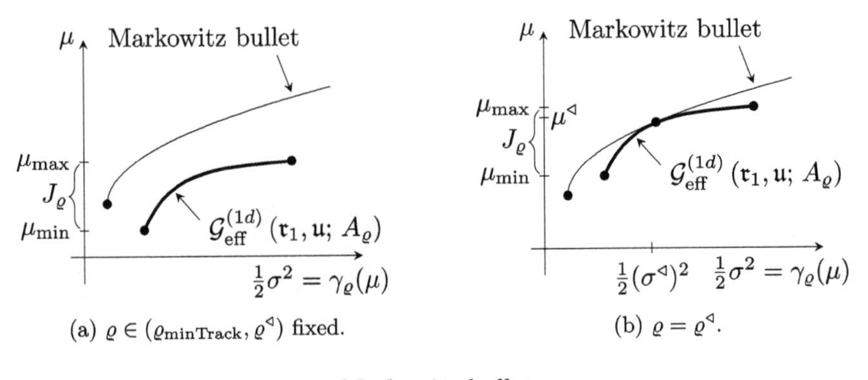

(a) $\varrho \in (\varrho_{\mathrm{minTrack}}, \varrho^\triangleleft)$ fixed.

(b) $\varrho = \varrho^\triangleleft$.

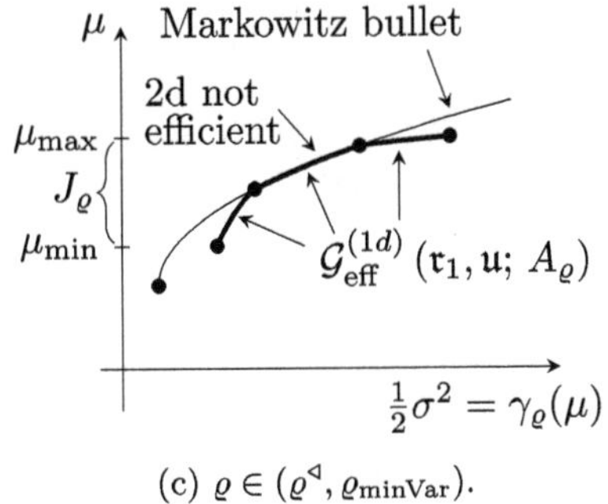

(c) $\varrho \in (\varrho^\triangleleft, \varrho_{\mathrm{minVar}})$.

Fig. 3.12 Illustration for Example 3.50; graphs of γ_ϱ

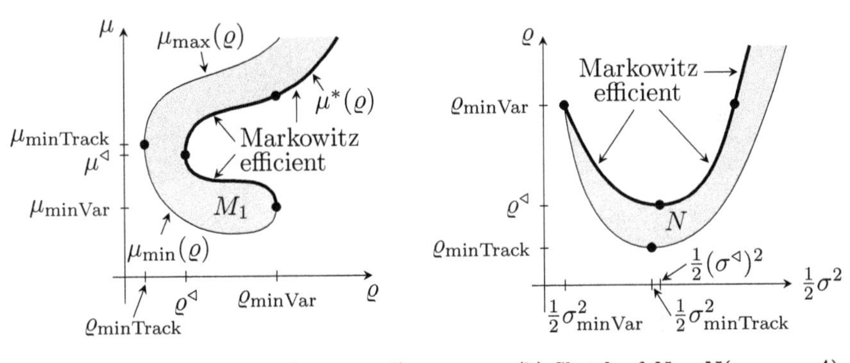

(a) Sketch of $M_1 = M_1(\mathfrak{r}_1, \mathfrak{r}_2, \mathfrak{u}; A)$.

(b) Sketch of $N = N(\mathfrak{r}_1, \mathfrak{r}_2, \mathfrak{u}; A)$.

Fig. 3.13 Illustration for Example 3.50

Chapter 4
Application Examples

The theory that we have developed in the previous chapters can be applied to several management problems, coming from corporations, individuals, or governments. The manager has to balance profit with more than one risk. We present a few situations where multiple risks arise.

Example 4.1 (Investment Funds with Illiquid Assets) *Many funds, such as real estate, hedge funds, infrastructure, or private equity, invest a considerable amount on illiquid assets. If a fund manager allocates insufficient cash to meet all redemptions, both expected and unexpected, the fund may face bankruptcy. On the other hand, the assets in these funds are subject to the risk of devaluations. Therefore, such fund managers must carefully manage the balance between profitability, valuation risk on the assets, and liquidity risks from having too little cash.*

Example 4.2 (Multiple Risks Arising in Corporations) *An oil company has risks in fluctuating oil prices but needs to make sure that it has enough cash to cover expected or unexpected payments. Therefore, it must balance oil price risk and liquidity risk.*

A global hotel corporation that operates in several countries will balance profitability with the several risks that arise. Legal risks stem from different legal jurisdictions. The corporation will also be subject to liquidity risk. It will need to ensure that it has enough cash to compensate for any expected or unexpected cash outflows, from payments to suppliers, workers, or debt payments. Other risks include a drop in occupancy rates or a significant decrease in real estate prices.

Example 4.3 (Household Finance Risks) *If a household decides to buy a house resorting to a mortgage, it will have to balance the cost on the mortgage with the incurring risks. On the one hand, typically, the household's primary asset will be real estate, so the household's net worth will be highly impacted*

S. Maier-Paape et al., *Scalar and Vector Risk in the General Framework of Portfolio Theory*, CMS/CAIMS Books in Mathematics 9,
https://doi.org/10.1007/978-3-031-33321-7_4

by real estate prices. On the other hand, it should ensure that it has enough funds in deposit accounts to meet unexpected needs, for example, from bills. The first risk is real estate risk; the second risk is liquidity risk.

Example 4.4 (Liability Management for Governments) *A government that is issuing debt typically has to balance the cost of the debt with the maturity structure of the debt, i.e., find the optimal debt maturity structure. Short-term debt issuance is typically cheaper than long-term issuance but entails higher risks, as the government has to refinance its debt frequently. Frequent refinancing leaves the government exposed to changes in interest rates, and the government debt service costs will increase if interest rates rise. Therefore, the government will have considerable interest rate risk. Furthermore, a government that is excessively reliant on short-term issuance is also more exposed to a credit crunch if it doesn't have enough liquid assets or a cash buffer. The reason is that, in case of a credit crunch, the investors holding the government's debt may not refinance it and ask for their funds back. Since the government will not have enough liquidity to pay back to the investors, it will face bankruptcy. This example shows an instance of liquidity risk.*

Single metrics can aggregate some types of risk. For example, in an investment portfolio, value at risk or expected shortfall can aggregate foreign exchange risk and stock price risk. But other risks are difficult to aggregate due to the lack of data to calibrate the correlation matrix and the heterogeneous nature of different risks. For example, liquidity risk and interest rate risk are intrinsically different. The first risk assesses if the corporation has enough cash to meet unexpected payments; the second evaluates if the company has enough capital to accommodate potential losses occurring from considerable variations in interest rates.

In practice, even if corporations aggregate different risks in their risk appetite frameworks, they will typically enforce individual limits on various risks. So, in practice, one always needs to deal with multiple risk measures and limits.

We will focus on the case of banks. Banks are subject to multiple risks, such as credit risk, liquidity risk, and interest rate risk. In the first section of this chapter, we will introduce balance sheet management problems and the risks associated with banks; the second section develops solutions of a linear-quadratic model necessary to solve the balance sheet management problem. Finally, the third section solves several balance sheet models for banks.

4.1 Bank Balance Sheet Management Problems

In this section, we address the balance sheet problem for banks. In Sect. 4.1.1, we review financial intermediation and the balance sheet. Section 4.1.2 describes the risks involved in the balance sheet and their measurement.

4.1.1 Bank Balance Sheet Problems

Banks are actively involved in financial intermediation, raising funds from creditors and shareholders, and investing these funds. Creditors include retail and corporate depositors, holders of bonds issued by the bank, or other banks involved in the money market or repurchase agreements (repos). By lending money to the bank, creditors receive interest. Shareholders invest cash in the bank's equity or bank capital and expect to receive dividends, depending on the bank's performance.

With the money from creditors, banks make investments in several asset classes. Typical investments include mortgages, consumer credit loans, corporate loans, treasury securities, short-term deposits at other banks, and deposits at the central bank. Figure 4.1 shows a schematic example of financial intermediation.

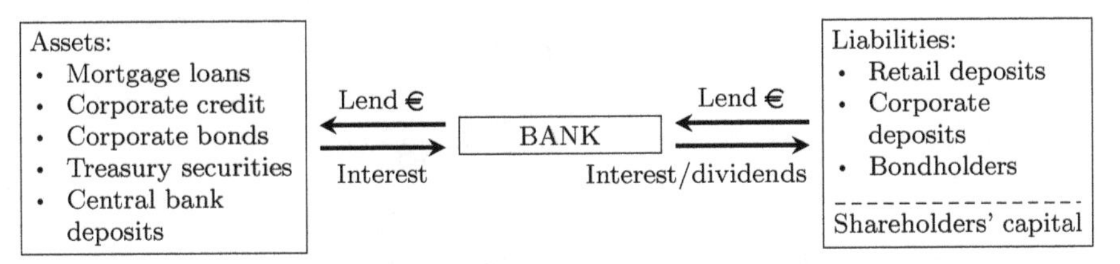

Fig. 4.1 An example of financial intermediation. The bank obtains the funds from equity and liabilities and makes investments in loans or securities. On the one hand, it receives a return from assets. On the other hand, it pays interest to liabilities, and dividends to shareholders' capital, depending upon the bank's performance

The balance sheet records assets and liabilities. We give a simplified example of a bank balance sheet in Table 4.1. The fundamental accounting equation applies, i.e., assets are equal to liabilities plus shareholder capital. In other words, shareholder capital, which is the net wealth of the bank, is given by assets minus liabilities.

Assets		**Liabilities**	
Central bank deposits	200	Retail deposits	600
Mortgage loans	500	Corporate deposits	200
Corporate credit	100	Repo funding	100
Treasury bonds	200		
		Capital	
		Shareholders' capital	100

Table 4.1 An example of a bank balance sheet. The fundamental accounting equation applies, i.e., assets are equal to liabilities plus shareholders' capital

In our discussion, we include $m \in \mathbb{N}$ assets and $k \in \mathbb{N}$ liabilities, as well as one cash market position, which we assume to be central bank overnight deposits. Central bank deposits are also an asset class, but we differentiate them as the riskless assets class.

Definition 4.5 (Bank Balance Sheet) *A bank balance sheet is a vector given by*

$$x = (x_0, x_1, \ldots, x_m, x_{m+1}, \ldots, x_{m+k})^\top \in \mathbb{R}^{1+m+k},$$

where x_0 is the risk-free investment (overnight central bank deposits), $\widehat{x^a} = (x_1, \ldots, x_m)^\top$ is the vector of the amounts invested in assets, and $\widehat{x^\ell} = (x_{m+1}, \ldots, x_{m+k})^\top$ is the vector of liabilities. We assume that x_0, $\widehat{x^a}$ are non-negative and that $\widehat{x^\ell}$ is non-positive. In total, we have $M := m + k$ risky positions. As before, $\widehat{x} = (x_1, \ldots, x_M)^\top = (x_1^a, \ldots, x_m^a, x_{m+1}^\ell, \ldots, x_{m+k}^\ell)^\top = (\widehat{x^a}^\top, \widehat{x^\ell}^\top)^\top \in \mathbb{R}^M$ denotes the risky part of the portfolio. Note that the superscripts "a" and "ℓ" may be occasionally used when emphasizing the asset and liability character of the balance sheet positions.

As we will see below, the balance sheet problem deals with a vector-valued risk $\mathfrak{r} = (\mathfrak{r}_1, \mathfrak{r}_2, \mathfrak{r}_3)$ as a function of $x = (x_0, \widehat{x^a}^\top, \widehat{x^\ell}^\top)^\top$. In our notation, $\mathfrak{r}_1$ will be interest rate risk, $\mathfrak{r}_2$ will be credit risk, and $\mathfrak{r}_3$ will be liquidity risk.

Notation 4.6 *We normalize $S_0 = (S_0^0, \widehat{S_0^a}^\top, \widehat{S_0^\ell}^\top)^\top \in \mathbb{R}^{1+M}$ in our setting to be the vector $\mathbf{1} = (1, \ldots, 1)^\top$ of length $1 + M = 1 + m + k$, corresponding to the central bank deposits, m assets, and k liabilities. Consistently with the previous notation, we have $\widehat{S_0} = (\widehat{S_0^a}^\top, \widehat{S_0^\ell}^\top)^\top$, $\widehat{S_0^a} = (S_0^1, \ldots, S_0^m)^\top$, and $\widehat{S_0^\ell} = (S^{m+1}, \ldots, S_0^{m+k})^\top$.*

As developed in the previous chapters for a one-period financial market (Definition 2.3), we assume that the assets and liabilities have outcomes given by $S_1 = (S_1^0, \widehat{S_1}^\top)^\top = (S_1^0, \widehat{S_1^a}^\top, \widehat{S_1^\ell}^\top)^\top$, which is a random vector of length $1 + m + k$, such that $\widehat{S_1^a} = (S_1^1, \ldots, S_1^m)^\top$ and $\widehat{S_1^\ell} = (S_1^{m+1}, \ldots, S_1^{m+k})^\top$ are the valuations for assets and liabilities at $t = 1$, respectively.

Knowing that the fundamental accounting equation states that shareholders' capital is equal to the difference between assets and liabilities, we obtain due to our sign constraints $x_0, \widehat{x^a} \geq 0$, and $\widehat{x^\ell} \leq 0$, that $S_t^\top x$ is the shareholders' capital at time $t \in \{0, 1\}$. Recall that we normalize $S_0^\top x = 1$, similarly to the unit initial cost portfolio assumption in Notation 2.5. Note that $|x_i|$, $i \in \{1, \ldots, M\}$, represents the fraction of position i relative to the shareholders' capital, but not relative to the total amount invested, i.e., to $S_0^T |x| \geq 1$.

Assumption 4.7 *Set $\mathcal{I}_a := \{1, \ldots, m\}$ and $\mathcal{I}_\ell := \{m + 1, \ldots, m + k\}$. We assume that $S_1^0 = (1 + R_0)$. $\mathrm{E}(S_1^i) = (1 + R_i)$, for $i \in \mathcal{I}_a$, represent the expected valuations on assets, whereas $\mathrm{E}(S_1^i) = (1 + R_i)$ for $i \in \mathcal{I}_\ell$ represent the expected valuations on liabilities.*

In this setting, the restriction $S_0^\top x = 1$ is the same as $\sum_{i=0}^{M} x_i = 1$. We now want to compute the expected payoff for the balance sheet, separating assets from liabilities. Thus, we compute

$$\begin{aligned}
\mathrm{E}(S_1^\top x) &= \mathrm{E}(x_0 S_1^0 + \widehat{S_1^a}^\top \widehat{x^a} + \widehat{S_1^\ell}^\top \widehat{x^\ell}) \\
&= \mathrm{E}(x_0 S_1^0 + \sum_{i\in\mathcal{I}_a} x_i S_1^i + \sum_{j\in\mathcal{I}_\ell} x_j S_1^j) \\
&= x_0(1+R_0) + \sum_{i\in\mathcal{I}_a} x_i(1+R_i) + \sum_{j\in\mathcal{I}_\ell} x_j(1+R_j) \\
&= x_0 R_0 + \sum_{i\in\mathcal{I}_a} x_i R_i + \sum_{j\in\mathcal{I}_\ell} x_j R_j + x_0 + \sum_{i\in\mathcal{I}_a} x_i + \sum_{j\in\mathcal{I}_\ell} x_j \\
&= x_0 R_0 + \widehat{R^a}^\top \widehat{x^a} + \widehat{R^\ell}^\top \widehat{x^\ell} + 1,
\end{aligned}$$

where $R = (R_0, \widehat{R^a}^\top, \widehat{R^\ell}^\top)^\top$, with $\widehat{R^a} = (R_1, \ldots, R_m)^\top$ and $\widehat{R^\ell} = (R_{m+1}, \ldots, R_{m+k})^\top$. Therefore, maximizing the payoff is the same as maximizing $x_0 R_0 + \widehat{R^a}^\top \widehat{x^a} + \widehat{R^\ell}^\top \widehat{x^\ell}$.

4.1.2 Risks and Their Measurements

Banks' balance sheets are subject to multiple risks. The majority of the balance sheet for many commercial banks comprises the banking book, i.e., activities unrelated to trading. We will focus on the banking book, but one can also extend the framework to incorporate market risk arising from trading activities. Some of the main risks for the banking book include credit risk, interest rate risk, and (funding) liquidity risk.

We will develop precise explanations on each risk, but, as a summary, credit risk arises essentially from potential losses on defaults from loans. Interest rate risk is the risk of impacts of the interest rate curve on the balance sheet, particularly from the mismatch in maturities between fixed-rate assets and liabilities. Liquidity risk arises when there is a lack of liquid securities that the bank can sell to mitigate the impact of potential withdrawals or outflows of funds from customers or institutional investors.

Many financial crises and bank failures were linked to these risks, as we document below in the following examples.

Example 4.8 (Savings and Loan Crisis) *Many believe that interest rate risk was the main factor for the savings and loans crisis in the early 1980s. Several savings and loan institutions faced losses when interest rates rose in the late 1970s, as they had previously issued a significant amount of fixed-rate long-term loans at lower rates.*

Example 4.9 (Liquidity Risk Failures) *More recent failures, such as Bear Stearns and Lehman Brothers in the USA or Northern Rock in the UK, are linked to liquidity risk. Many authors believe that excessive reliance on short-term wholesale funding was the chief culprit of the collapse of these institutions. Short-term wholesale funding such as repo markets, by nature, is less stable than traditional retail funding such as customer deposits.*

Example 4.10 (Credit Risk Crises) *Credit risk episodes have also caused the failure or near-failure of several European banks recently. Several banks failed or were bailed out during the last few years due to multiple defaults on debts after excessive risks taken by customers. These episodes occurred in several jurisdictions in the European Union.*

We will not cover all the means to measure each type of risk, but we give here a brief overview of each risk and its measurement. The reader interested in knowing more can consult several books on bank risk management [14, 61, 62, 83, 97] or asset-liability management [2, 27, 37, 74].

Interest Rate Risk

Let us first address interest rate risk. Interest rate risk is the risk due to the mismatch in maturities between fixed-rate assets and liabilities. To illustrate this risk, let us give an extreme example of a balance sheet with 1000 in 30-year fixed-rate mortgages as assets receiving 3% interest and 1000 in overnight funding paying 1% interest as a liability.

The net interest margin (i.e., the difference between interest earned on assets and paid on liabilities) is equal to $(3\% \cdot 1000) - (1\% \cdot 1000) = 20$. But what if the short-term interest rates increase by 4%? Then the bank still receives 3% on the mortgages (because they are fixed-rate) and pays 5% on overnight funding. In this scenario, the bank will have a negative interest margin of $(3\% \cdot 1000) - (5\% \cdot 1000) = -20$. Therefore, the bank has transitioned from a profit to a loss due to the rise in interest rates. The mortgage pays more than overnight funding at the beginning due to the compensation for interest rate risk.

We now exemplify two standard ways of computing interest rate risk: change in the economic value of equity (EVE) and normal value at risk (VaR). Of course, many other measures are possible. However, these two types are standard and allow us to solve the bank balance sheet problem in a tractable fashion. These measures are also described in the textbooks on asset-liability management and risk management cited above.

Change in Economic Value of Equity (EVE)

Duration-based measures are the starting point for assessing interest rate risk for a balance sheet. Using the previous notation, let us assume that τ_i, for $i \in \{1, \ldots, M\}$, is the maturity for asset or liability class i.

The most basic way of measuring interest rate risk in the balance sheet is to address the sensitivity of the economic value of equity (EVE) to an upward parallel interest rate risk shock. One could also assess downward interest rate risk shocks or other types of interest rate risk shocks. Since banks typically fund themselves with short-dated liabilities to invest in long-dated assets (maturity transformation), they lose with upward shifts in the interest rate curve.

The standard way to compute the change in EVE is to map all the assets and liabilities into maturity buckets. The interest rate risk is calculated for each maturity bucket by resorting to the modified duration for that maturity. Although assets with the same maturity can have different modified durations, for aggregation purposes, one assumes that assets with the same maturity have the same duration. We will develop the notation in more detail. For simplicity, we assume that maturities are within $\mathcal{M} = \{1, 2, 3, 5, 7, 10, 20, 30\}$; otherwise, we round to the nearest maturity. For each $i \in \{1, \ldots, M\} = \{1, \ldots, m+k\}$, we assume that the maturity of asset or liability x_i is $\tau_i \in \mathcal{M}$. In practice, this is not the case, but for the purpose of measuring interest rate risk, each asset or liability is typically mapped to the nearest maturity within a certain discrete set of time buckets such as $\mathcal{M}$.

For each maturity $\tau \in \mathcal{M}$, the modified duration $D_\tau \in \mathbb{R}_{>0}$ is calculated based on the duration of a par Treasury bond which has the same maturity. Specifically, let

$$P_\tau(y, \gamma) := \sum_{i=1}^{\tau} \frac{\gamma}{(1+y)^i} + \frac{1}{(1+y)^\tau} \tag{4.1.1}$$

$$= \frac{\gamma}{y} + \frac{1}{(1+y)^\tau}\left(1 - \frac{\gamma}{y}\right) \tag{4.1.2}$$

be the price of the bond with coupon γ, yield to maturity y, and repayment at maturity at par, assuming that $y \neq 0$.

From formula (4.1.2), we can see that, for general values of $y > -1$ and $y \neq 0$, the bond is priced at par ($P_\tau(y, \gamma) = 1$) if and only if

$$\frac{\gamma}{y}\left(1 - \frac{1}{(1+y)^\tau}\right) = 1 - \frac{1}{(1+y)^\tau}, \tag{4.1.3}$$

which is equivalent to $y = \gamma$. In particular, $P_\tau(y, y) = 1$.

Then the modified duration for the par Treasury bond with maturity τ is defined by

$$D_\tau := -\frac{dP_\tau}{dy}(y_\tau, y_\tau) \tag{4.1.4}$$
$$= \frac{1}{y_\tau} - \frac{1}{y_\tau(y_\tau+1)^\tau} > 0, \tag{4.1.5}$$

where y_τ is the Treasury yield for maturity $\tau \in \mathcal{M}$.

Definition 4.11 (Change in Economic Value of Equity) *Let*

$$x = (x_0, x_1, \ldots, x_m, x_{m+1}, \ldots, x_{m+k})^\top \in \mathbb{R}^{1+m+k}$$

be a balance sheet as defined in Definition 4.5. Then the change in economic value of equity is a risk measure

$$\mathfrak{r}_1(x) = \mathfrak{r}_{\text{EVE}}(x) = \sum_{i=1}^{m+k} x_i D_{\tau_i} \sigma_{\tau_i}, \tag{4.1.6}$$

where D_{τ_i} is the modified duration for maturity τ_i and $\sigma_{\tau_i} \in \mathbb{R}$ is an interest rate shock for maturity τ_i.

σ_τ can take several forms that are standard when measuring interest rate risk and represent a shock to interest rates. For example, parallel shifts to the interest rate risk curve are equal to a constant $\sigma_\tau = \alpha$. A slope shift is typically given by $\sigma_\tau = \alpha + \beta \cdot \tau$. Figure 4.2 illustrates parallel shifts and slope shifts to the yield curve.

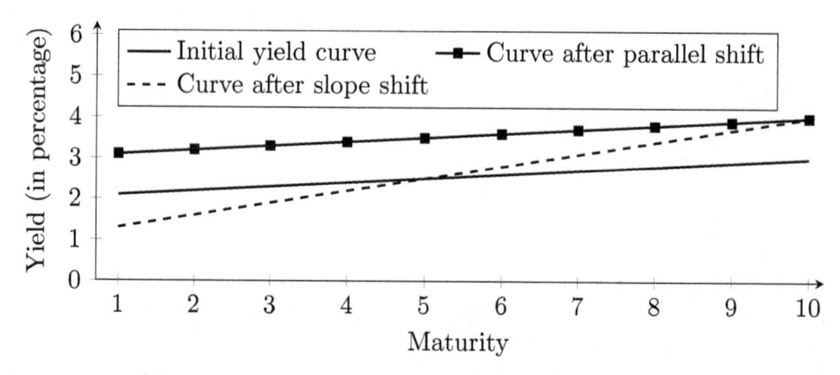

Fig. 4.2 A simplified example of shocks to the yield curve. Parallel shifts move the yield curve upward or downward in a parallel manner. Slope shifts change the slope of the yield curve

We will focus on parallel upward interest rate risk shocks. In this case, σ_τ is a positive constant, independent of $\tau > 0$. But the change in EVE can be computed using several other interest rate risk shocks (e.g., downward parallel shifts, changes in the yield curve slope, etc.). In practice, to determine the interest upward shock, one can consider that interest rate variations follow a normal distribution with a standard deviation σ, so that we can compute the worst possible increment within a confidence level of 99%. In this case, $\sigma_T = \mathcal{N}^{-1}(0.99) \cdot \sigma$, where $\mathcal{N}$ is the standard normal distribution, so in this particular instance, the interest rate risk shock does not depend on the maturity. The interest rate risk $\mathfrak{r}_1$ is the market value loss for an adverse interest rate shock of $\mathcal{N}^{-1}(0.99) \cdot \sigma$. This amounts to a one-factor value at risk measure, assuming a parallel shift in interest rates.

Normal Value at Risk

For each maturity $\tau \in \mathcal{M}$, let B_τ be the net amount invested in assets of maturity τ. Given $x \in \mathbb{R}^{1+m+k}$, this can be calculated as

$$B_\tau = B_\tau(x) = \sum_{\tau_i = \tau} x_i, \tag{4.1.7}$$

since we assumed $S_0 = (1, \dots, 1)^\top$. If $B_\tau > 0$, this means that in total, there are more investments in assets than funding from liabilities for that maturity. Otherwise, the bank has more liabilities than assets in that maturity.

We denote i the indices for assets and j the indices for liabilities. To calculate *interest rate risk*, again, we use a simple approximation from bond math. Suppose that the Treasury yield curve, which contains the Treasury yields for the different maturities, is given by $(y_T)_{T \in \mathcal{M}}$. Consider that this curve is subject to a shock and is equal to $(\widetilde{y}_T)_{T \in \mathcal{M}}$, after the shock. Then the change in the yield for maturity τ due to this shock is given by $\Delta y_T = \widetilde{y}_T - y_T$. The change in the value of fixed-income assets for a given maturity in (4.1.7) is approximately given by the modified duration times the change in the yield, i.e.,

$$\Delta W_\tau := -B_\tau \cdot D_\tau \cdot \Delta y_T, \tag{4.1.8}$$

where we use the approximation for the modified duration (4.1.5).

Now, let us consider the potential change of value in a balance sheet of net assets and liabilities for all the given maturities together:

$$\sum_{\tau \in \mathcal{M}} \Delta W_\tau.$$

We assume that $(\sigma_T)_{T \in \mathcal{M}}$ is the volatility vector, where $\sigma_\tau \in \mathbb{R}_{>0}$ is the volatility of the τ-maturity interest rate and $(\rho_{Ts})_{T,s \in \mathcal{M}}$ is the correlation matrix between the τ-maturity rate and the s-maturity rate. $(\sigma_{Ts})_{T,s \in \mathcal{M}}$ is the covariance between the τ-maturity and the s-maturity interest rates, i.e., $\sigma_{Ts} = \mathrm{Cov}(\Delta y_T, \Delta y_s) = \rho_{Ts}\sigma_T\sigma_s$. Notice that these can be obtained by low-dimensionality reduction such as principal component analysis (PCA), which is a standard procedure.

Using the duration approximation, we can compute the variance of $\sum_{\tau\in\mathcal{M}} \Delta W_\tau$:

$$\begin{aligned}\operatorname{Var}\left(\sum_{\tau\in\mathcal{M}} \Delta W_\tau\right) &= \operatorname{Cov}(\sum_{\tau\in\mathcal{M}} \Delta W_\tau, \sum_{s\in\mathcal{M}} \Delta W_s)\\ &= \sum_{\tau\in\mathcal{M}}\sum_{s\in\mathcal{M}} B_\tau B_s D_\tau D_s \operatorname{Cov}(\Delta y_\tau, \Delta y_s)\\ &= \sum_{\tau\in\mathcal{M}}\sum_{s\in\mathcal{M}} B_\tau B_s D_\tau D_s \sigma_{\tau s}\\ &= \sum_{\tau\in\mathcal{M}}\sum_{s\in\mathcal{M}} B_\tau B_s D_\tau D_s \rho_{\tau s}\sigma_\tau\sigma_s.\end{aligned}$$

The normal value at risk for interest rate risk with a confidence level of α is then based on the standard deviation:

$$\mathfrak{r}_{\mathrm{NVaR}}(x) := \widehat{\mathfrak{r}}_{\mathrm{NVaR}}\left(\widehat{x^a}\right) = \mathcal{N}^{-1}(\alpha)\sqrt{\sum_{\tau\in\mathcal{M}}\sum_{s\in\mathcal{M}} B_\tau B_s D_\tau D_s \sigma_{\tau s}},$$

where B_τ is defined by (4.1.7) and D_τ is defined by (4.1.5). For convenience, we will actually use the equivalent quadratic function $\mathfrak{r}_{\mathrm{intSqr}}(x) := (\mathfrak{r}_{\mathrm{NVaR}}(x))^2/(2(\mathcal{N}^{-1}(\alpha))^2)$. To get a standard vector representation of this quadratic function, define $\beta_\tau = (\beta_\tau^1, \ldots, \beta_\tau^M)^\top$, where $\beta_\tau^i = 1$ if the ith asset is a bond with duration τ and $\beta_\tau^i = 0$ otherwise. Then we can represent $B_\tau = \widehat{x}^\top \beta_\tau$. It follows that

$$B_\tau B_s = \widehat{x}^\top \beta_\tau \beta_s^\top \widehat{x}.$$

Define

$$\Theta := \sum_\tau \sum_s \beta_\tau \beta_s^\top D_\tau D_s \sigma_{\tau s}. \tag{4.1.9}$$

Then Θ is a symmetric, positive semi-definite matrix, and we can represent the risk $\mathfrak{r}_1$ as

$$\mathfrak{r}_1(x) = \mathfrak{r}_{\mathrm{intSqr}}(x) := \frac{1}{2}\widehat{x}^\top \Theta\, \widehat{x}. \tag{4.1.10}$$

Credit Risk

Credit risk is the risk associated with potential defaults from the borrowers of the bank. Thus, credit risk only focuses solely on the asset side, not on the liability side. Specific credit risk measures include credit value at risk, and several models are available. We will not be able to discuss all the models in detail. Standard references include Vasicek [110] model or CreditMetrics

[54], but many other models are available. In addition, the interested reader can refer to the several books and articles published on credit risk ([15, 17, 34, 53, 100, 117]).

For our book, we resort to credit risk measures that are tractable when computing optimal balance sheets. Therefore, we focus on linear and quadratic credit risk measures. Linear credit risk measures are standard and used for capital requirements. Quadratic measures, although less standard, can still give good approximations to credit risk, allow for diversification effects, and enable the tractability of the computation of optimal balance sheets.

Linear Credit Risk

Linear measures of credit risk are used in practice. For example, credit risk capital requirements under Basel II and III are a linear function of the allocations to each asset class. Assuming that we have different asset classes indicated by the indices $\mathcal{I}_a = \{1, \ldots, m\}$ as in Definition 4.5, linear measures of credit risk are of the form

$$\mathfrak{r}_2(x) = \widehat{\mathfrak{r}}_2(\widehat{x^a}) = \mathfrak{r}_{\text{CredLin}}(x) := \sum_{i \in \mathcal{I}_a} x_i c_i.$$

We now show how to determine the credit risk $c_i \in \mathbb{R}_{\geq 0}$ for each asset class. For ease of exposition, we will suppress the index $i \in \mathcal{I}_a$ in the following paragraphs and focus on how to calculate the credit risk for a particular asset class i. Some assets can have negligible credit risk, such as certain Treasury bonds and cash held at central banks, so $c_i = 0$ is allowed. For these assets, the calculation of credit risk is trivial, as it is just zero. The following discussion applies to assets with credit risk, which have a nonzero probability of default.

The Vasicek [110] model became the basis for capital standards under the advanced-IRB approach in Basel II. Although the model is more prevalent in its book-value form, assuming incurred losses, we use its mark-to-market counterpart to be consistent with interest rate risk measurement.

The Vasicek paper calculates mark-to-market credit portfolio loss distributions. Let H be the investment horizon (which we assume equal to one year), T the maturity of the credits, L the credit portfolio loss, ρ the credit correlation within the same asset class, and r the discount rate. The loss distribution is given by

$$P\left[L \leq \chi\right] = F\left(\frac{\chi}{a}; PD; \frac{\rho H}{T}\right), \tag{4.1.11}$$

where

$$F(\widetilde{\chi}, \widetilde{p}, \widetilde{\rho}) := \mathcal{N}\left(\frac{\sqrt{1-\widetilde{\rho}}\mathcal{N}^{-1}(\widetilde{\chi}) - \mathcal{N}^{-1}(\widetilde{p})}{\sqrt{\widetilde{\rho}}}\right), \tag{4.1.12}$$

and

$$a := LGD \cdot \exp(-r(T-H)). \tag{4.1.13}$$

LGD denotes the loss given default and PD the T-year probability of default. The marginal probability of default MP can be used to infer PD:

$$PD = 1 - \exp(-MP \cdot T).$$

From the loss distribution, we can compute the credit value at risk measure of credit risk. The worst credit portfolio loss c with a confidence level of α is given by

$$P[L \leq c] = \alpha. \tag{4.1.14}$$

Solving (4.1.14) for c, we get the credit risk factor

$$c = a\mathcal{N}\left(\frac{\sqrt{\frac{\rho}{T}}\mathcal{N}^{-1}(\alpha) + \mathcal{N}^{-1}(PD)}{\sqrt{1-\frac{\rho}{T}}}\right) > 0. \tag{4.1.15}$$

Quadratic Credit Risk

Many credit risk models have been developed over the years. Some of the most well-known models include Vasicek [110], CreditMetrics [54], CreditRisk+ [116], or copula models. In comparison to other risk models such as interest rate risk or market risk, credit risk models have to incorporate the highly skewed nature of credit risk.

Although one can apply the theory to such models, we will focus here on quadratic models. First, quadratic models ensure the tractability of the solutions. Second, quadratic models can still incorporate credit correlations among different asset classes and are an improvement compared to linear models. Third, it is still possible to ensure skewness of the loss distribution if the marginal distributions for risk on each asset class are skewed. We detail the formulation below.

For this, we follow the procedure in Rosenberg and Schuermann [95], which consists of two steps:

- Step 1 – determine the credit risk for each of the asset classes based on the skewed. distributions.
- Step 2 – mix those risks using the correlation matrix between these asset classes.

This procedure has been used in the context of risk aggregation purposes.

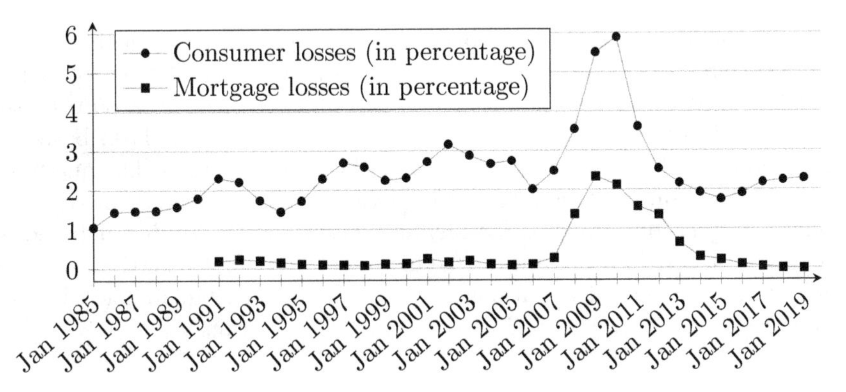

Fig. 4.3 Example of historical credit losses for consumer credit and mortgage loans, for US data. The x axis represents the dates, whereas the y axis represents the yearly credit losses as a percentage of the credits

Let us first look into step 1. For each asset class $i \in \{1, \ldots, m\}$, consider the historical credit losses. An example of historical credit losses can be observed in Fig. 4.3. Based on the historical losses, we can calculate the worst possible loss within a certain confidence interval. Let c_i denote this worst possible loss. Notice that the risk number calculated for this asset class uses the historical losses and therefore is not normal due to the skewed shape of the historical distribution.

Then we proceed to step 2, mixing the risks under the correlation matrix between these asset classes. Given the historical series of credit losses, we may calculate the correlation between these asset classes. In the example given in Fig. 4.3, the correlation between the two series is 0.83.

Let us denote the credit risk correlation by $(\rho^C_{i_1 i_2})_{i_1,i_2 \in \{1,\ldots,m\}}$. Then we can calculate the credit risk for the balance sheet, which is only influenced by assets:

$$\begin{aligned}\mathfrak{r}_2(x) &= \mathfrak{r}_{\text{CredSqr}}(x) = \widehat{\mathfrak{r}}_{\text{CredSqr}}\left(\widehat{x^a}\right) \\ &:= \sqrt{\sum_{i_1,i_2 \in \{1,\ldots,m\}} x^a_{i_1} x^a_{i_2} c_{i_1} c_{i_2} \rho^C_{i_1 i_2}} = \sqrt{(\widehat{x^a})^\top \Theta_{\text{cred}}\, \widehat{x^a}}. \end{aligned} \tag{4.1.16}$$

Remark 4.12 *We now show that the matrix $\Theta_{\text{cred}} \in \mathbb{R}^{m \times m}$ is positive semi-definite, so the square root in formula (4.1.16) is well-defined. Let $c = (c_i)_{i \in \{1,\ldots,m\}}$ and $\odot$ denote componentwise multiplication for vectors. For every $\widetilde{x} \in \mathbb{R}^m$,*

$$\widetilde{x}^\top \Theta_{\text{cred}}\, \widetilde{x} = (c \odot \widetilde{x})^\top \rho^C\, (c \odot \widetilde{x}) \geq 0, \tag{4.1.17}$$

by the positive semi-definiteness of the correlation matrix $(\rho^C_{i_1 i_2})_{i_1,i_2 \in \{1,\ldots,m\}}$.

We point out the differences between the quadratic risk calculation and the linear credit risk measure in the previous section. The linear credit risk model assumes that the credit risk of the balance sheet is a sum of the credit risk in each asset class, given by the Vasicek model [110], based on the parametric distribution (4.1.11) for credit portfolio losses. The quadratic credit risk measure does not sum the individual risks; instead, it mixes them under a correlation measure. For each asset class, credit risk is not based on a parametric distribution, but on the historical probability distribution. The choices of individual distributions, even for the linear or quadratic case, are many; we focus on two possibilities for ease of exposition.

Liquidity Risk

Liquidity risk is the risk associated with the bankruptcy for the bank whose balance sheet problem we are addressing. Ultimately, banks fail because of not having enough liquid assets to sell to face payment demands from creditors.

In Table 4.2, we give an example of a liquidity dry-up. In that simple example of a bank, we assume that, due to systemic problems or problems due to the bank, the funding through repurchase agreements (repos) dries up. In the example in the table, the bank does not have enough liquid assets (central bank deposits and marketable Treasury securities) to compensate for the withdrawals and thus faces bankruptcy.

This simple example illustrates the problems of overreliance on short-term wholesale funding and not having enough liquid securities to compensate for the volatility of outflows associated with that type of funding. Many of the most important bank failures (Bear Stearns, Lehman Brothers, Northern Rock) are thought to be related to this type of imbalance in the balance sheet. In a nutshell, this example illustrates liquidity risk.

Linear Liquidity Risk

Many liquidity stress tests can give rise to linear liquidity risk measures. Typically, these stress tests assess whether the bank has enough liquid resources to compensate the potential withdrawals (also known as runoffs) from liabilities after a certain period.

Let us consider then a balance sheet as in Definition 4.5. We want to assess the liquidity shortfall that the balance sheet will have within a specific horizon, i.e., whether there are enough liquid assets at the end of the horizon to compensate withdrawals from creditors.

The liquid value of assets in a stress scenario at the end of the horizon is determined by haircuts $l_i \in [0, 1]$, $i = 0, \ldots, m$. Central bank deposits have no liquidity risk, so we assume that their haircut is $l_0 = 0$. These haircuts determine the proportion of $1 - l_i$ realizable in cash under a stress

Balance sheet		
Assets	$t=0$	$t=1$
Central bank deposits	100	100
Treasury securities	100	80
Mortgage loans	800	800
Liabilities and capital		
Retail deposits	500	500
Repo funding	400	0
Shareholder capital	100	100

Table 4.2 An example of a liquidity dry-up. Between time $t = 0$ and $t = 1$, we assume that the repo market dries up so that the bank cannot rely on that funding. We presume that Treasury securities also have a devaluation of 20% between $t = 0$ and $t = 1$. At time $t = 1$, the bank has only 180 in liquid assets (80 in securities plus 100 in central bank deposits). The liquidity shortfall at time $t = 1$, given by the outflows in the liabilities (400) and the stock of liquid assets (180), is 220. Since the withdrawals on repos are much higher than liquid assets, the bank will face bankruptcy

event. Short-term deposits typically have zero haircuts, as they serve as cash to compensate withdrawals from creditors. Liquid securities typically have haircuts close to zero, which correspond to the potential devaluations on securities; for example, a 10% haircut on a security indicates that, in a stress event, the security can be sold at 90% of its face value to compensate runoffs. Less liquid securities, such as mortgages or corporate credit, have haircuts close to 1, indicating that realizing cash on these securities is difficult.

The potential outflows on liabilities are also given by the runoff values $l_i \in [0, 1]$, $i = m+1, \ldots, m+k$. In this context, shareholders' equity or long-term debt have zero outflows, as shareholders or bondholders cannot simply ask for their money back from the bank in the near future. Retail deposits typically have very low runoff values; although customers can typically claim their money back, statistically, the outflows on these assets tend to be low (although bank runs, i.e., depositors herding to claim their money back, are possible). Short-term institutional or wholesale funding, such as repo funding, tends to be more unstable, as creditors may not refinance these credits in times of crises. The example in Table 4.2 illustrates such episodes. As a consequence, wholesale funding receives runoff values l_i closer to 1.

The liquidity shortfall, given by the difference in the liabilities outflows and the realizable values of assets, is shown in the following expression:

Definition 4.13 (Linear Liquidity Shortfall) *Suppose $x \in \mathbb{R}^{1+m+k}$ is a balance sheet as in Definition 4.5, $l_i \in [0, 1]$, $i = 1, \ldots, m$, are the haircuts for assets, and $l_i \in [0, 1]$, $i = m+1, \ldots, m+k$, are the runoffs for liabilities. Then the liquidity shortfall for this balance sheet is given by*

$$\mathfrak{r}_3(x) = \mathfrak{r}_{\text{LiqLin}}(x) := \max\left(- \sum_{j=m+1}^{m+k} l_j x_j - \sum_{i=0}^{m} (1 - l_i) x_i, 0 \right). \tag{4.1.18}$$

Remark 4.14 *The expression inside the* max *in (4.1.18) shows two terms:* $-\sum_{j=m+1}^{m+k} l_j x_j$ *is the total runoff in liabilities, whereas* $\sum_{i=0}^{m}(1-l_i)x_i$ *are the total assets realizable in cash. There will be a shortfall in liquidity if the total runoff in liabilities exceeds the realizable cash from assets; otherwise, there is no shortfall.*

Quadratic Liquidity Risk

For *liquidity risk* measurement, we can replicate the procedure that we used in credit risk. Again step 1 is to calculate the individual risks for each asset and liability class; step 2 mixes the risk using the correlation matrix. Notice that when we use step 1, we will use the historical distribution, and thus the risk does not stem from a normal distribution.

Let us divide the index $i \in \{0, \ldots, m\}$ on the asset side (including central bank deposits) into the liquid and illiquid assets: $i \in L \cup \bar{L}$, where L is the set of indices for liquid assets (essentially central bank deposits, Treasury bonds, and corporate bonds) and $\bar{L}$ the set for illiquid assets such as mortgages and corporate credit.

Let us assume that the initial value of liquid assets (which are sellable) is given by

$$K = \sum_{i \in L} x_i. \tag{4.1.19}$$

Since $S_0 = (1, \ldots, 1)^\top$, K is essentially the market value of liquid securities at time $t = 0$ in the balance sheet, which depends on the composition of the balance sheet.

Now, what are the events that decrease liquidity? For liquid assets, liquidity shrinks by a market devaluation of the assets. If a liquid asset reduces its value by 10 million euros, the bank will have 10 million less liquidity.

For each liquid asset class (securities), the individual liquidity risk is the haircut, i.e., the worst possible decrease in market value within a certain level of confidence based on historical data. For asset class $i \in L$, we assume that this haircut is given by $l_i \in [0, 1]$. For example, for a Treasury bond, this number is the worst market loss within a confidence level.

Runoffs on liabilities dictate their loss in liquidity. For example, if depositors take away 10 million euros in deposits, the bank will have 10 million less liquidity. For each liability class, $j \in \{m+1, \ldots, m+k\}$, we assume that the individual liquidity risk is the worst possible outflow within a certain confidence level. Let us call this $l_j \in [0, 1]$. For example, in the case of deposits, we take the worst decrease in deposit volumes within a certain confidence level.

In step 2, we need to mix the risks using the correlation matrices. This correlation matrix $\left(\rho^{\mathrm{Liq}}_{i_1,i_2}\right)_{i_1,i_2\in\{1,\ldots,m+k\}}$ uses the historical data series of market prices on securities and redemptions on liabilities.

Combining the potential market devaluations with the potential redemptions, and using the correlation matrix, we get the combined liquidity shock for the balance sheet, which we call liquidity at risk (LaR), as proposed by some authors, such as Cont et al. [30].

Definition 4.15 (Quadratic Liquidity at Risk) *Suppose $x \in \mathbb{R}^{1+m+k}$ is a balance sheet as in Definition 4.5 and $L \subset \{0,\ldots,m\}$ is the set of liquid assets. Let $l_i \in [0,1]$, $i \in L$, be the haircuts for liquid assets and $l_j \in [0,1]$, $j = m+1,\ldots,m+k$, the runoffs for liabilities. Then the quadratic liquidity at risk (LaR) for this balance sheet is given by*

$$
\begin{aligned}
\mathfrak{r}_{\mathrm{LaR}}(x) = \Bigg[&\sum_{i_1,i_2\in L} x^a_{i_1} x^a_{i_2} l_{i_1} l_{i_2} \rho^{\mathrm{Liq}}_{i_1,i_2} + \sum_{j_1,j_2\in\{m+1,\ldots,m+k\}} x^\ell_{j_1} x^\ell_{j_2} l_{j_1} l_{j_2} \rho^{\mathrm{Liq}}_{j_1,j_2} \\
&- 2 \sum_{i\in L, j\in\{m+1,\ldots,m+k\}} x^a_i x^\ell_j l_i l_j \rho^{\mathrm{Liq}}_{i,j} \Bigg]^{1/2} \\
=: &\sqrt{\widehat{x}^\top \Theta_{\mathrm{LaR}}\, \widehat{x}}.
\end{aligned}
$$

Remark 4.16 $\Theta_{\mathrm{LaR}} \in \mathbb{R}^{(m+k)\times(m+k)}$ *is also positive semi-definite. The proof follows the same lines as Remark 4.12.*

Liquidity at risk essentially computes the liquidity drain associated with a balance sheet. It is the analogue of value at risk [62] for liquidity. Whereas value at risk measures market risk, liquidity at risk measures funding liquidity risk. In the value at risk framework, banks assess their financial health by comparing the level of capital to the output of the value at risk measure. If capital is higher, the bank has enough capital to withstand losses; otherwise, it does not. In liquidity at risk, one needs to compare the stock of liquid assets with the potential liquidity drain (LaR). Similarly to Definition 4.13, we then compute the liquidity shortfall.

Definition 4.17 (Quadratic Liquidity Shortfall) *Suppose we have the same setting as in Definition 4.15. Then the liquidity shortfall is given by*

$$
\mathfrak{r}_3(x) = \mathfrak{r}_{\mathrm{LaRShort}}(x) := \max\left(\mathfrak{r}_{\mathrm{LaR}}(x) - \sum_{i\in L} x^a_i, 0\right). \tag{4.1.20}
$$

If $\beta = (\beta_0,\ldots,\beta_m)^\top \in \mathbb{R}^{1+m}$ is the vector such that $\beta_i = 1$ if $i \in L$ and $\beta_i = 0$ otherwise, then the previous expression can be written as

$$\mathfrak{r}_3(x) = \max\left(\sqrt{\widehat{x}^\top \Theta_{\text{LaR}}\, \widehat{x}} - (x_0, \widehat{x^a}^\top) \cdot \beta, 0\right). \tag{4.1.21}$$

Remark 4.18 *As in Definition 4.13, Eq. (4.1.20) compares two quantities inside the maximum. If the stock of total liquid assets $\sum_{i \in L} x_i^a$ exceeds the potential liquidity drain $\mathfrak{r}_{\text{LaR}}(x)$, then there is no liquidity shortfall; otherwise, the shortfall is equal to the difference between LaR and the stock of liquid assets.*

We are interested in liquidity shortfall more so than liquidity at risk because we want to include the hedging power of liquid securities as a mitigant to liquidity risk. Liquidity at risk is only concerned with the potential drains in liquidity and does not include the initial stock of liquid assets $\sum_{i \in L} x_i^a$. When facing balance sheet choices, it is essential to include these liquid assets in the liquidity risk metric. For example, think of two balance sheets with the same liquidity at risk of 100, but the first has no liquid securities, while the second has 200 in liquid assets. The first balance sheet will have a shortfall of 100, meaning that it will likely face bankruptcy as it has no liquid assets to compensate runoffs. But the second balance sheet has a zero liquidity shortfall, meaning that it has enough liquid assets to withstand outflows. It is therefore crucial to incorporate the hedging power of liquid securities in the liquidity risk measure when assessing optimal balance sheet choices.

Other Risks

In banking, several other risks exist, which we will not highlight in detail here. Some examples include market risk, operational risk, legal risk, and climate risk, whose measurement has tremendous developments at the time of writing. One can extend our vector risk framework to incorporate all these risks.

4.2 Bank Balance Sheet Problems Involving Linear Interest Rate and Credit Risk

Bank balance sheet problems are used to illustrate the application of the theories developed in the previous chapters. These problems are examined using models adopting various different risk measures with real-world data.

In this section, a bank balance sheet problem that consists of general investable assets and bonds is modeled with linear risk measures following the discussion in Júdice and Zhu [65]. The risks of the items in the balance sheets are characterized with a coefficient representing risk per unit of investment. We will distinguish assets and liabilities with superscripts a and ℓ, respectively. Notation of those coefficients is summarized in Table 4.3.

Balance sheet items	Allocation	Returns	Interest Rate risk	Credit Risk	Range of indices
General assets	$x_j^a \geq 0$	R_j^a	$I_j^a > 0$	$c_j > 0$	$j = 1, \ldots, m_{gA}$
Long bonds	$x_j^a \geq 0$	R_j^a	$I_j^a > 0$	None	$j = m_{gA} + 1, \ldots, m_{gA} + m_{lB}$
Short bonds	$x_s^\ell \leq 0$	R_s^ℓ	$I_s^\ell > 0$	None	$s = 1, \ldots, k$
Central bank funds	$x_0 \in \mathbb{R}$	R_0	$I_0 = 0$	None	$\{0\}$

Table 4.3 Notation (see also [65, Table 1])

We assume that there are $k \in \mathbb{N}$ short bonds, $m_{lB} \in \mathbb{N}$ long bonds, and $m_{gA} \in \mathbb{N}$ general assets distinguished with subscripts. The total number of assets is then $m = m_{gA} + m_{lB}$. Subscript 0 is reserved for the overnight central bank funding which has no interest rate risk and may be long or short. The returns can take both positive and negative values (negative interest rate is a reality and is allowed in our model). The only assumption we make is that the central bank funding has the lowest return, i.e.,

$$R_s^\ell, R_j^a > R_0, \tag{4.2.1}$$

which amounts to say the yield curve is not inverted.

4.2.1 Linear Programming Model

Denote the vector of shares in various assets in the balance sheet by $x = (x_0, \widehat{x^a}^\top, \widehat{x^\ell}^\top)^\top \in \mathbb{R}^{1+m+k}$. The linear programming model for the bank balance sheet is our primal problem:

$$p = \max_{x \in \mathbb{R}^{1+m+k}} \sum_{j=1}^{m} R_j^a x_j^a + \sum_{s=1}^{k} R_s^\ell x_s^\ell + R_0 x_0 \tag{4.2.2}$$

subject to

$$\begin{aligned} &\sum_{j=1}^{m} I_j^a x_j^a + \sum_{s=1}^{k} I_s^\ell x_s^\ell \leq r^i \\ &\sum_{j=1}^{m_{gA}} c_j x_j^a \leq r^c \\ &\widehat{x^a} \geq 0, \text{and } \widehat{x^\ell} \leq 0 \\ &\sum_{j=1}^{m} x_j^a + \sum_{s=1}^{k} x_s^\ell + x_0 = e. \end{aligned} \tag{4.2.3}$$

where $r^i > 0$ and $r^c > 0$ are limits for the interest and credit risks, respectively, and e is the equity. In our model, there is no limitation on leveraging the balance sheet. As a result, the equity does not impact the optimal solution. Thus, we assume $e = 0$ to simplify the notation. Using Eq. (4.2.3) to eliminate x_0, the linear programming problem becomes

$$\mu = p = \max_{\widehat{x} \in \mathbb{R}^{m+k}} \mathfrak{u}(\widehat{x}) := \sum_{j=1}^{m} (R_j^a - R_0) x_j^a + \sum_{s=1}^{k} (R_s^\ell - R_0) x_s^\ell \tag{4.2.4}$$

subject to

$$\lambda^i : \quad \mathfrak{r}_1(\widehat{x}) := \sum_{j=1}^{m} I_j^a x_j^a + \sum_{s=1}^{k} I_s^\ell x_s^\ell \leq r^i$$

$$\lambda^c : \quad \mathfrak{r}_2(\widehat{x}) := \sum_{j=1}^{m_{gA}} c_j x_j^a \leq r^c$$

$$\widehat{x^a} \geq 0, \text{ and } \widehat{x^\ell} \leq 0.$$

The first two constraints in (4.2.4) correspond to the dual variables λ^i, λ^c, representing the sensitivity with respect to changes in the limits for interest rate risk r^i and credit risk r^c, respectively.

4.2.2 The Dual Problem

The linear programming problem (4.2.4) has a dual problem for the dual variable $\lambda = (\lambda^i, \lambda^c)$ (the corresponding primal variables are marked in front of the constraints):

$$d = \min_{\lambda \in \mathbb{R}^2} \quad r^i \lambda^i + r^c \lambda^c \tag{4.2.5}$$

subject to

$$x_j^a : \quad I_j^a \lambda^i + c_j \lambda^c \geq R_j^a - R_0, \; j = 1, \ldots, m_{gA}$$

$$x_j^a : \quad I_j^a \lambda^i \geq (R_j^a - R_0), \; j = m_{gA} + 1, \ldots, m$$

$$x_s^\ell : \quad I_s^\ell \lambda^i \leq (R_s^\ell - R_0), \; s = 1, \ldots, k$$

$$\lambda^i, \lambda^c \geq 0.$$

The constraints corresponding to x_j^a, $j = m_{gA}+1, \ldots, m$ and x_s^ℓ, $s = 1, \ldots, k$ in (4.2.5) lead to

$$u := \min_{s=1,\ldots,k} \frac{R_s^\ell - R_0}{I_s^\ell} \geq \lambda^i \geq \max_{j=m_{gA}+1,\ldots,m} \frac{R_j^a - R_0}{I_j^a} := l > 0. \tag{4.2.6}$$

Let α be an index among $\{1, ..., k\}$ such that $u = (R^\ell_\alpha - R_0)/I^\ell_\alpha$. Then the short bond represented by α has the lowest cost per unit of interest risk. Similarly, let β be an index among $\{m_{gA}+1, \ldots, m\}$ such that $(R^a_\beta - R_0)/I^a_\beta = l$, and then the long bond represented by β has the highest return per unit of interest risk.

The two dual variables are actually related. Given λ^i, the constraints corresponding to $x^a_j, j = 1, \ldots, m_{gA}$ in (4.2.5) tell us that

$$\lambda^c \geq \max_{j=1,\ldots,m_{gA}} \frac{R^a_j - R_0 - \lambda^i I^a_j}{c_j}. \tag{4.2.7}$$

If, for a solution of (4.2.4), at least one of $x^a_j, j = 1, ..., m_{gA}$ is positive, then the complimentary slackness condition implies that (4.2.7) is actually an equality. Thus, we can replace λ^c with the right-hand side of (4.2.7) and simplify the dual problem to the following form:

Lemma 4.19 *Suppose that at least one of $x^a_j, j = 1, ..., m_{gA}$ is positive in the optimal balance sheet of* (4.2.4). *Then the dual problem in* (4.2.5) *can be represented as a one variable optimization problem:*

$$d = \min_{\lambda^i \in [l,u]} \phi(\lambda^i) \tag{4.2.8}$$

where $\phi(\lambda^i) := \max_{j=0,1,\ldots,m_{gA}} L_j(\lambda^i)$, $L_0(\lambda^i) = r^i \lambda^i$ *and*

$$L_j(\lambda^i) := \frac{r^c(R^a_j - R_0)}{c_j} + \frac{r^i c_j - r^c I^a_j}{c_j} \lambda^i, \quad j = 1, ..., m_{gA}. \tag{4.2.9}$$

The function ϕ is convex and polyhedral, i.e., locally linear, and its possible behaviors are illustrated in Fig. 4.4.

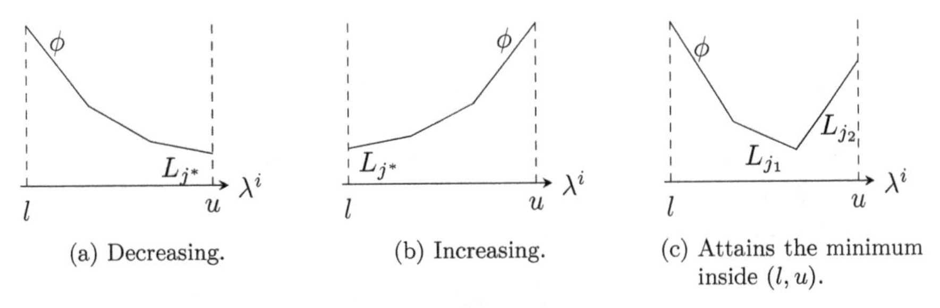

Fig. 4.4 Illustration of ϕ; see Theorem 4.20 (cf. also [65, Figures 1, 2, and 3])

4.2.3 Optimal Balance Sheet

Finding the optimal balance sheet and the companion dual solutions corresponding to each scenario are simple algebraic exercises. We summarize the results in the next theorem without proof. The optimal values of the variables are indicated with bars.

Theorem 4.20 (Optimal Balance Sheet [65]) *Let the setting of Table 4.3 be given with additionally* (4.2.1). *We consider the primal and dual linear programming problems* (4.2.4) *and* (4.2.5) *with given constants* $r^i > 0$ *and* $r^c > 0$. *Suppose furthermore that at least one of* $x_j^a, j = 1, ..., m_{gA}$ *is positive in the optimal balance sheet of* (4.2.4). *The optimal balance sheet of* (4.2.4) *can then be determined explicitly according to the behavior of* ϕ *as below:*
Case 1. *Suppose that the function* ϕ *is monotone decreasing on* $[l, u]$ *as in Fig. 4.4a. We select some* j^* *such that* L_{j^*} *is among all* $j \in \{1, \ldots, m_{gA}\}$ *satisfying* $L_j(u) = \phi(u) = d = p$ *that coincide with* ϕ *near* u *and consider two cases:*

(a) *If* $(R_{j^*}^a - R_0)/I_{j^*}^a > u$, *then the optimal balance sheet is* $\bar{x}_{j^*}^a = r^c/c_{j^*}$, $\bar{x}_\alpha^\ell = -(I_{j^*}^a r^c - c_{j^*} r^i)/c_{j^*} I_\alpha^\ell \leq 0$ *and* $\bar{x}_0 = -\bar{x}_{j^*}^a - \bar{x}_\alpha^\ell$, *The corresponding dual solution of* (4.2.5) *is*

$$(\bar{\lambda}^i, \bar{\lambda}^c) = \left(u, \left(\frac{R_{j^*}^a - R_0}{I_{j^*}^a} - u \right) \frac{I_{j^*}^a}{c_{j^*}} \right).$$

(b) *Otherwise, it holds* $(R_{j^*}^a - R_0)/I_{j^*}^a = u$ *and the optimal balance sheet is* $\bar{x}_{j^*}^a = r^i/I_{j^*}^a$, *and* $\bar{x}_0 = -\bar{x}_{j^*}^a$, *with the dual solution*

$$(\bar{\lambda}^i, \bar{\lambda}^c) = (u, 0).$$

Case 2. *Suppose that the function* ϕ *is monotone increasing on* $[l, u]$ *as in Fig. 4.4b. We select some* j^* *such that* L_{j^*} *is among all* $j \in \{0, 1, \ldots, m_{gA}\}$ *satisfying* $L_j(l) = \phi(l) = d = p$ *that coincide with* ϕ *near* l *and consider two cases:*

(a) *If* $j^* \neq 0$ *and* $(R_{j^*}^a - R_0)/I_{j^*}^a > l$, *then the optimal balance sheet is* $\bar{x}_{j^*}^a = r^c/c_{j^*}$, $\bar{x}_\beta^a = (c_{j^*} r^i - I_{j^*}^a r^c)/c_{j^*} I_\beta^a \geq 0$, *and* $\bar{x}_0 = -\bar{x}_{j^*}^a - \bar{x}_\beta^a$, *and the corresponding dual solution is*

$$(\bar{\lambda}^i, \bar{\lambda}^c) = \left(l, \left(\frac{R_{j^*}^a - R_0}{I_{j^*}^a} - l \right) \frac{I_{j^*}^a}{c_{j^*}} \right).$$

(b) *Otherwise, it holds* $(R_{j^*}^a - R_0)/I_{j^*}^a = l$ *or* $j^* = 0$ *and the optimal balance sheet is* $\bar{x}_\beta^a = r^i/I_\beta^a$ *and* $\bar{x}_0 = -\bar{x}_\beta^a$, *with the dual solution*

$$(\bar{\lambda}^i, \bar{\lambda}^c) = (l, 0).$$

Case 3. *Suppose that the function ϕ attains the unique minimum inside (l, u); then there are two possible cases:*

(a) *The function ϕ attains the unique minimum at the intersection of the graphs of L_{j_1} and L_{j_2}, where $j_1, j_2 \in \{1, \ldots, m_{gA}\}$. Then $(r^i c_{j_1} - r^c I^a_{j_1}) < 0$ and $(r^i c_{j_2} - r^c I^a_{j_2}) > 0$ and the optimal balance sheet is $\bar{x}_0 = -\bar{x}^a_{j_1} - \bar{x}^a_{j_2}$, and*

$$(\bar{x}^a_{j_1}, \bar{x}^a_{j_2}) = \left(\frac{\begin{vmatrix} r^i & I^a_{j_2} \\ r^c & c_{j_2} \end{vmatrix}}{\begin{vmatrix} I^a_{j_1} & I^a_{j_2} \\ c_{j_1} & c_{j_2} \end{vmatrix}}, \frac{\begin{vmatrix} I^a_{j_1} & r^i \\ c_{j_1} & r^c \end{vmatrix}}{\begin{vmatrix} I^a_{j_1} & I^a_{j_2} \\ c_{j_1} & c_{j_2} \end{vmatrix}} \right) \in \mathbb{R}^2_{>0}. \tag{4.2.10}$$

The dual solution is

$$(\bar{\lambda}^i, \bar{\lambda}^c) = \left(\frac{\begin{vmatrix} c_{j_1} & R^a_{j_1} - R_0 \\ c_{j_2} & R^a_{j_2} - R_0 \end{vmatrix}}{\begin{vmatrix} c_{j_1} & I^a_{j_1} \\ c_{j_2} & I^a_{j_2} \end{vmatrix}}, \frac{\begin{vmatrix} R^a_{j_1} - R_0 & I^a_{j_1} \\ R^a_{j_2} - R_0 & I^a_{j_2} \end{vmatrix}}{\begin{vmatrix} c_{j_1} & I^a_{j_1} \\ c_{j_2} & I^a_{j_2} \end{vmatrix}} \right) \in \mathbb{R}^2_{>0} \tag{4.2.11}$$

and the common primal and dual value is

$$p = d = \frac{\begin{vmatrix} r^i c_{j_1} - r^c I^a_{j_1} & R^a_{j_1} - R_0 \\ r^i c_{j_2} - r^c I^a_{j_2} & R^a_{j_2} - R_0 \end{vmatrix}}{\begin{vmatrix} c_{j_1} & I^a_{j_1} \\ c_{j_2} & I^a_{j_2} \end{vmatrix}} > 0. \tag{4.2.12}$$

(b) *The function ϕ attains the unique minimum at the intersection of the graphs of L_{j*} and $r^i \lambda^i$. Then $(r^i c_{j*} - r^c I^a_{j*}) < 0$ and the optimal balance sheet is $\bar{x}_0 = -\bar{x}^a_{j*} = -r^i / I^a_{j*}$. The dual solution is*

$$(\bar{\lambda}^i, \bar{\lambda}^c) = ((R^a_{j*} - R_0)/I^a_{j*}, 0),$$

and the common primal and dual value is $p = d = r^i (R^a_{j*} - R_0)/I^a_{j*}$.

4.2.4 Efficient Frontier

In a linear model, the efficient frontier can always be represented by a polyhedral function. Using the dual solution and $p = d$ for (4.2.4) and (4.2.5), locally the efficient frontier can be represented by

$$\mu = \bar{\lambda}^i r^i + \bar{\lambda}^c r^c. \tag{4.2.13}$$

However, varying r^i and r^c can also bring a change of the dual solution, and therefore in general, this efficient frontier may be quite complicated. We illustrate this by working out the explicit representation of the efficient frontier in the simple case when there are only two general assets involved, labeled by indices j_1 and j_2. Moreover, we focus on the interesting cases when the risk-adjusted returns of these two assets are better than the minimal bond return u, i.e.,

$$\frac{R^a_{j_k} - R_0}{I^a_{j_k}} > u > l, \text{ for } k = 1, 2.$$

We already know that locally the efficient frontier will be represented by (4.2.13) with dual solution determined by Theorem 4.20. The concrete form of the dual solution is determined by the behavior of the function ϕ in Lemma 4.19. In the case when only two general assets j_1, j_2 are involved, the behavior of ϕ is determined by the slopes of L_{j_1}, L_{j_2}. We assume without loss of generality that

$$0 < \frac{c_{j_1}}{I^a_{j_1}} < \frac{c_{j_2}}{I^a_{j_2}}.$$

Then the two lines $r^i c_{j_k} - r^c I^a_{j_k} = 0, \ \ k = 1, 2$ separate the first quadrant $\mathbb{R}^2_{>0} := \{(r^i, r^c) : r^i, r^c > 0\}$ of the (r^i, r^c)-plane into three regions:

$$\begin{aligned} A &:= \{(r^i, r^c) : r^i c_{j_1} - r^c I^a_{j_1} \geq 0\} \cap \mathbb{R}^2_{>0}, \\ B &:= \{(r^i, r^c) : r^i c_{j_1} - r^c I^a_{j_1} < 0 \ \&\ r^i c_{j_2} - r^c I^a_{j_2} > 0\} \cap \mathbb{R}^2_{>0}, \\ C &:= \{(r^i, r^c) : r^i c_{j_2} - r^c I^a_{j_2} \leq 0\} \cap \mathbb{R}^2_{>0}. \end{aligned} \tag{4.2.14}$$

We can see that ϕ is increasing in A and decreasing in C and attains a minimum in (l, u) in B. Thus, by Theorem 4.20 the efficient frontier can be represented by

$$\mu = \nu(r^i, r^c) = \begin{cases} \dfrac{l(r^i c_{j_1} - r^c I^a_{j_1}) + (R^a_{j_1} - R_0) r^c}{c_{j_1}} & \text{if } (r^i, r^c) \in A, \\[2ex] \dfrac{\begin{vmatrix} r^i c_{j_1} - r^c I^a_{j_1} & R^a_{j_1} - R_0 \\ r^i c_{j_2} - r^c I^a_{j_2} & R^a_{j_2} - R_0 \end{vmatrix}}{\begin{vmatrix} c_{j_1} & I^a_{j_1} \\ c_{j_2} & I^a_{j_2} \end{vmatrix}} & \text{if } (r^i, r^c) \in B, \\[2ex] \dfrac{u(r^i c_{j_2} - r^c I^a_{j_2}) + (R^a_{j_2} - R_0) r^c}{c_{j_2}} & \text{if } (r^i, r^c) \in C. \end{cases} \tag{4.2.15}$$

Remark 4.21 *Note that on the common boundary of A and B, we have $r^i c_{j_1} - r^c I^a_{j_1} = 0$ and the representations of the efficient frontier in* (4.2.15) *coincide. Similarly, on the common boundary of B and C, we have $r^i c_{j_2} - r^c I^a_{j_2} = 0$ and, again, the representations of the efficient frontier in* (4.2.15) *coincide. By Proposition 3.12, $\nu(r^i, r^c)$ is a concave function. Thus, in this example, the efficient frontier is represented by a continuous concave polyhedral function.*

4.2.5 Financial Meanings

The results derived in the previous section have very intuitive financial meanings that are also described in Júdice and Zhu [65]. As the authors show, the dual problem (4.2.5) finds the lowest price of the combined insurance against interest rate risk and credit risk $r^i\lambda^i + r^c\lambda^c$. λ^i and λ^c are the interest rate and credit risk insurance prices for the balance sheet. The restrictions in (4.2.5) state that the value on insurance needs to be higher than the individual risk-adjusted returns on assets and liabilities. The dual value is equal to the minimum of such possible prices.

In Eq. (4.2.6), u is the lowest interest rate risk-adjusted cost on liabilities, l is the highest interest rate risk-adjusted return on long bonds, and $l \leq u$. If $l > u$, then there is an arbitrage. We will illustrate this in the concrete example below.

Equation (4.2.9) also has an important financial meaning. For each asset $j = 1, \ldots, m_{gA}$, the price of credit risk results from removing the return from interest rate risk

$$\frac{R_j^a - R_0 - \lambda^i I_j^a}{c_j}.$$

It is shown in [65] that for these prices actually holds

$$\lambda^c = \max_{j=1,\ldots,m_{gA}} \frac{R_j^a - R_0 - \lambda^i I_j^a}{c_j},$$

in case the right-hand side is non-negative. As a consequence, given λ^i, the combined interest rate and credit risk insurance price is

$$r^c\lambda^c + r^i\lambda^i = r^c \max_{j=1,\ldots,m_{gA}} \frac{R_j^a - R_0 - \lambda^i I_j^a}{c_j} + r^i\lambda^i,$$

or, in other words,

$$\phi(\lambda^i) = \max_{j=0,1,\ldots,m_{gA}} L_j(\lambda^i). \tag{4.2.16}$$

Figure 4.4 shows the possibilities for ϕ, a polyhedral convex function. It is shown in [65] that the arbitrage-free price of insurance is the minimum of $\phi(\lambda^i)$ over $\lambda^i \in [l, u]$.

The function $\phi(\lambda^i)$ is locally equal to $L_j(\lambda^i)$, for some j. Ignoring the case of $j = 0$, the function is increasing or decreasing depending on two opposite effects: the insurance price for interest rate risk, $r^i\lambda^i$, increasing in λ_i, and the credit risk insurance price $r^c \frac{R_j^a - R_0 - \lambda^i I_j^a}{c_j}$, decreasing in λ^i. So, the slope of $L_j(\lambda^i)$ will depend upon the rate of change to λ^i in the interest rate risk component – equal to r^i – and in the credit risk component, equal to $r^c I_j^a / c_j$. As a consequence, the combined price is increasing if $r^i > r^c I_j^a / c_j$, or equivalently

$$r^i / I_j^a > r^c / c_j, \tag{4.2.17}$$

and nonincreasing otherwise. The first situation will happen when credit risk is exhausted first; the second situation arises when $r^i / I_j^a \leq r^c / c_j$, and in this case, interest rate risk is exhausted first.

Theorem 4.20 illustrates the three possible situations. In Case 1, ϕ is a decreasing function, attaining a minimum at u. With the exception of the pathological case (b) in the theorem, $r^i / I_j^a < r^c / c_j$, for $j = 1, \ldots, m_{gA}$, which means that the interest rate risk is exhausted first when compared to credit risk. In the general case (a), the balance sheet will invest in a general asset j^*. Since interest rate risk is exhausted first, to consume all the credit risk, the bank manager will need to violate the interest rate budget, but he has a quick fix for this: he simply needs to hedge the interest rate risk with the least expensive short bond (after adjusting for risk). Thus, the optimal balance sheet consists of a general asset and a short bond; the net funds from this pair will be invested or financed from the central bank funds, depending on whether the general asset balance exceeds the financing from the short bonds.

The pathological Case 1 (b) occurs when no risk-adjusted return on general assets outperforms the risk-adjusted return on bonds, so it doesn't pay off to invest in general assets. In this case, the balance sheet will only have one asset, a long bond, financed by the central bank. The credit risk restriction is not binding (hence $\bar{\lambda}^c = 0$, since no credit risk is actually consumed in this case.

When ϕ is an increasing function, as in Case 2, it attains a minimum at l. In the general case (a), the balance sheet will include a general asset j^*. Locally, we have $r^i / I_j^a > r^c / c_j$, meaning that credit risk is exhausted first. So the bank manager will invest all the resources in the general asset j^*, but she/he will have slack in interest rate risk, which she/he will fill with the long bond with the highest returning yield (after adjusting for risk). The balance sheet thus is comprised of a general asset and a long bond, funded by the central bank. Again, there is a pathological Case 2 (b), similar to Case 1 (b).

Case 3 in Theorem 4.20 occurs when locally we have both situations, i.e., for some j's, the interest rate risk is exhausted first, and for other j's, credit risk is consumed first. In this case, the solution is attained in the interior of $[l, u]$. In the general case, the optimal balance sheet will be comprised of two general assets that bind the restrictions on credit and interest rate risk, funded by the central bank. Again, there is a pathological Case 3 (b), where the balance sheet is a long bond, funded by the central bank.

In all cases, we see that the optimal balance sheet has not more than three instruments, including deposits or funding from the central bank. Therefore, to make this approach manageable in practice, Júdice and Zhu propose using diversification constraints described in their paper [65] .

4.2.6 Linear Bank Balance Sheet Example with Real Data

Following [65, Section 5] we illustrate the application of the solutions with real data, taken from databases that report yields, interest rates, default rates, and losses given defaults. The data is organized and described in Table 4.4, which showcase a simplified version of a real bank balance sheet risk allocation problem. We refer to [65] for details of the source of the data. We assume that the upper limits of both interest rate and credit risks are 50, i.e., $r^i = r^c = 50$. To simplify the discussion, we view central bank funding as the short of liquidity.

Interest Rate Risk Arbitrage

Examining Table 4.4, we notice that R^ℓ_6, R^ℓ_7 are smaller than R_0, violating Assumption (4.2.1). As a result, $u = (R^\ell_7 - R_0)/I^\ell_7 = -0.78 < 0 < 0.0451 = (R^a_5 - R_0)/I^a_5 = l$, which indicates an interest rate risk arbitrage. In practice, however, we know that certificate of deposit funding is limited. Therefore,

Asset class	Allocation	Returns	Interest rate risk	Credit risk
15-year mortgages	x^a_1	$R^a_1 = 3.22\%$	$I^a_1 = 10.35\%$	$c_1 = 9.6\%$
30-year mortgages	x^a_2	$R^a_2 = 3.76\%$	$I^a_2 = 26.99\%$	$c_2 = 9.9\%$
BBB corporate bonds	x^a_3	$R^a_3 = 5.02\%$	$I^a_3 = 32.04\%$	$c_3 = 1.6\%$
Consumer credit	x^a_4	$R^a_4 = 7.68\%$	$I^a_4 = 3.96\%$	$c_4 = 14.3\%$
AAA corporate bonds	x^a_5	$R^a_5 = 4.02\%$	$I^a_5 = 35.90\%$	None
10-year treasury bonds	x^a_6	$R^a_6 = 2.69\%$	$I^a_6 = 19.96\%$	None
5-year treasury bonds	x^a_7	$R^a_7 = 2.51\%$	$I^a_3 = 10.70\%$	None
2-year treasury bonds	x^a_8	$R^a_8 = 2.48\%$	$I^a_8 = 4.44\%$	None
Liquidity	x_0	$R_0 = 2.40\%$	$I_0 = 0.00\%$	None
Liability class	**Allocation**	**Returns**	**Interest rate risk**	**Credit risk**
Central bank funding	x_0	$R_0 = 2.40\%$	$I_0 = 0.00\%$	None
1-month commercial paper	x^ℓ_1	$R^\ell_1 = 2.42\%$	$I^\ell_1 = 0.19\%$	None
3-month commercial paper	x^ℓ_2	$R^\ell_2 = 2.67\%$	$I^\ell_2 = 0.57\%$	None
2-year financial bonds	x^ℓ_3	$R^\ell_3 = 3.29\%$	$I^\ell_3 = 4.39\%$	None
5-year financial bonds	x^ℓ_4	$R^\ell_4 = 3.53\%$	$I^\ell_4 = 10.39\%$	None
10-year financial bonds	x^ℓ_5	$R^\ell_5 = 3.93\%$	$I^\ell_5 = 18.75\%$	None
24-month deposit	x^ℓ_6	$R^\ell_6 = 0.81\%$	$I^\ell_6 = 4.55\%$	None
12-month deposit	x^ℓ_7	$R^\ell_7 = 0.61\%$	$I^\ell_7 = 2.29\%$	None

Table 4.4 Returns for the different asset and liability classes with $m_{gA} = m_{lB} = 4$, $m = m_{gA} + m_{lB} = 8$, and $k = 7$ (cf. [65, Table 2])

the practical implication is that the bank can construct such interest rate risk arbitrages as a standalone position to the extent that the certificate of deposit funding is exhausted. Thus, in practice, balance sheet manager will focus on the assets in Table 4.4 excluding these assets.

Graphic Illustration of Optimal Balance Sheets

Excluding the 12- and 24-month deposits from Table 4.4, the rest of the assets do satisfy Assumption (4.2.1), with

$$u = (R_5^\ell - R_0)/I_5^\ell = 0.0816 > 0.0451 = (R_5^a - R_0)/I_5^a = l > 0.$$

Let us graphically illustrate the optimal balance sheet by drawing the four lines $L_j, j = 1, \ldots, m_{gA} = 4$ representing the four general assets along with L_0 in the interval $[l, u]$ in Fig. 4.5.

Clearly, $\phi = \max(L_3, L_4)$ and the optimal dual solution $\bar{\lambda}^i$ is at the intersection of L_3, L_4. This fits Case 3 (a) in Theorem 4.20. Using (4.2.10), (4.2.11), and (4.2.12), we can find the primal and dual solutions to be

$$(\bar{x}_3^a, \bar{x}_4^a) = (114, 336), \text{ and } \bar{x}_0 = -450,$$

and

$$(\bar{\lambda}^i, \bar{\lambda}^c) = (0.0638, 0.3515).$$

Moreover, the expected return corresponding to this optimal portfolio is

$$\mu = p = d = 20.77.$$

The solution suggests that the optimal balance sheet should only contain consumer credit and BBB corporate bonds. Moreover, from Fig. 4.5, we can also see that the lines representing these two assets are way above the others which indicates that they produce much better risk-adjusted returns. It is no surprise that asset managers are well aware of the advantage of such assets. An indirect evidence is the many flyers we receive enticing us to take on consumer credit.

Since consumer credit is competed by many, next let's assume that it is no longer available. Removing L_4, we get a situation illustrated in Fig. 4.5.

Now we can see that $\phi = L_3$ and BBB corporate bonds is the only general asset class in the optimal balance sheet. Since ϕ is decreasing, we have the situation described in Case 1 (a) of Theorem 4.20. Thus, the optimal balance sheet consists of

$$\bar{x}_3^a = 3125, \ \bar{x}_5^\ell = -5073, \text{ and } \bar{x}_0 = 1948.$$

The dual solution is

$$(\bar{\lambda}^i, \bar{\lambda}^c) = (0.0816, 0.0035),$$

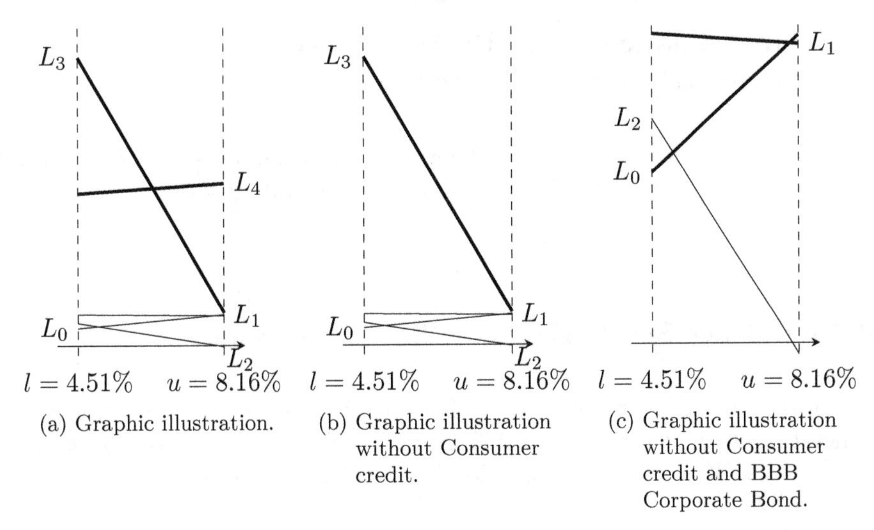

(a) Graphic illustration.

(b) Graphic illustration without Consumer credit.

(c) Graphic illustration without Consumer credit and BBB Corporate Bond.

Fig. 4.5 Graphic illustration (see also [65, Figures 5, 6, and 9])

and the corresponding expected return is $\mu = p = d = 4.25$.

What if both consumer credit and BBB corporate bonds are not available? Then we get Fig. 4.5. We can see that in this case, $\phi = \max(L_0, L_1)$ and the minimum is attained inside (l, u). This fits Case 3 (b) of Theorem 4.20. Moreover, if the 15-year mortgage represented by L_1 is also not available, then $\phi = \max(L_0, L_2)$, and the minimum is again attained inside (l, u), so we have a similar situation. Under the environment (determined by the yield curve), a bank balance sheet manager should prefer the 15-year mortgage since L_1 is above L_2 in the entire interval of $[l, u]$. Of course, if the environment changes, e.g., if the interval $[l, u]$ were shifted to the left, this preference could change. Finally, if there are no general assets available, then $\phi = L_0$. This fits Case 2 (b) of Theorem 4.20. The optimal balance sheet contains only a long bond financed by the central bank funding.

The examples here illustrate that the various scenarios discussed in Theorem 4.20 do arise naturally in practical situations, and this theorem provides useful guide to bank balance sheet managers. Of course, practical balance sheet management is more involved. Several important aspects that are not discussed here are issue of diversification and computation procedure for real-world problems that involve a large number of assets and sensitivities. We refer the reader to Júdice and Zhu [65] for further discussions of these issues.

4.3 Bank Balance Sheet Problem with Quadratic Interest Rate and Linear Credit Risk

This section discusses the bank balance sheet problem of Sect. 4.2 with the same linear credit risk but with a quadratic interest rate risk that was already discussed in Sect. 4.1.2.

4.3.1 Balance Sheets Enforcing Sign Constraints

We again assume that there are k short bonds, m_{lB} long bonds, and m_{gA} general assets as described in Table 4.3 except for the column regarding the interest risk. Following the discussion in Sect. 4.1.1, the interest risk is given instead by

$$\widehat{\mathfrak{r}}_1(\widehat{x}) = \widehat{\mathfrak{r}}_1\left(\widehat{x^a}, \widehat{x^\ell}\right) = \frac{1}{2}\widehat{x}^\top \Theta \widehat{x},$$

where Θ is a positive definite matrix of size $m + k$ with $m = m_{lB} + m_{gA}$ (cf. (4.1.10)). We aim to give a tractable solution to the balance sheet management problem while enriching the structure that we presented in the linear risk case. Unlike linear risks, quadratic risks allow nontrivial covariance matrices, so the diversification effect is taken into account. To make the notation concise, we define $\widehat{1} = (1, \dots, 1)^\top$ and $\widehat{c} := (c_1, \dots, c_{m_gA}, 0 \dots, 0)^\top \in \mathbb{R}^{m+k} \backslash \{0\}$. Again, using $r^i > 0$ and $r^c > 0$ to denote the limits on the interest and credit risk, respectively, the balance sheet optimization problem becomes

$$\max_{x \in \mathbb{R}^{1+m+k}} \quad x_0 R_0 + \langle \widehat{R}, \widehat{x} \rangle \tag{4.3.1}$$

subject to

$$\begin{aligned}
&\frac{1}{2}\widehat{x}^\top \Theta \widehat{x} \le r^i \\
&\langle \widehat{c}, \widehat{x} \rangle \le r^c \\
&x_0 + \langle \widehat{1}, \widehat{x} \rangle = 0 \\
&x_0 \in \mathbb{R}, \widehat{x^a} \ge 0, \widehat{x^\ell} \le 0.
\end{aligned}$$

The objective function corresponds to the return on the balance sheet. The first and second restrictions are the constraints on quadratic interest risk and linear credit risk. The third restriction is similar to constraint (4.2.3) assuming equity is 0. The last line of the restrictions states that amounts invested in general assets and long bonds assume a positive sign, the funding amounts from liabilities (short bonds) take a negative sign, whereas the central bank funding can either be long or short.

Note that the constraints on credit and interest risks do not involve x_0. Thus, we can eliminate x_0 using $x_0 = -\langle \widehat{1}, \widehat{x}\rangle$ so that the problem becomes

$$\mu := \max_{\widehat{x}\in\mathbb{R}^{m+k}} \langle \widehat{R} - R_0\widehat{1}, \widehat{x}\rangle \tag{4.3.2}$$
$$\text{subject to}$$
$$\lambda^i: \quad \frac{1}{2}\widehat{x}^\top \Theta \widehat{x} \le r^i$$
$$\lambda^c: \quad \langle \widehat{c}, \widehat{x}\rangle \le r^c$$
$$\widehat{\lambda^a}: \quad \widehat{x^a} \ge 0$$
$$\widehat{\lambda^\ell}: \quad \widehat{x^\ell} \le 0.$$

Here λ^i, λ^c, $\widehat{\lambda^a}$, and $\widehat{\lambda^\ell}$ are dual variables. For this problem, it turns out that the alternative formulation below is more convenient.

Example 4.22 *Consider the companion problem*

$$r^i := \min_{\widehat{x}\in\mathbb{R}^{m+k}} \frac{1}{2}\widehat{x}^\top \Theta \widehat{x} \tag{4.3.3}$$
$$\text{subject to}$$
$$\lambda^\mu: \quad -\langle \widehat{R} - R_0\widehat{1}, \widehat{x}\rangle \le -\mu$$
$$\lambda^c: \quad \langle \widehat{c}, \widehat{x}\rangle \le r^c$$
$$\widehat{\lambda^a}: \quad \widehat{x^a} \ge 0$$
$$\widehat{\lambda^\ell}: \quad \widehat{x^\ell} \le 0.$$

We assume that both linear constrains are binding. The corresponding Lagrangian is

$$L(\widehat{x},\lambda) = \frac{1}{2}\widehat{x}^\top \Theta \widehat{x} + \lambda^\mu(\mu - \langle \widehat{R} - R_0\widehat{1}, \widehat{x}\rangle) + \lambda^c(\langle \widehat{c}, \widehat{x}\rangle - r^c)$$
$$- \langle \widehat{\lambda^a}, \widehat{x^a}\rangle + \langle \widehat{\lambda^\ell}, \widehat{x^\ell}\rangle, \tag{4.3.4}$$

where $\lambda = [\lambda^\mu, \lambda^c, \widehat{\lambda^a}^\top, \widehat{\lambda^\ell}^\top]^\top \in \mathbb{R}^{2+m+k}$. *By Lagrangian strong duality in Theorem A.31, we have*

$$r^i = \inf_{\widehat{x}} \sup_{\lambda\ge 0} L(\widehat{x},\lambda) = \sup_{\lambda\ge 0} \inf_{\widehat{x}} L(\widehat{x},\lambda). \tag{4.3.5}$$

To calculate $\inf_{\widehat{x}} L(\widehat{x},\lambda)$, *we solve*

$$0 = \nabla_{\widehat{x}} L(\widehat{x},\lambda) = \Theta\widehat{x} - \lambda^\mu(\widehat{R} - R_0\widehat{1}) + \lambda^c\widehat{c} - [\widehat{\lambda^a}^\top, -\widehat{\lambda^\ell}^\top]^\top. \tag{4.3.6}$$

It follows that the optimal solution is given by

$$\widehat{x} = \Theta^{-1}(\lambda^\mu(\widehat{R} - R_0\widehat{1}) - \lambda^c\widehat{c} + [\widehat{\lambda^a}^\top, -\widehat{\lambda^\ell}^\top]^\top). \tag{4.3.7}$$

Thus, the dual problem is

$$r^i = \sup_{\lambda \geq 0} \quad \lambda^\mu \mu - \lambda^c r^c - \frac{1}{2}\Big\langle \lambda^\mu(\widehat{R} - R_0\widehat{1}) - \lambda^c\widehat{c} + [\widehat{\lambda^a}^\top, -\widehat{\lambda^\ell}^\top]^\top,$$
$$\Theta^{-1}(\lambda^\mu(\widehat{R} - R_0\widehat{1}) - \lambda^c\widehat{c} + [\widehat{\lambda^a}^\top, -\widehat{\lambda^\ell}^\top]^\top)\Big\rangle. \tag{4.3.8}$$

Denote the dual solution by $\bar{\lambda}$. Then for a problem that both the expected return and the credit risk are relevant, we must have $\bar{\lambda}^\mu, \bar{\lambda}^c > 0$, so that they are determined by equations

$$\mu = \Big\langle \bar{\lambda}^\mu(\widehat{R} - R_0\widehat{1}) - \bar{\lambda}^c\widehat{c} + [\overline{\lambda^a}^\top, -\overline{\lambda^\ell}^\top]^\top, \Theta^{-1}(\widehat{R} - R_0\widehat{1})\Big\rangle. \tag{4.3.9}$$

and

$$r^c = \Big\langle \bar{\lambda}^\mu(\widehat{R} - R_0\widehat{1}) - \bar{\lambda}^c\widehat{c} + [\overline{\lambda^a}^\top, -\overline{\lambda^\ell}^\top]^\top, \Theta^{-1}\widehat{c}\Big\rangle. \tag{4.3.10}$$

We now turn to $[\overline{\lambda^a}^\top, -\overline{\lambda^\ell}^\top]^\top \in \mathbb{R}^{m+k}$. Decomposing the Euclidean space, they belong to $\mathbb{R}^{m+k} = U \times V$, where $U = \mathbb{R}^t$ is the subspace in which the components of $[\overline{\lambda^a}^\top, -\overline{\lambda^\ell}^\top]^\top$ are nonzero numbers (assuming there are t such nonzero components) and $V = \mathbb{R}^{m+k-t}$ is the subspace in which the components of $[\overline{\lambda^a}^\top, -\overline{\lambda^\ell}^\top]^\top$ are zeros. Denote the projection operators onto U and V by P_U and P_V, respectively. We will use the same notation for their corresponding representing matrices. To have a clearer picture of these representing matrices, let us assume without loss of generality that the t nonzero components of $[\overline{\lambda^a}^\top, -\overline{\lambda^\ell}^\top]^\top$ are the first t components. Then $P_U = (I_t, 0) \in \mathbb{R}^{t\times(m+k)}$ and $P_V = (0, I_{m+k-t}) \in \mathbb{R}^{(m+k-t)\times(m+k)}$ so that $I_{m+k} = \begin{bmatrix} P_U \\ P_V \end{bmatrix} \in \mathbb{R}^{(m+k)\times(m+k)}$ is the identity matrix. We can also decompose Θ^{-1} according to this by

$$\Theta^{-1} = \begin{bmatrix} P_U \\ P_V \end{bmatrix} \Theta^{-1}[P_U^\top, P_V^\top] = \begin{bmatrix} P_U\Theta^{-1}P_U^\top & P_U\Theta^{-1}P_V^\top \\ P_V\Theta^{-1}P_U^\top & P_V\Theta^{-1}P_V^\top \end{bmatrix}, \tag{4.3.11}$$

where $P_U\Theta^{-1}P_U^\top \in \mathbb{R}^{t\times t}, P_V\Theta^{-1}P_V^\top \in \mathbb{R}^{(m+k-t)\times(m+k-t)}$ are invertible, and $P_U\Theta^{-1}P_V^\top \in \mathbb{R}^{t\times(m+k-t)}$ is the transpose matrix of $P_V\Theta^{-1}P_U^\top \in \mathbb{R}^{(m+k-t)\times t}$. Defining

$$u = P_U\left(\lambda^\mu(\widehat{R} - R_0\widehat{1}) - \lambda^c\widehat{c} + [\widehat{\lambda^a}^\top, -\widehat{\lambda^\ell}^\top]^\top\right) \in \mathbb{R}^t$$

and

$$v = P_V\left(\lambda^\mu(\widehat{R} - R_0\widehat{1}) - \lambda^c\widehat{c} + [\widehat{\lambda^a}^\top, -\widehat{\lambda^\ell}^\top]^\top\right)$$
$$= P_V\left(\lambda^\mu(\widehat{R} - R_0\widehat{1}) - \lambda^c\widehat{c}\right) \in \mathbb{R}^{m+k-t},$$

we can write (4.3.8) *as*

$$r^i = \sup_{\lambda\geq 0} \lambda^\mu\mu - \lambda^c r^c - \frac{1}{2}\langle \begin{bmatrix} u \\ v \end{bmatrix}, \Theta^{-1}\begin{bmatrix} u \\ v \end{bmatrix}\rangle \tag{4.3.12}$$
$$= \sup_{\lambda\geq 0} \lambda^\mu\mu - \lambda^c r^c - \frac{1}{2}\langle u, P_U\Theta^{-1}P_U^\top u\rangle$$
$$- \langle u, P_U\Theta^{-1}P_V^\top v\rangle - \frac{1}{2}\langle v, P_V\Theta^{-1}P_V^\top v\rangle.$$

Note that v is independent of $[\widehat{\lambda^a}^\top, -\widehat{\lambda^\ell}^\top]^\top$ and u is related to the nonzero components of $[\widehat{\lambda^a}^\top, -\widehat{\lambda^\ell}^\top]^\top$. Thus, the dual solution can be derived by setting the derivative of the quadratic function in (4.3.12) *with respect to the nonzero components of $[\overline{\lambda^a}^\top, -\overline{\lambda^\ell}^\top]^\top$ or, equivalently, with respect to u to zero. This yields the equation $P_U\Theta^{-1}P_U^\top\bar{u} = P_U\Theta^{-1}P_V^\top\bar{v}$ or*

$$\bar{u} := P_U\left(\bar{\lambda}^\mu(\widehat{R} - R_0\widehat{1}) - \bar{\lambda}^c\widehat{c} + [\overline{\lambda^a}^\top, -\overline{\lambda^\ell}^\top]^\top\right) \tag{4.3.13}$$
$$= \left(P_U\Theta^{-1}P_U^\top\right)^{-1} P_U\Theta^{-1}P_V^\top\bar{v} \in \mathbb{R}^t, \tag{4.3.14}$$

where $\bar{u}$ is the vector u corresponding to the dual solution and

$$\bar{v} := P_V\left(\bar{\lambda}^\mu(\widehat{R} - R_0\widehat{1}) - \bar{\lambda}^c\widehat{c}\right) \in \mathbb{R}^{m+k-t}.$$

It follows that

$$P_V[\overline{\lambda^a}^\top, -\overline{\lambda^\ell}^\top]^\top = 0 \in \mathbb{R}^{m+k-t} \tag{4.3.15}$$

and from (4.3.13) *we get*

$$P_U[\overline{\lambda^a}^\top, -\overline{\lambda^\ell}^\top]^\top \tag{4.3.16}$$
$$= (P_U\Theta^{-1}P_U^\top)^{-1}P_U\Theta^{-1}P_V^\top\bar{v} - P_U(\bar{\lambda}^\mu(\widehat{R} - R_0\widehat{1}) - \bar{\lambda}^c\widehat{c})$$
$$= ((P_U\Theta^{-1}P_U^\top)^{-1}P_U\Theta^{-1}P_V^\top P_V - P_U)(\bar{\lambda}^\mu(\widehat{R} - R_0\widehat{1}) - \bar{\lambda}^c\widehat{c}) \in \mathbb{R}^t.$$

To ease the notation, we denote

$$Q_{U,V} := I_{m+k} + \begin{bmatrix} (P_U\Theta^{-1}P_U^\top)^{-1}P_U\Theta^{-1}P_V^\top P_V - P_U \\ 0 \end{bmatrix} \in \mathbb{R}^{(m+k)\times(m+k)}.$$

Then Eqs. (4.3.9) *and* (4.3.10) *can be represented as*

$$\mu = \langle Q_{U,V}(\bar{\lambda}^\mu(\widehat{R} - R_0\widehat{1}) - \bar{\lambda}^c\widehat{c}), \Theta^{-1}(\widehat{R} - R_0\widehat{1})\rangle. \tag{4.3.17}$$

and

$$r^c = \langle Q_{U,V}(\bar{\lambda}^\mu(\widehat{R} - R_0\widehat{1}) - \bar{\lambda}^c\widehat{c}), \Theta^{-1}\widehat{c}\rangle. \tag{4.3.18}$$

Denoting $\alpha := \langle Q_{U,V}(\widehat{R}-R_0\widehat{1}), \Theta^{-1}(\widehat{R}-R_0\widehat{1})\rangle$, $\beta_1 := \langle \Theta^{-1}(\widehat{R}-R_0\widehat{1}), Q_{U,V}\widehat{c}\rangle$, $\beta_2 := \langle Q_{U,V}(\widehat{R} - R_0\widehat{1}), \Theta^{-1}\widehat{c}\rangle$, *and* $\gamma := \langle Q_{U,V}\widehat{c}, \Theta^{-1}\widehat{c}\rangle$, *Eqs.* (4.3.17) *and* (4.3.18) *become*

$$\mu = \alpha\bar{\lambda}^\mu - \beta_1\bar{\lambda}^c, \quad r^c = \beta_2\bar{\lambda}^\mu - \gamma\bar{\lambda}^c. \tag{4.3.19}$$

Since we assume that in Problem 4.22, both linear constrains are binding, the system of Eqs. (4.3.19) *has a unique solution. This implies that* $\alpha\gamma - \beta_1\beta_2 \neq 0$. *Solving* (4.3.19) *and combining with* (4.3.15) *and* (4.3.16), *it follows that the dual solution is*

$$\begin{aligned}
\bar{\lambda}^\mu &= \frac{\gamma\mu - \beta_1 r^c}{\alpha\gamma - \beta_1\beta_2} \\
\bar{\lambda}^c &= \frac{\beta_2\mu - \alpha r^c}{\alpha\gamma - \beta_1\beta_2} \\
\begin{bmatrix} \overline{\lambda^a} \\ -\overline{\lambda^\ell} \end{bmatrix} &= (Q_{U,V} - I_{m+k})\left[\frac{\gamma\mu - \beta_1 r^c}{\alpha\gamma - \beta_1\beta_2}(\widehat{R} - R_0\widehat{1}) - \frac{\beta_2\mu - \alpha r^c}{\alpha\gamma - \beta_1\beta_2}\widehat{c}\right].
\end{aligned} \tag{4.3.20}$$

By Eq. (4.3.7), *the primal solution is then*

$$\bar{x} = \Theta^{-1}Q_{U,V}\left[\frac{\gamma\mu - \beta_1 r^c}{\alpha\gamma - \beta_1\beta_2}(\widehat{R} - R_0\widehat{1}) - \frac{\beta_2\mu - \alpha r^c}{\alpha\gamma - \beta_1\beta_2}\widehat{c}\right]. \tag{4.3.21}$$

Note that the derivation here assumes that we know which components of the dual solution are nonzero. In practice, of course, we don't know that before deriving the dual solution. Therefore, the solution derived above merely provided us with the pattern of the solution. To derive the solution, theoretically, we can test all the possible decompositions $\mathbb{R}^{m+k} = U \times V$ and the dual solution calculated by (4.3.20) is the one that maximizes (4.3.12). In practice, this works only for small-scale problems. For a problem involving a large number of assets and liabilities, we need to use numerical methods.

Remark 4.23 *The solution formula in* (4.3.21) *indicates that we may have a property similar to Markowitz efficient portfolios: since* $\bar{x}$ *is linear in the problem parameters, one can use linear combinations of the efficient portfolios to move between points on the efficient frontier whose dual solutions correspond to the same decomposition* $\mathbb{R}^{m+k} = U \times V$.

Example 4.24 *The method discussed above also applies to problem* (4.3.2) *directly. However, the computation is more involved. Using the Lagrangian*

$$L(\widehat{x},\lambda) = \langle \widehat{R} - R_0\widehat{1}, \widehat{x}\rangle + \lambda^i\left(r^i - \frac{1}{2}\widehat{x}^\top \Theta \widehat{x}\right) + \lambda^c(r^c - \langle \widehat{c}, \widehat{x}\rangle) + \langle \widehat{\lambda^a}, \widehat{x^a}\rangle - \langle \widehat{\lambda^\ell}, \widehat{x^\ell}\rangle, \quad (4.3.22)$$

by Lagrangian strong duality in Theorem A.31, we have

$$\mu = \sup_{\widehat{x}} \inf_{\lambda \geq 0} L(\widehat{x},\lambda) = \inf_{\lambda \geq 0} \sup_{\widehat{x}} L(\widehat{x},\lambda). \quad (4.3.23)$$

To calculate $\sup_{\widehat{x}} L(\widehat{x},\lambda)$*, we solve*

$$0 = \nabla_{\widehat{x}} L(\widehat{x},\lambda) = \widehat{R} - R_0\widehat{1} - \lambda^c\widehat{c} + \begin{bmatrix} \widehat{\lambda^a} \\ -\widehat{\lambda^\ell} \end{bmatrix} - \lambda^i \Theta \widehat{x} \quad (4.3.24)$$

to derive the relationship between the optimal balance sheet $\widehat{x}$ *and the dual variables* $\lambda = (\lambda^i, \lambda^c, \widehat{\lambda^a}^\top, \widehat{\lambda^\ell}^\top)^\top \in \mathbb{R}^{2+m+k}$,

$$\widehat{x} = \frac{1}{\lambda^i}\Theta^{-1}\left(\widehat{R} - R_0\widehat{1} - \lambda^c\widehat{c} + \begin{bmatrix} \widehat{\lambda^a} \\ -\widehat{\lambda^\ell} \end{bmatrix}\right) =: \frac{1}{\lambda^i}\Theta^{-1} z_\lambda. \quad (4.3.25)$$

It follows that the dual problem of (4.3.2) *is*

$$\mu = \inf_{\lambda \geq 0} \left[r^c\lambda^c + r^i\lambda^i + \frac{1}{2\lambda^i} z_\lambda^\top \Theta^{-1} z_\lambda \right]. \quad (4.3.26)$$

We can see that the dual solution can be determined using a method similar to that of Example 4.22. However, the equations for $\overline{\lambda^i}$ *and* $\overline{\lambda^c}$ *are more involved than those for* $\overline{\lambda^\mu}$ *and* $\overline{\lambda^c}$ *in Example 4.22. We leave the details to the readers.*

4.3.2 An Approach to Bank Balance Sheets Using the Tracking Error

Instead of enforcing sign constraints for $\widehat{x^a}$ and $\widehat{x^\ell}$ directly like in (4.3.1), one might also use the tracking error to induce the sign constraints indirectly as in the example below. This example is related to Example 2.85, where we discussed Markowitz-efficient portfolios with a tracking error constraint, but here the unit cost condition (2.4.35) is replaced by the credit risk constraint (4.3.29). Nevertheless, similar ideas apply.

Example 4.25 *As we have seen before, the problem described in Example 4.22 becomes difficult to solve analytically and leads to corner solutions, hindering diversification. A more practical and tractable approach specifies that the optimal balance sheet* $\widehat{x}$ *should be within a certain distance* r^b

to a benchmark balance sheet $\widehat{x}^*$*, while satisfying the minimum return requirement and the credit risk limit specified by the bank's risk appetite. For* $R_0 \in \mathbb{R}, \widehat{R}, \widehat{c} \in \mathbb{R}^{m+k}$ *given by the market data and* $\widehat{x}^*$ *prescribed, we consider the three-parameter problem* ($\mu, r^c \in \mathbb{R}, r^b > 0$ *are given):*

$$\frac{1}{2}\sigma^2 := \min_{\widehat{x} \in \mathbb{R}^{m+k}} \frac{1}{2}\widehat{x}^\top \Theta \widehat{x} \tag{4.3.27}$$

subject to

$$\lambda^\mu : \quad \langle \widehat{R} - R_0 \widehat{1}, \widehat{x} \rangle \geq \mu \tag{4.3.28}$$

$$\lambda^c : \quad \langle \widehat{c}, \widehat{x} \rangle \leq r^c \tag{4.3.29}$$

$$\lambda^b : \quad \|\widehat{x} - \widehat{x}^*\|_\Theta \leq r^b, \tag{4.3.30}$$

where $\Theta \in \mathbb{R}^{(m+k)\times(m+k)}$ *is symmetric, positive definite and the norm* $\|\cdot\|_\Theta$ *is specified by* $\|\widehat{x}\|_\Theta = \sqrt{\widehat{x}^\top \, \Theta \, \widehat{x}}$*, which is equivalent to the Euclidean norm, and therefore induces the same topology.*

The benchmark balance sheet $\widehat{x}^*$ *and distance* r^b *should be chosen so the ball specified by* $B^\Theta_{r^b}(\widehat{x}^*) = \{\widehat{x} \in \mathbb{R}^{m+k} : \|\widehat{x} - \widehat{x}^*\|_\Theta \leq r^b\}$ *lies in the interior of the set that specifies non-negativity for assets and non-positivity for liabilities:*

$$\widehat{x^a} \geq 0, \tag{4.3.31}$$

$$\widehat{x^\ell} \leq 0. \tag{4.3.32}$$

One can find such a ball, when the benchmark $\widehat{x}^*$ *lies inside the interior of the set given by* (4.3.31)–(4.3.32)*, i.e.,* $\{\widehat{x} \in \mathbb{R}^{m+k} : \widehat{x^a} > 0 \text{ and } \widehat{x^\ell} < 0\}$.

The solution to the problem is given by the Karush-Kuhn-Tucker (KKT), or first-order, conditions. We start by forming the Lagrangian. The tracking error constraint (4.3.30) *can be rephrased as* $\frac{1}{2}(\widehat{x} - \widehat{x}^*)^\top \Theta (\widehat{x} - \widehat{x}^*) \leq \frac{1}{2}(r^b)^2$*. Hence, the Lagrangian with Lagrange multipliers* $(\lambda^\mu, \lambda^c, \lambda^b) \in \mathbb{R}^3$ *can be formed as*

$$\begin{aligned} L(x, \lambda) = {} & \frac{1}{2}\widehat{x}^\top \Theta \widehat{x} + \lambda^\mu \left[\mu - \langle \widehat{R} - R_0 \widehat{1}, \widehat{x} \rangle\right] + \lambda^c \left[\langle \widehat{c}, \widehat{x} \rangle - r^c\right] \\ & + \frac{\lambda^b}{2} \left[(\widehat{x} - \widehat{x}^*)^\top \Theta (\widehat{x} - \widehat{x}^*) - (r^b)^2\right]. \end{aligned} \tag{4.3.33}$$

The first-order conditions are then given by

$$\Theta\widehat{x} - \lambda^{\mu}(\widehat{R} - R_0\widehat{1}) + \lambda^c\widehat{c} + \lambda^b\Theta(\widehat{x} - \widehat{x}^*) = 0, \tag{4.3.34}$$

$$\lambda^{\mu}\left[\mu - \langle\widehat{R} - R_0\widehat{1}, \widehat{x}\rangle\right] = 0, \tag{4.3.35}$$

$$\lambda^c\left[\langle\widehat{c}, \widehat{x}\rangle - r^c\right] = 0, \tag{4.3.36}$$

$$\lambda^b\left[(\widehat{x} - \widehat{x}^*)^\top\Theta(\widehat{x} - \widehat{x}^*) - (r^b)^2\right] = 0, \tag{4.3.37}$$

$$\lambda^{\mu},\ \lambda^c,\ \lambda^b \geq 0, \tag{4.3.38}$$

$$\langle\widehat{R} - R_0\widehat{1}, \widehat{x}\rangle \geq \mu, \tag{4.3.39}$$

$$\langle\widehat{c}, \widehat{x}\rangle \leq r^c, \tag{4.3.40}$$

$$\|\widehat{x} - \widehat{x}^*\|_{\Theta} \leq r^b. \tag{4.3.41}$$

The condition (4.3.34) *sets the gradient of the Lagrangian to zero, while* (4.3.35)–(4.3.37) *are the complementary slackness conditions. In* (4.3.38)*, the Lagrange multipliers are non-negative, due to the fact that the optimization problem has inequality constraints; finally, the conditions* (4.3.39)–(4.3.41) *are the feasibility constraints.*

In the above formulation, the Lagrange multipliers λ^{μ}*,* λ^c*, and* λ^b *can correspond to binding constraints (positive multipliers) or nonbinding (multipliers equal to zero). To make a complete analysis of the solution, we would have to consider all the seven different cases: exactly one constraint binding (three cases), exactly two constraints binding (three cases), and exactly three constraints binding (one case). The case of having all the constraints nonbinding (with the corresponding Lagrange multipliers equal to zero) is of no interest, since by Eq.* (4.3.34) *this would lead to the trivial solution* $\widehat{x} = \widehat{0}$*. For ease of exposition, we will not consider all the seven cases mentioned above, but focus on the case of all the constraints* (4.3.39)–(4.3.41) *binding. Therefore, we assume that*

$$\lambda^{\mu} > 0, \lambda^c > 0, \lambda^b > 0. \tag{4.3.42}$$

Let us first start with Eq. (4.3.34)*. Solving for* $\widehat{x}$*, we get*

$$\widehat{x} = \frac{1}{1+\lambda^b}\,\Theta^{-1}\,(\lambda^{\mu}(\widehat{R} - R_0\widehat{1}) - \lambda^c\widehat{c} + \lambda^b\,\Theta\,\widehat{x}^*). \tag{4.3.43}$$

Now we multiply this equation from the left by $(\widehat{R} - R_0\widehat{1})^\top$ *and use that* $(\widehat{R} - R_0\widehat{1})^\top\,\widehat{x} \geq \mu$ *is binding to get*

$$\mu = \frac{1}{1+\lambda^b}\left(\lambda^{\mu}A - \lambda^cB + \lambda^bC\right), \tag{4.3.44}$$

where $A := (\widehat{R} - R_0\widehat{1})^\top\,\Theta^{-1}\,(\widehat{R} - R_0\widehat{1})$*,* $B := (\widehat{R} - R_0\widehat{1})^\top\,\Theta^{-1}\,\widehat{c}$*, and* $C := (\widehat{R} - R_0\widehat{1})^\top\,\widehat{x}^*$*. This equation is the same as*

$$\lambda^{\mu}A - \lambda^cB = \mu - \lambda^b(C - \mu), \tag{4.3.45}$$

so it is linear in λ.

On the other hand, we can multiply Eq. (4.3.43) *from the left by* $\widehat{c}^\top$ *and use the fact that* $\widehat{c}^\top\widehat{x} \leq r^c$ *is binding to get*

$$r^c = \frac{1}{1+\lambda^b}\,(\lambda^\mu B - \lambda^c D + \lambda^b \mathrm{E}), \tag{4.3.46}$$

where $D := \widehat{c}^\top \Theta^{-1}\widehat{c}$ *and* $E := \widehat{c}^\top\widehat{x}^*$. *This equation is equivalent to the following linear equation:*

$$\lambda^\mu B - \lambda^c D = r^c - \lambda^b(E - r^c). \tag{4.3.47}$$

Therefore, Eqs. (4.3.45) *and* (4.3.47) *form a linear system given by*

$$\begin{bmatrix} A & -B \\ -B & D \end{bmatrix} \begin{bmatrix} \lambda^\mu \\ \lambda^c \end{bmatrix} = \begin{bmatrix} \mu - \lambda^b(C-\mu) \\ \lambda^b(E-r^c) - r^c \end{bmatrix}. \tag{4.3.48}$$

The matrix on the left-hand side is invertible, provided that $\widehat{R} - R_0\widehat{1}$ *and* $\widehat{c}$ *are linearly independent. If that were not the case, then either one of the columns would be zero, which cannot happen if we assume* $\widehat{R} - R_0\widehat{1}$ *and* $\widehat{c}$ *are nonzero, or*

$$\begin{bmatrix} A \\ -B \end{bmatrix} = \alpha \begin{bmatrix} -B \\ D \end{bmatrix}, \tag{4.3.49}$$

where α *is a nonzero constant. But if that were the case, then we would have*

$$(A + \alpha B) + \alpha(B + \alpha D) = 0, \tag{4.3.50}$$

which is equivalent to

$$(\widehat{R} - R_0\widehat{1} + \alpha\widehat{c})^\top\, \Theta^{-1}\,(\widehat{R} - R_0\widehat{1} + \alpha\widehat{c}) = 0. \tag{4.3.51}$$

We would arrive at a contradiction, by the positive-definiteness of Θ *and the linear independence of* $\widehat{R} - R_0\widehat{1}$ *and* $\widehat{c}$.

Since the left-hand side matrix in Eq. (4.3.48) *is invertible,* λ^μ *and* λ^c *can be expressed as an affine function of* λ^b:

$$\lambda^\mu = a_1 + b_1\lambda^b, \tag{4.3.52}$$

$$\lambda^c = a_2 + b_2\lambda^b, \tag{4.3.53}$$

for suitable $a_i, b_i \in \mathbb{R}$, $i \in 1, 2$. *The coefficients are easily determined by Cramer's rule:*

$$a_1 = \Delta^{-1}[\mu D - r^c B], \tag{4.3.54}$$

$$a_2 = \Delta^{-1}[\mu B - r^c A], \tag{4.3.55}$$

$$b_1 = \Delta^{-1}[B(E - r^c) - D(C - \mu)], \tag{4.3.56}$$

$$b_2 = \Delta^{-1}[A(E - r^c) - B(C - \mu)], \tag{4.3.57}$$

where $\Delta = AD - B^2$.

Now we plug in Eqs. (4.3.52)–(4.3.53) *into Eq.* (4.3.43) *to get*

$$\widehat{x} = \frac{1}{1 + \lambda^b}(a_3 + \lambda^b b_3), \tag{4.3.58}$$

where $a_3 := a_1 \Theta^{-1}(\widehat{R} - R_0 \widehat{1}) - a_2 \Theta^{-1}\widehat{c} \in \mathbb{R}^{m+k}$ *and* $b_3 := b_1 \Theta^{-1}(\widehat{R} - R_0 \widehat{1}) - b_2 \Theta^{-1}\widehat{c} + \widehat{x}^* \in \mathbb{R}^{m+k}$.

Plugging Eq. (4.3.58) *into the binding constraint* (4.3.41) *and multiplying by* $(1 + \lambda_b)$, *we get the equation*

$$(a_3 + \lambda^b b_3 - (1+\lambda^b)\widehat{x}^*)^\top \Theta (a_3 + \lambda^b b_3 - (1+\lambda^b)\widehat{x}^*) = (1+\lambda_b)^2 (r^b)^2. \tag{4.3.59}$$

This is a quadratic equation in λ^b, *which can be solved when the discriminant is non-negative. Finally, in order to compute* (4.3.58) *with this* λ^b, *one has to check whether the calculated Lagrange multipliers indeed satisfy* (4.3.42).

4.3.3 Efficient Frontier for the Tracking Error Approach

Here we describe the efficient frontier associated with problem (4.3.27)–(4.3.30), assuming that the radius of the ellipsoid in (4.3.30) is fixed. We have a similar reasoning to Example 2.85, considering two cases. Case 1 refers to optimal balance sheets located in the interior of the ellipsoid. Case 2 refers to optimal balance sheets located on the boundary of the ellipsoid, i.e., when (4.3.30) is binding.

Let us first tackle Case 1, i.e., $\|\widehat{x} - \widehat{x}^*\|_\Theta < r^b$. Then $\lambda^\mu > 0$, $\lambda^c > 0$, and $\lambda^b = 0$. Equations (4.3.52)–(4.3.53) give

$$\lambda^\mu = \Delta^{-1}[\mu D - r^c B], \tag{4.3.60}$$

$$\lambda^c = \Delta^{-1}[\mu B - r^c A], \tag{4.3.61}$$

Multiplying Eq. (4.3.34) by $\widehat{x}$ from the left, we get $\sigma^2 = \widehat{x}^\top \Theta \widehat{x} = \lambda^\mu \mu - \lambda^c r^c \geq 0$ since Θ is positive definite. After rearranging, this yields

$$\sigma = \sqrt{\lambda^\mu \mu - \lambda^c r^c}. \tag{4.3.62}$$

Plugging this into the expressions obtained in (4.3.60)–(4.3.61), we get

$$\sigma = (\sqrt{\Delta})^{-1}\sqrt{[\mu D - r^c B]\mu - [\mu B - r^c A]r^c}$$
$$= (\sqrt{\Delta})^{-1}\sqrt{D\mu^2 - 2B\mu r^c + A(r^c)^2}. \tag{4.3.63}$$

Also, notice that by (4.3.58), and since $\lambda^b = 0$, the optimal balance sheets are given by

$$\widehat{x} = a_1\Theta^{-1}(\widehat{R} - R_0\widehat{1}) - a_2\Theta^{-1}\widehat{c}$$
$$= \Delta^{-1}[\mu D - r^c B]\Theta^{-1}(\widehat{R} - R_0\widehat{1}) - \Delta^{-1}[\mu B - r^c A]\Theta^{-1}\widehat{c}$$
$$= \mu\Delta^{-1}[D\Theta^{-1}(\widehat{R} - R_0\widehat{1}) - B\Theta^{-1}\widehat{c}] + r^c\Delta^{-1}[A\Theta^{-1}\widehat{c} - B\Theta^{-1}(\widehat{R} - R_0\widehat{1})],$$

so that they lie in a 2d plane and inside the ball $\{\widehat{x} : \|\widehat{x} - \widehat{x}^*\|_\Theta < r^b\}$.

Let us now go into Case 2, i.e., $\|\widehat{x} - \widehat{x}^*\|_\Theta = r^b$, corresponding to $\lambda^\mu, \lambda^c, \lambda^b > 0$. In this case, the optimal balance sheets are given by Eq. (4.3.58):

$$\widehat{x} = \frac{1}{1 + \lambda^b}\,(a_3 + \lambda^b b_3),$$

where $a_3 = a_3(\mu, r^c)$ and $b_3 = b_3(\mu, r^c)$ are both linear in μ and r^c, but independent of r^b. Also, by Eq. (4.3.59), $\lambda^b = \lambda^b(\mu, r^c, r^b)$. Notice that given r^b, μ, and r^c, there can be at most two values of λ^b that solve the quadratic equation and which also satisfy the positivity condition $\lambda^b > 0$; if there are two possible values of λ^b, one should choose the value that minimizes the objective function in (4.3.27).

Next, we left-multiply Eq. (4.3.58) by $\widehat{x}^\top\Theta$ to get

$$\sigma^2 = \widehat{x}^\top\Theta\widehat{x} = \frac{1}{1 + \lambda^b}\,(\widehat{x}^\top\Theta a_3 + \lambda^b\widehat{x}^\top\Theta b_3) \geq 0. \tag{4.3.64}$$

Now,

$$\widehat{x}^\top\Theta a_3 = a_1\mu - a_2 r^c. \tag{4.3.65}$$

Also, $\widehat{x}^\top\Theta b_3 = b_1\mu - b_2 r^c + \widehat{x}^\top\Theta\widehat{x}^*$. Using that $(\widehat{x} - \widehat{x}^*)^\top\Theta(\widehat{x} - \widehat{x}^*) = (r^b)^2$, we have $\widehat{x}^\top\Theta\widehat{x}^* = \frac{1}{2}[\sigma^2 + F - (r^b)^2]$, where $F := (\widehat{x}^*)^\top\Theta\widehat{x}^*$. Therefore,

$$\widehat{x}^\top\Theta b_3 = b_1\mu - b_2 r^c + \frac{1}{2}[\sigma^2 + F - (r^b)^2]. \tag{4.3.66}$$

Plugging Eqs. (4.3.65) and (4.3.66) into (4.3.64), we get

$$\sigma^2\left(\frac{2 + \lambda^b}{2(1 + \lambda^b)}\right) = \frac{1}{1 + \lambda^b}\left(a_1\mu - a_2 r^c + \lambda^b(b_1\mu - b_2 r^c + \frac{1}{2}[F - (r^b)^2])\right). \tag{4.3.67}$$

Rearranging, we get the optimal σ for $\widehat{x} \in \partial B^\Theta_{r^b}(\widehat{x}^*)$, i.e., at the boundary of the ellipsoid:

$$\sigma = \sqrt{\frac{2}{2+\lambda^b}\left(a_1\mu - a_2 r^c + \lambda^b(b_1\mu - b_2 r^c + \frac{1}{2}[F - (r^b)^2])\right)}, \tag{4.3.68}$$

where the dependency of λ^b, a_1, a_2, b_1, and b_2 on μ, r^b and r^c is given by Eqs. (4.3.59) and (4.3.54)–(4.3.57).

It remains to discuss which points on the graph of σ are indeed efficient. Having three risk functions and one utility function in (4.3.27)–(4.3.30), the (vector risk, utility) space is four dimensional and the efficient frontier generically will be a three-dimensional hypersurface in $\mathbb{R}^4$ contained in the graph of $\frac{1}{2}\sigma^2$. But not all these optimal points are necessarily efficient. To discuss this problem in full detail would be extremely laborious and certainly beyond the scope of this book. However, we can give partial answers on efficiency by freezing one of the linear constraints (4.3.28) or (4.3.29). Let us, for instance, prescribe the credit risk constraint, i.e., fix r^c = const and plug this constraint into the portfolio set:

$$A = A_{r^c} := \left\{x = \left(0, \widehat{x}^\top\right)^\top \in \mathbb{R}^{1+m+k} : \langle \widehat{c}, \widehat{x}\rangle = r^c\right\}^{\cdot} \tag{4.3.69}$$

Our new problem now becomes

$$\frac{1}{2}\sigma^2 = \min_{(0,\widehat{x}^\top)^\top \in A} \frac{1}{2}\widehat{x}^\top \Theta \,\widehat{x} \tag{4.3.70}$$

$$\text{subject to } \langle \widehat{R} - R_0\,\widehat{1}, \widehat{x}\rangle \geq \mu \tag{4.3.71}$$

$$\frac{1}{2}\,\|\widehat{x} - \widehat{x}^*\|_\Theta^2 \leq \frac{1}{2}(r^b)^2 \tag{4.3.72}$$

where besides the utility constraint (4.3.71) now only two risk functions

$$\begin{aligned} \mathfrak{r}_1(x) = \mathfrak{r}_{\text{Var}}(x) &:= \frac{1}{2}\widehat{x}^\top \Theta \widehat{x} \qquad \text{and} \\ \mathfrak{r}_2(x) = \mathfrak{r}_{\text{Track}}(x) &:= \frac{1}{2}(\widehat{x} - \widehat{x}^*)^\top \Theta(\widehat{x} - \widehat{x}^*). \end{aligned} \tag{4.3.73}$$

are involved. The efficient frontier of (4.3.70)–(4.3.72) lies in the three-dimensional (vector risk, utility) space. Fortunately, this problem is exactly in the form (3.3.17)–(3.3.19), with (4.3.69) in a similar form as (3.3.14), for which the efficient frontier/efficient portfolios in the sense of Definitions 3.7 and 3.8 are already discussed in Example 3.50. So transferring these results will give all efficient points in the three-dimensional (vector risk, utility) space for (4.3.70)–(4.3.72) as well.

Remark 4.26 *Note that a similar approach will work when we prescribe the utility constraint, i.e., fix* μ = const *and use the portfolio set:*

$$A = A_\mu := \left\{x = \left(0, \widehat{x}^\top\right)^\top \in \mathbb{R}^{1+m+k} : \langle \widehat{R} - R_0\,\widehat{1}, \widehat{x}\rangle = \mu\right\}.$$

The new problem here is

$$\frac{1}{2}\sigma^2 = \min_{x \in A} \frac{1}{2}\widehat{x}^\top \Theta \widehat{x} \tag{4.3.74}$$

$$\text{subject to } \mathfrak{u}(\widehat{x}) := -\langle \widehat{c}, \widehat{x} \rangle \geq -r^c \tag{4.3.75}$$

$$\frac{1}{2}\|\widehat{x} - \widehat{x}^*\|_\Theta^2 \leq \frac{1}{2}(r^b)^2, \tag{4.3.76}$$

where (4.3.75) may be viewed as "utility" constraint and the risk functions are the same as in (4.3.73). Again, transferring Example 3.50 provides all efficient points in the three-dimensional (vector risk, utility) space for (4.3.74)–(4.3.76).

4.3.4 Determining the Radius of the Ball r^b to Use in the Tracking Error Problem

An important question, as we will see when we apply this solution to real-world data, is to determine if, given a point $\widehat{x}^*$ and a radius r^b, the ball $B^\Theta_{r^b}(\widehat{x}^*) = \{\widehat{x} \in \mathbb{R}^{m+k} : \|\widehat{x} - \widehat{x}^*\|_\Theta \leq r^b\}$ lies within the set $\{\widehat{x} \in \mathbb{R}^{m+k} : \widehat{x^a} \geq 0 \text{ and } \widehat{x^\ell} \leq 0\}$, so that we can guarantee the non-negativity of the asset vector $\widehat{x^a}$ and the non-positivity of the liability vector $\widehat{x}^\ell$. The question at hand now is different than before: we know that such $\widehat{x}^*$ and r^b exist, by the fact that $\{\widehat{x} \in \mathbb{R}^{m+k} : \widehat{x^a} \geq 0 \text{ and } \widehat{x^\ell} \leq 0\}$ has a non-empty interior $\{\widehat{x} \in \mathbb{R}^{m+k} : \widehat{x^a} > 0 \text{ and } \widehat{x^\ell} < 0\}$; now we want to look for conditions under which $B^\Theta_{r^b}(\widehat{x}^*)$ for specific $\widehat{x}^*$ and r^b lies in that set.

We assume that $\widehat{x}^*$ lies within the interior of the region specified by the non-negativity and non-positivity constraints (inequalities (4.3.31)–(4.3.32)). Now, in order to determine if the ball lies within that region, we solve m optimization problems for assets $i = 1, \ldots, m$

$$d_i := \min_{\widehat{x} \in \mathbb{R}^{m+k}} x_i \tag{4.3.77}$$

subject to

$$(\widehat{x} - \widehat{x}^*)^\top \Theta (\widehat{x} - \widehat{x}^*) = (r^b)^2, \tag{4.3.78}$$

and k optimization problems for liabilities $i = m+1, \ldots, m+k$

$$d_i := \max_{\widehat{x} \in \mathbb{R}^{m+k}} x_i \tag{4.3.79}$$

subject to

$$(\widehat{x} - \widehat{x}^*)^\top \Theta (\widehat{x} - \widehat{x}^*) = (r^b)^2, \tag{4.3.80}$$

in order to find the conditions under which d_i is non-negative for assets and non-positive for liabilities.

Let us start by looking into the problem (4.3.77) and (4.3.78). The solution is easily obtainable using the first-order conditions:

$$\widehat{e}^i - 2\lambda\,\Theta(\widehat{x} - \widehat{x}^*) = 0, \quad \lambda \in \mathbb{R}, \tag{4.3.81}$$

where $\widehat{e}^i \in \mathbb{R}^{m+k}$ is i-th vector of the canonical basis. We can readily see that $\lambda \neq 0$; otherwise, we get to a contradiction $\widehat{e}^i = 0$. Solving equation (4.3.81) for $\widehat{x}$, we get

$$\widehat{x} - \widehat{x}^* = \frac{1}{2\lambda}\Theta^{-1}\widehat{e}^i. \tag{4.3.82}$$

Plugging this into Eq. (4.3.78), we get

$$\frac{1}{(2\lambda)^2}(\widehat{e}^i)^\top \Theta^{-1}\widehat{e}^i = (r^b)^2, \tag{4.3.83}$$

which is the same as

$$\frac{1}{2\lambda} = \pm\frac{r^b}{\sqrt{(\widehat{e}^i)^\top \Theta^{-1}\widehat{e}^i}}. \tag{4.3.84}$$

Since we want to find the minimum, and the numerator and denominator on the right-hand side are non-negative (the denominator is actually positive by the positive-definiteness of Θ), we have

$$\frac{1}{2\lambda} = -\frac{r^b}{\sqrt{(\widehat{e}^i)^\top \Theta^{-1}\widehat{e}^i}}. \tag{4.3.85}$$

Plug this into Eq. (4.3.82) to get

$$\widehat{x} = \widehat{x}^* - \frac{r^b}{\sqrt{(\widehat{e}^i)^\top \Theta^{-1}\widehat{e}^i}}\Theta^{-1}\widehat{e}^i. \tag{4.3.86}$$

Therefore, the solution to (4.3.77)–(4.3.78) is given by the ith entry of the previous vector:

$$d_i = \left[\widehat{x}^* - \frac{r^b}{\sqrt{(\widehat{e}^i)^\top \Theta^{-1}\widehat{e}^i}}\Theta^{-1}\widehat{e}^i\right]_i, \quad i = 1, \ldots, m. \tag{4.3.87}$$

Solving the problem (4.3.79)–(4.3.80) for liabilities $i = m+1, \ldots, m+k$ is similar and yields

$$d_i = \left[\widehat{x}^* + \frac{r^b}{\sqrt{(\widehat{e}^i)^\top \Theta^{-1}\widehat{e}^i}}\Theta^{-1}\widehat{e}^i\right]_i. \tag{4.3.88}$$

Therefore, d_i is non-negative for assets $i = 1, \ldots, m$ if and only if

$$\widehat{x}_i^* \geq \left[\frac{r^b}{\sqrt{(\widehat{e}^i)^\top \Theta^{-1} \widehat{e}^i}} \Theta^{-1} \widehat{e}^i \right]_i, \tag{4.3.89}$$

and d_i is non-positive for liabilities $i = m+1, \ldots, m+k$ if and only if

$$-\widehat{x}_i^* \geq \left[\frac{r^b}{\sqrt{(\widehat{e}^i)^\top \Theta^{-1} \widehat{e}^i}} \Theta^{-1} \widehat{e}^i \right]_i. \tag{4.3.90}$$

As a conclusion, given r^b, if we take

$$\widehat{x}_i^* := \left[\frac{r^b}{\sqrt{(\widehat{e}^i)^\top \Theta^{-1} \widehat{e}^i}} \Theta^{-1} \widehat{e}^i \right]_i, \quad \text{for} \quad i = 1, \ldots, m, \tag{4.3.91}$$

$$\widehat{x}_i^* := -\left[\frac{r^b}{\sqrt{(\widehat{e}^i)^\top \Theta^{-1} \widehat{e}^i}} \Theta^{-1} \widehat{e}^i \right]_i, \quad \text{for} \quad i = m+1, \ldots, m+k, \tag{4.3.92}$$

then the ball $B^{\Theta}_{r^b}(\widehat{x}^*) = \{\widehat{x} \in \mathbb{R}^{m+k} : \|\widehat{x} - \widehat{x}^*\|_\Theta \leq r^b\}$ is contained in the set $\{\widehat{x} \in \mathbb{R}^{m+k} : \widehat{x}^a \geq 0 \text{ and } \widehat{x}^\ell \leq 0\}$. On the other hand, when the benchmark $\widehat{x}^* \in \mathbb{R}^{m+k}$ is prescribed and lies in the interior of the region described by (4.3.31) and (4.3.32), then using (4.3.89), (4.3.90) we can easily determine a maximal radius $r^b > 0$ such that $B^{\Theta}_{r^b}(\widehat{x}^*)$ is contained in $\{\widehat{x} \in \mathbb{R}^{m+k} : \widehat{x}^a \geq 0 \text{ and } \widehat{x}^\ell \leq 0\}$.

4.3.5 Example Using Real-World Data

We now apply the previous results to real-world data. We use the same setting as in Sect. 4.2.6, with the exception that we now consider a quadratic interest rate risk measure, described in (4.1.10). In Eq. (4.1.9), we defined the interest rate risk (IRR) covariance matrix underlying this risk measure,

$$\Theta = \sum_\tau \sum_s \beta_\tau \beta_s^\top D_\tau D_s \sigma_{\tau s}, \tag{4.3.93}$$

which is not guaranteed to be invertible. Therefore, we borrow the idea of shrinkage estimators which are used in the context of portfolio optimization (see, for instance, the chapter on shrinkage estimation in Cornuéjols et al. [31]).

Shrinkage estimators typically modify the covariance matrix by using a convex combination of the estimated covariance matrix and one other more structured matrix. In our case, we will modify the covariance matrix Θ in the following fashion:

$$\Theta^{mod} := (1-\delta) \cdot \Theta + \delta \cdot \Theta^0, \tag{4.3.94}$$

where Θ^0 is a shrinkage target estimator and $\delta \in [0,1]$ is the shrinkage parameter. Cornuéjols, Peña, and Tütüncü point out that $\delta = 1/2$ is commonly used, so we will use the same parameter here. Moreover, we will choose Θ^0 as the matrix with the diagonal entries of Θ, i.e., $\Theta^0_{i,j} = \Theta_{i,j}$ if $i = j$ and $\Theta^0_{i,j} = 0$ otherwise, which ensures that the individual variance of assets and liabilities is not changed.

We use the data in Table 4.4 for the returns and the credit risk factors, as we still consider the problem to be linear with regard to the target return and credit risk restrictions. However, we will use a quadratic interest rate risk objective for the interest rate risk, deploying the asset-liability covariance matrix Θ in Eq. (4.3.93), which is based on the interest rate covariance matrix $(\sigma_{\tau s})_{\tau,s\in\mathcal{M}}$ given in Table 4.5.

	1 year	2 year	3 year	5 year	7 year	10 year	20 year	30 year
1 year	0.1611	0.1427	0.1280	0.1034	0.0907	0.0785	0.0565	0.0560
2 year	0.1427	0.1396	0.1315	0.1144	0.1042	0.0927	0.0703	0.0664
3 year	0.1280	0.1315	0.1278	0.1170	0.1093	0.0993	0.0780	0.0731
5 year	0.1034	0.1144	0.1170	0.1171	0.1148	0.1085	0.0904	0.0849
7 year	0.0907	0.1042	0.1093	0.1148	0.1158	0.1118	0.0962	0.0913
10 year	0.0785	0.0927	0.0993	0.1085	0.1118	0.1103	0.0972	0.0934
20 year	0.0565	0.0703	0.0780	0.0904	0.0962	0.0972	0.0895	0.0863
30 year	0.0560	0.0664	0.0731	0.0849	0.0913	0.0934	0.0863	0.0897

Table 4.5 Matrix $(\sigma_{\tau s})_{\tau,s\in\mathcal{M}}$ for interest rate covariance multiplied by factor 10^3. In this case, $\mathcal{M} = \{1,2,3,5,7,10,20,30\}$, as defined in Sect. 4.1.2. The covariance was calculated using annual data for US Treasury yields from 1995 to 2018

Using Table 4.6, we can then compute the covariance matrix Θ from (4.3.93). The result is given in Table 4.7. This matrix has rank 5, so we need to modify it according to Eq. (4.3.94). The resulting matrix Θ^{mod}, for $\delta = 1/2$, is shown in Table 4.8. Comparing Θ^{mod} with Θ, we see that the diagonal elements are preserved, meaning that the variance for each asset and liability is the same, which is a desirable property of the modification. The condition number for Θ^{mod} is equal to 569, improving sharply when compared to Θ, whose condition number is $+\infty$. Moreover, the modulus of the eigenvalue of Θ^{mod} that is closest to zero equals 7.65×10^{-5}, meaning that Θ^{mod} is invertible.

Now we apply the solution that we obtained in Example 4.25. We assume that the returns and credit risk factors are the same as in Table 4.4. As in Sect. 4.2.6, we don't consider deposit funding.

We start by defining the radius of the ellipsoid as $r^b = 0.5$. In Table 4.9, we show how we set the benchmark balance sheet using the expressions in (4.3.91)–(4.3.92) as a first guess. These equations yield a benchmark balance sheet that overconcentrates on short-term liabilities, which is risky from the liquidity standpoint. Even though we are not considering liquidity risk directly in the problem, we still would like to start from a benchmark balance sheet that is balanced, so we adjust the liability side of the balance sheet.

We now set the remaining limits in the optimization problem (4.3.27)–(4.3.30), taking $r^c = 5.5$ and $\mu = 1.23$, which ensure the feasibility of the benchmark balance sheet $\widehat{x}^*$. With all the parameters set, we can now solve the optimization problem. The solution procedure is applied in the next steps.

We compute the constants in Example 4.25:

Asset class	Rounded maturity
15-year mortgages	5
30-year mortgages	10
BBB corporate bonds	20
Consumer credit	2
AAA corporate bonds	20
10-year Treasury bonds	10
5-year Treasury bonds	5
2-year Treasury bonds	2
Liability class	**Rounded maturity**
1-month commercial paper	1
3-month commercial paper	1
2-year financial bonds	2
5-year financial bonds	5
10-year financial bonds	10

Table 4.6 Rounded maturities to calculate the matrix $(\beta_{i,j})_{i,j\in\{1,\cdots,M\}}$, used to compute the asset-liability IRR covariance matrix (4.3.93), where $M = m + k$ is the number of risky positions as in Definition 4.5. Recall that, according to Sect. 4.1.2, $\beta^i_\tau = 1$ if the ith asset/liability is a bond with maturity τ and $\beta^i_\tau = 0$ otherwise. Since 15-year and 30-year mortgages have amortizing payments, for the purpose of interest rate risk, we considered the rounded maturity as the closest to the duration of these asset classes

0.25	0.44	0.63	0.10	0.63	0.44	0.25	0.10	0.05	0.05	0.10	0.25	0.44
0.44	0.83	1.27	0.15	1.27	0.83	0.44	0.15	0.07	0.07	0.15	0.44	0.83
0.63	1.27	2.03	0.20	2.03	1.27	0.63	0.20	0.08	0.08	0.20	0.63	1.27
0.10	0.15	0.20	0.05	0.20	0.15	0.10	0.05	0.03	0.03	0.05	0.10	0.15
0.63	1.27	2.03	0.20	2.03	1.27	0.63	0.20	0.08	0.08	0.20	0.63	1.27
0.44	0.83	1.27	0.15	1.27	0.83	0.44	0.15	0.07	0.07	0.15	0.44	0.83
0.25	0.44	0.63	0.10	0.63	0.44	0.25	0.10	0.05	0.05	0.10	0.25	0.44
0.10	0.15	0.20	0.05	0.20	0.15	0.10	0.05	0.03	0.03	0.05	0.10	0.15
0.05	0.07	0.08	0.03	0.08	0.07	0.05	0.03	0.02	0.02	0.03	0.05	0.07
0.05	0.07	0.08	0.03	0.08	0.07	0.05	0.03	0.02	0.02	0.03	0.05	0.07
0.10	0.15	0.20	0.05	0.20	0.15	0.10	0.05	0.03	0.03	0.05	0.10	0.15
0.25	0.44	0.63	0.10	0.63	0.44	0.25	0.10	0.05	0.05	0.10	0.25	0.44
0.44	0.83	1.27	0.15	1.27	0.83	0.44	0.15	0.07	0.07	0.15	0.44	0.83

Table 4.7 Matrix Θ multiplied by 10^2, IRR asset-liability covariance

0.25	0.22	0.32	0.05	0.32	0.22	0.13	0.05	0.02	0.02	0.05	0.13	0.22
0.22	0.83	0.63	0.08	0.63	0.41	0.22	0.08	0.03	0.03	0.08	0.22	0.41
0.32	0.63	2.03	0.10	1.01	0.63	0.32	0.10	0.04	0.04	0.10	0.32	0.63
0.05	0.08	0.10	0.05	0.10	0.08	0.05	0.03	0.01	0.01	0.03	0.05	0.08
0.32	0.63	1.01	0.10	2.03	0.63	0.32	0.10	0.04	0.04	0.10	0.32	0.63
0.22	0.41	0.63	0.08	0.63	0.83	0.22	0.08	0.03	0.03	0.08	0.22	0.41
0.13	0.22	0.32	0.05	0.32	0.22	0.25	0.05	0.02	0.02	0.05	0.13	0.22
0.05	0.08	0.10	0.03	0.10	0.08	0.05	0.05	0.01	0.01	0.03	0.05	0.08
0.02	0.03	0.04	0.01	0.04	0.03	0.02	0.01	0.02	0.01	0.01	0.02	0.03
0.02	0.03	0.04	0.01	0.04	0.03	0.02	0.01	0.01	0.02	0.01	0.02	0.03
0.05	0.08	0.10	0.03	0.10	0.08	0.05	0.03	0.01	0.01	0.05	0.05	0.08
0.13	0.22	0.32	0.05	0.32	0.22	0.13	0.05	0.02	0.02	0.05	0.25	0.22
0.22	0.41	0.63	0.08	0.63	0.41	0.22	0.08	0.03	0.03	0.08	0.22	0.83

Table 4.8 Matrix Θ^{mod} multiplied by 10^2, IRR asset-liability covariance after applying a shrinkage parameter $\delta = 1/2$

Asset class	Balance sheet (4.3.91)–(4.3.92)	Benchmark $\widehat{x}^*$ used
15-year mortgages	13.4	13.4
30-year mortgages	7.3	7.3
BBB corporate Bonds	4.6	4.6
Consumer credit	29.0	29.0
AAA corporate Bonds	4.6	4.6
10-year Treasury Bonds	7.3	7.3
5-year Treasury Bonds	13.4	13.4
2-year Treasury Bonds	29.0	29.0
Liability class	**Balance sheet (4.3.91)–(4.3.92)**	**Benchmark $\widehat{x}^*$ used**
1-month commercial paper	−51.5	−20
3-month commercial paper	−51.5	−20
2-year financial bonds	−29.0	−20
5-year financial bonds	−13.4	−20
10-year financial bonds	−7.3	−20

Table 4.9 Setting the benchmark balance sheet $\widehat{x}^*$. We use expressions (4.3.91)–(4.3.92) for a first guess of the benchmark balance sheet, using the value of $r^b = 0.5$. However, as we can see from the middle column, the calculated benchmark overweighs short-term liabilities, making the balance sheet risky from the liquidity standpoint (even though we are not considering liquidity risk in the optimization problem). Therefore, we adjust the liability amounts in the benchmark balance sheet $\widehat{x}^*$

$$
\begin{aligned}
A &= (\widehat{R} - R_0\widehat{1})^\top (\Theta^{mod})^{-1} (\widehat{R} - R_0\widehat{1}) = 8.60 \\
B &= (\widehat{R} - R_0\widehat{1})^\top (\Theta^{mod})^{-1} \widehat{c} = 23.33 \\
C &= (\widehat{R} - R_0\widehat{1})^\top \widehat{x}^* = 1.23 \\
D &= \widehat{c}^\top (\Theta^{mod})^{-1} \widehat{c} = 72.59 \\
E &= \widehat{c}^\top \widehat{x}^* = 6.23.
\end{aligned}
$$

Then, we compute the values in (4.3.54)–(4.3.57):

$$
\begin{aligned}
a_1 &= \Delta^{-1}[\mu D - r^c B)] = -0.49 \\
a_2 &= \Delta^{-1}[\mu B - r^c A)] = -0.23 \\
b_1 &= \Delta^{-1}[B(E - r^c) - D(C - \mu)] = 0.21 \\
b_2 &= \Delta^{-1}[A(E - r^c) - B(C - \mu)] = 0.08.
\end{aligned}
$$

Next, we compute the values of a_3 and b_3:

$$
\begin{aligned}
a_3 &= a_1(\Theta^{mod})^{-1}(\widehat{R} - R_0\widehat{1}) - a_2(\Theta^{mod})^{-1}\widehat{c} \\
&= (13.5, 3.6, -1.0, 27.0, -0.9, -0.7, -1.5, -3.3, \\
&- 2.1, -18.3, -18.6, -5.4, -2.2)^\top \\
b_3 &= b_1(\Theta^{mod})^{-1}(\widehat{R} - R_0\widehat{1}) - b_2(\Theta^{mod})^{-1}\widehat{c} + \widehat{x}^* \\
&= (8.5, 6.1, 5.0, 28.0, 4.9, 7.4, 13.3, 28.5, \\
&- 22.3, -15.3, -13.8, -18.3, -19.3)^\top.
\end{aligned}
$$

We have everything in place to determine λ^b in Eq. (4.3.59). We perform a change of variable $\tilde{\lambda} = 1 + \lambda^b$ to make the equation more manageable:

$$
((a_3 - b_3) + \tilde{\lambda}(b_3 - \widehat{x}^*))^\top \Theta^{mod}((a_3 - b_3) + \tilde{\lambda}(b_3 - \widehat{x}^*)) = (\tilde{\lambda})^2(r^b)^2.
$$

This equation in $\tilde{\lambda}$ can be rewritten in standard form:

$$
\eta_1\tilde{\lambda}^2 + \eta_2\tilde{\lambda} + \eta_3 = 0,
$$

where

$$
\begin{aligned}
\eta_1 &= (b_3 - \widehat{x}^*)^\top \Theta^{mod}(b_3 - \widehat{x}^*) - (r^b)^2 = -0.19 \\
\eta_2 &= 2(a_3 - b_3)^\top \Theta^{mod}(b_3 - \widehat{x}^*) = 0.00 \\
\eta_3 &= (a_3 - b_3)^\top \Theta^{mod}(a_3 - b_3) = 4.08.
\end{aligned}
$$

Thus, we can compute the roots, which are 4.61 and −4.61. Switching back to the original variable λ^b, and enforcing $\lambda^b > 0$, we get that $\lambda^b = 3.61$.

Using Eq. (4.3.58)

$$
\widehat{x} = \frac{1}{1 + \lambda^b}\,(a_3 + \lambda^b b_3),
$$

we can compute the optimal balance sheet for the problem (4.3.27)–(4.3.30). The result is given in Table 4.10. We can readily observe that the optimal balance sheet improves credit risk and interest rate risk when compared to the benchmark, maintaining the target return. Furthermore, the balance sheet deleverages considerably, and preserves diversification, a most useful feature in practice.

Asset class	Benchmark $\widehat{x}^*$	Optimal balance sheet
15-year mortgages	13.4	9.6
30-year mortgages	7.3	5.5
BBB corporate bonds	4.6	3.7
Consumer credit	29.0	27.8
AAA corporate Bonds	4.6	3.6
10-year Treasury bonds	7.3	5.6
5-year Treasury bonds	13.4	10.1
2-year Treasury bonds	29.0	21.6
Liability class	**Benchmark $\widehat{x}^*$**	**Optimal balance sheet**
1-month commercial paper	−20	−17.9
3-month commercial paper	−20	−15.9
2-year financial bonds	−20	−14.9
5-year financial bonds	−20	−15.5
10-year financial bonds	−20	−15.6
Measures	**Benchmark $\widehat{x}^*$**	**Optimal balance sheet**
Return	1.23	1.23
Credit risk	6.23	5.50
Interest rate risk σ	2.27	1.79

Table 4.10 Computation of the optimal balance sheet in problem (4.3.27)–(4.3.30), within the ellipsoid region determined by the benchmark balance sheet and the radius r^b, using the expression (4.3.58). The difference between the shown assets and liabilities is equal to 7.8, which represents the overnight funding that is needed to compensate that difference. Compared to the benchmark, the optimal balance sheet achieves lower credit risk and interest rate risk while maintaining the same target return. The optimal balance sheet deleverages considerably when compared to the benchmark balance sheet, and avoids corner solutions, a critical feature when computing optimal balance sheets in practice

4.4 Comments and Further Developments

The patterns observed in Sects. 4.2 and 4.3 largely persist in more general settings. Of course, there are additional technical difficulties. We will briefly discuss these generalizations and point to interesting directions for further research while appropriate. Furthermore, we will give a sophisticated log drawdown and log TWR application.

4.4.1 Linear Model Involving More Than Two Kinds of Risks

In the example illustrated in Sect. 4.2, we considered only two kinds of risks: the interest risk and the credit risk. As we have seen in Sect. 4.1, there are many more kinds of risks involved in bank balance sheet management problems. What happens if we extend the linear model discussed in Sect. 4.2 to situations involving more than two risks? The discussion below follows that of Júdice et al. [64]. If we use positive signs for the portfolio weights of both assets and liabilities, then a general linear model of the bank balance sheet optimization problem can be represented by a linear programming problem:

$$\max_{x \in \mathbb{R}^N} \langle c, x \rangle \quad \text{subject to } Ax \leq b, x \geq 0 \tag{4.4.1}$$

where $x \in \mathbb{R}^N$ is the vector of portfolio weights, $c \in \mathbb{R}^N$ represents the adjusted expected return vector, the vector $b \in \mathbb{R}^M$ represents the limits on M different risks, and $A = [a_1, a_2, \dots, a_N] \in \mathbb{R}^{M \times N}$ is a matrix whose column $a_n \in \mathbb{R}^M$ describes the rates which asset n generates in the M different risks. By Example A.33 in the appendix, the dual problem of (4.4.1) is

$$\min_{\lambda \in \mathbb{R}^M} \langle b, \lambda \rangle \quad \text{subject to } A^\top \lambda \geq c, \lambda \geq 0. \tag{4.4.2}$$

In the dual problem, the constraint $a_n^\top \lambda \geq c_n$ corresponds to the asset n. Due to the complementary slackness condition, asset n is involved in the optimal portfolio only when $a_n^\top \lambda = c_n$. It is well-known in linear programming that the optimal solution occurs on the corner points of the feasible region characterized by

$$a_{n_k}^\top \lambda = c_{n_k}, k = 1, 2, \dots, K, \tag{4.4.3}$$

and

$$b \in \text{cone}\{a_{n_k}, k = 1, 2, \dots, K\}. \tag{4.4.4}$$

Here cone(S) signals the cone generated by the set S. Recall

Theorem 4.27 (Carathéodory's Theorem for the Conical Hull) *If a point $b \in \mathbb{R}^M$ lies in the conical hull of a set S, then b can be written as the conical combination of at most M points in S.*

Thus, to determine the dual solution, at most M equations among those in (4.4.3) suffices, i.e., $K \leq M$. That is to say, in the optimal portfolio, we need at most M assets. This is similar to the conclusion we have seen in Sect. 4.2. However, the discussion on the financial meaning of the conditions characterizing the optimal solution is much more involved. We refer to [64] for details.

4.4.2 Linear-Quadratic Model Involving More Than Two Kinds of Risks

We have seen in the last subsection that when only linear risks are considered, the number of assets in the optimal balance sheet will not need to exceed the number of risks. This causes unrealistic concentration of the balance sheet. The reason is that assuming linear risk amounts to say the risks generated by different assets are perfectly correlated which is, of course, not what happens in practice. Thus, a more realistic model needs to consider correlation between risks generated by different assets. This leads to a linear-quadratic model in which linear risks and quadratic risks are mixed. A simple case involving two kinds of risks has been discussed in Sect. 4.3. Here, we briefly discuss the technical complications when more than two risks are involved. The situation below in Example 4.28 arises, e.g., when besides a linear (utility) return constraint, there are $M-1$ linear and K quadratic risk constraints and furthermore one quadratic risk function to be minimized.

Example 4.28 *In general, we will have the following model:*

$$\begin{aligned} p = \min_{x\in\mathbb{R}^N} & \frac{1}{2}\langle x, \Theta x\rangle \\ \text{subject to } & Ax \le b^{(1)} = (b_1,\dots,b_M)^\top \in \mathbb{R}^M, \\ & \frac{1}{2}\langle x, \Psi_k x\rangle \le b_{M+k},\ k=1,\dots,K. \\ & x \ge 0. \end{aligned} \tag{4.4.5}$$

Here $x\in\mathbb{R}^N$, $A\in\mathbb{R}^{M\times N}$, with $M<N$ has rank M, and $\Psi_k\in\mathbb{R}^{N\times N}, k=1,\dots,K$ are symmetric, positive semidefinite matrices and Θ is symmetric, positive definite as well. To simplify the notation, we separate a vector $v\in\mathbb{R}^{M+K}$ into two blocks: $v=((v^{(1)})^\top,(v^{(2)})^\top)^\top$, where $v^{(1)}=(v_1,\dots,v_M)^\top$ and $v^{(2)}=(v_{M+1},\dots,v_{M+K})^\top$. We denote $\Psi_{v^{(2)}}:=\sum_{k=1}^K v_{M+k}\Psi_k$ and introduce the Lagrangian:

$$\begin{aligned} L(x,\lambda) = \frac{1}{2}\langle x, \Theta x\rangle + \langle \lambda^{(1)}, Ax-b^{(1)}\rangle + \frac{1}{2}\langle x, \Psi_{\lambda^{(2)}} x\rangle \\ - \langle \lambda^{(2)}, b^{(2)}\rangle - \langle \lambda^{(3)}, x\rangle. \end{aligned} \tag{4.4.6}$$

By Lagrangian duality, we have

$$p = \inf_x \sup_{\lambda\ge 0} L(x,\lambda) = \sup_{\lambda\ge 0}\inf_x L(x,\lambda) = d. \tag{4.4.7}$$

To calculate $\inf_x L(x,\lambda)$, we solve

$$0 = \nabla_x L(x,\lambda) = (\Theta + \Psi_{\lambda^{(2)}})x + A^\top\lambda^{(1)} - \lambda^{(3)} \tag{4.4.8}$$

to derive the minimum of the map $x \mapsto L(x, \lambda)$ to be attained at

$$x = (\Theta + \Psi_{\lambda^{(2)}})^{-1}(\lambda^{(3)} - A^\top \lambda^{(1)}). \tag{4.4.9}$$

It follows that the dual problem of (4.4.5) is

$$\begin{aligned} d = \sup_{\lambda \geq 0} \quad & - \langle b^{(1)}, \lambda^{(1)} \rangle - \langle b^{(2)}, \lambda^{(2)} \rangle \\ & - \frac{1}{2} \langle \lambda^{(3)} - A^\top \lambda^{(1)}, (\Theta + \Psi_{\lambda^{(2)}})^{-1}(\lambda^{(3)} - A^\top \lambda^{(1)}) \rangle. \end{aligned} \tag{4.4.10}$$

Let $\overline{\lambda} = (\overline{\lambda^{(1)}}^\top, \overline{\lambda^{(2)}}^\top, \overline{\lambda^{(3)}}^\top)^\top \in \mathbb{R}^{M+K+N}$ denote the solution to the dual problem (4.4.10). For all the constraints to be relevant, we must have $\overline{\lambda^{(1)}}, \overline{\lambda^{(2)}} > 0$. Thus, these two components of the dual solution are determined by the following system of equations if we know $\overline{\lambda^{(3)}}$:

$$b^{(1)} = A(\Theta + \Psi_{\overline{\lambda^{(2)}}})^{-1}(\overline{\lambda^{(3)}} - A^\top \overline{\lambda^{(1)}}), \tag{4.4.11}$$

and, for $k = 1, \ldots, K$,

$$b_{M+k} = \frac{1}{2}(\overline{\lambda^{(3)}} - A^\top \overline{\lambda^{(1)}})^\top (\Theta + \Psi_{\overline{\lambda^{(2)}}})^{-1} \Psi_k (\Theta + \Psi_{\overline{\lambda^{(2)}}})^{-1} (\overline{\lambda^{(3)}} - A^\top \overline{\lambda^{(1)}}). \tag{4.4.12}$$

We turn to $\overline{\lambda^{(3)}}$ which can be determined using the same method as in Example 4.22. Decompose the Euclidean space $\mathbb{R}^N$ into $U \times V$ where $U = \mathbb{R}^t$ is the subspace in which the components of $\overline{\lambda^{(3)}}$ are nonzero numbers and $V = \mathbb{R}^{N-t}$ is the subspace in which the components of $\overline{\lambda^{(3)}}$ are zeros. We again assume without loss of generality that the t nonzero components of $\overline{\lambda^{(3)}}$ are the first t components. Denote the projection operators onto U and V by $P_U = (I_t, 0) \in \mathbb{R}^{t \times N}$ and $P_V = (0, I_{N-t}) \in \mathbb{R}^{(N-t) \times N}$, respectively, where we again use the same notation for their corresponding representing matrices. Set

$$Q_{U,V} := -I_N + \begin{bmatrix} (P_U(\Theta + \Psi_{\overline{\lambda^{(2)}}})^{-1} P_U^\top)^{-1} P_U (\Theta + \Psi_{\overline{\lambda^{(2)}}})^{-1} P_V^\top P_V - P_U \\ 0 \end{bmatrix} \in \mathbb{R}^{N \times N}.$$

Then using a computation similar to Example 4.22, we obtain

$$\overline{\lambda^{(3)}} = \begin{bmatrix} P_U \\ P_V \end{bmatrix} \overline{\lambda^{(3)}} = (I_N + Q_{U,V}) A^\top \overline{\lambda^{(1)}}. \tag{4.4.13}$$

Thus, $\overline{\lambda^{(1)}} \in \mathbb{R}^M, \overline{\lambda^{(2)}} \in R^K$ are determined by

$$b^{(1)} = A(\Theta + \Psi_{\overline{\lambda^{(2)}}})^{-1} Q_{U,V} A^{\top} \overline{\lambda^{(1)}}, \tag{4.4.14}$$

and, for $k = 1, \ldots, K$,

$$b_{M+k} = \frac{1}{2} \left\langle Q_{U,V} A^{\top} \overline{\lambda^{(1)}}, (\Theta + \Psi_{\overline{\lambda^{(2)}}})^{-1} \Psi_k (\Theta + \Psi_{\overline{\lambda^{(2)}}})^{-1} Q_{U,V} A^{\top} \overline{\lambda^{(1)}} \right\rangle. \tag{4.4.15}$$

It follows that the primal solution is given by (4.4.9)

$$\overline{x} = (\Theta + \Psi_{\overline{\lambda^{(2)}}})^{-1} (\overline{\lambda^{(3)}} - A^{\top} \overline{\lambda^{(1)}}), \tag{4.4.16}$$

with $\overline{\lambda^{(3)}}$ *from* (4.4.13). *Again, the nonzero components of* $\overline{\lambda^{(3)}}$ *are not given in general. Thus, to find the solution, we need to try all the decompositions of the form* $\mathbb{R}^N = U \times V$ *and find the one that corresponds to the maximum value for the dual problem* (4.4.10) *for* $\overline{\lambda^{(1)}}, \overline{\lambda^{(2)}}$ *calculated with* (4.4.14) *and* (4.4.15). *Of course, for reasonably large* N, *this exhaustive search is too complex to be practical and calculating the solution numerically is a more reasonable approach.*

Remark 4.29 *Although the method used here is similar to that of Example 4.22, due to the involvement of* $\overline{\lambda^{(2)}}$, *both the dual and primal solutions are no longer linearly dependent on the parameters* $b^{(1)}$ *and* $b^{(2)}$.

4.4.3 Diversification by Modifying a Benchmark

An illustration of this method has been discussed in Example 4.25 for the special case when the constraints corresponding to both the interest and the credit risks are not binding. There we use the $\|\cdot\|_{\Theta}$ distance to the benchmark. Below we discuss the solution to the general problem in which the $\|\cdot\|_{\Theta}$ distance is replaced by the $\|\widehat{x}\|_{\Psi} = \sqrt{\langle \widehat{x}, \Psi \widehat{x} \rangle}$ distance where Θ and Ψ may be different:

$$\mu := \max_{\widehat{x} \in \mathbb{R}^{n+m+s}} \quad \langle \widehat{R} - R_0 \widehat{1}, \widehat{x} \rangle \tag{4.4.17}$$

subject to

$$\begin{aligned}
\lambda^i &: \quad \frac{1}{2} \langle \widehat{x}, \Theta \widehat{x} \rangle \le r^i \\
\lambda^c &: \quad \langle \widehat{c}, \widehat{x} \rangle \le r^c \\
\lambda^b &: \quad \frac{1}{2} \langle \widehat{x} - \widehat{x}^*, \Psi(\widehat{x} - \widehat{x}^*) \rangle \le r^b.
\end{aligned}$$

Here $\Theta, \Psi \in \mathbb{R}^{(n+m+s)\times(n+m+s)}$ are symmetric, positive definite matrices and $\widehat{x}^*$ is a benchmark portfolio. The Lagrangian for this problem is

$$L(\widehat{x},\lambda) = \langle \widehat{R} - R_0\widehat{1}, \widehat{x}\rangle - \lambda^i\left(\frac{1}{2}\langle \widehat{x}, \Theta\widehat{x}\rangle - r^i\right) - \lambda^c\left(\langle \widehat{c}, \widehat{x}\rangle - r^c\right) \quad (4.4.18)$$
$$-\lambda^b\left(\frac{1}{2}\langle \widehat{x} - \widehat{x}^*, \Psi(\widehat{x} - \widehat{x}^*)\rangle - r^b\right),$$

where $\lambda = (\lambda^i, \lambda^c, \lambda^b)^\top \geq 0$. Then we can follow the familiar route of using Lagrangian duality:

$$p = \mu = \sup_{\widehat{x}} \inf_{\lambda \geq 0} L(\widehat{x},\lambda) = \inf_{\lambda \geq 0} \sup_{\widehat{x}} L(\widehat{x},\lambda) = d \quad (4.4.19)$$

to change our focus to the dual problem. We calculate $\sup_{\widehat{x}} L(\widehat{x},\lambda)$ by solving

$$0 = \nabla_{\widehat{x}} L(\widehat{x},\lambda) = \widehat{R} - R_0\widehat{1} - \lambda^c\widehat{c} + \lambda^b\Psi\widehat{x}^* - (\lambda^i\Theta + \lambda^b\Psi)\widehat{x} \quad (4.4.20)$$

to derive the maximum of the map $\widehat{x} \mapsto L(\widehat{x},\lambda)$ to be attained at

$$\widehat{x} = (\lambda^i\Theta + \lambda^b\Psi)^{-1}(\widehat{R} - R_0\widehat{1} - \lambda^c\widehat{c} + \lambda^b\Psi\widehat{x}^*) =: (\lambda^i\Theta + \lambda^b\Psi)^{-1}z_\lambda. \quad (4.4.21)$$

It follows that the dual problem of (4.4.17) is

$$d = \inf_{\lambda \geq 0} \lambda^i r^i + \lambda^c r^c + \lambda^b\left(r^b - \frac{\|\widehat{x}^*\|_\Psi^2}{2}\right) + \frac{1}{2}\left\langle z_\lambda, (\lambda^i\Theta + \lambda^b\Psi)^{-1}z_\lambda\right\rangle. \quad (4.4.22)$$

In the most general case when the three components of the dual solution are all nonzero, we get that the inequalities in (4.4.17) are binding. Thus, λ^i, λ^c and λ^b are all positive and may be determined by a system of three equations:

$$r^c - \langle \widehat{c}, (\lambda^i\Theta + \lambda^b\Psi)^{-1}z_\lambda\rangle = 0 \quad (4.4.23)$$

$$r^i - \frac{1}{2}\langle z_\lambda, (\lambda^i\Theta + \lambda^b\Psi)^{-1}\Theta(\lambda^i\Theta + \lambda^b\Psi)^{-1}z_\lambda\rangle = 0 \quad (4.4.24)$$

and

$$r^b - \frac{\|\widehat{x}^*\|_\Psi^2}{2} + \langle \Psi[\widehat{x}^* - \frac{1}{2}(\lambda^i\Theta + \lambda^b\Psi)^{-1}z_\lambda], (\lambda^i\Theta + \lambda^b\Psi)^{-1}z_\lambda\rangle = 0 \quad (4.4.25)$$

If some of the components of the dual solution are zero, then we can reduce the number of equations in the system. Note that Eqs. (4.4.23), (4.4.24), and (4.4.25) are highly nonlinear in λ^c, λ^i, and λ^b. Example 4.25 corresponds to the case when $\Psi = \Theta$ although with a different setting of minimizing the quadratic risk instead of maximizing the expected return. In that case, Eqs. (4.4.23), (4.4.24), and (4.4.25) simplify to

$$r^c(\lambda^i + \lambda^b) - \langle \widehat{c}, \Theta^{-1}z_\lambda\rangle = 0 \quad (4.4.26)$$

$$r^i(\lambda^i + \lambda^b)^2 - \frac{1}{2}\langle z_\lambda, \Theta^{-1} z_\lambda \rangle = 0 \tag{4.4.27}$$

and

$$\left(r^b - r^i - \frac{\|\widehat{x}^*\|_\Theta^2}{2} \right) (\lambda^i + \lambda^b) + \langle \widehat{x}^*, z_\lambda \rangle = 0 \tag{4.4.28}$$

The argument above also applies to the situation when there are multiple quadratic interest risks and linear credit risks as discussed in Example 4.28. The details are left for the readers.

4.4.4 Log Drawdown Risk and Log TWR Utility Application

The examples so far had all used the linear-quadratic structure of risk and utility function to calculate exact solutions. However, sometimes the models do not fit this tight setting.

For instance, in praxis many investors are sensitive to drawdowns of their portfolio equity. As a possible application for this, combined with a log TWR utility constraint, we discuss the following example:

Example 4.30 (Log Drawdown Risk and Log TWR Utility) *We use the setup of Sect. 2.2 with a d-period financial market and geometric returns G_t, $t = 1, \ldots, d$ (see Definition 2.35). Let $\mathfrak{u}_{\ln\mathrm{TWR}} \colon \mathcal{A}_1 \to \mathbb{R} \cup \{-\infty\}$ be the* log TWR *utility function of a d-period financial market as in* (2.2.18)–(2.2.20) *and $\mathfrak{r}_{\mathrm{drawdown}} \colon \mathcal{A}_1 \to \mathbb{R}_{\geq 0} \cup \{+\infty\}$ be the expected current* log drawdown *(see Definition 2.48), where*

$$\mathcal{A}_1 = \left\{ x = \left(x_0, \widehat{x}^\top\right)^\top \in \mathbb{R}^{M+1} : \sum_{j=0}^{M} x_j = 1 \right\} \tag{4.4.29}$$

was the set of fully invested portfolios proportion vectors (see (2.2.10)*). For a given $\mu \in \mathbb{R}$ and $\varrho \geq 0$, we intend to study*

$$\begin{aligned} \min_{x \in \mathcal{A}_1} \ & \mathfrak{r}_{\mathrm{drawdown}}(x) \\ \text{subject to } & \mathfrak{u}_{\ln\mathrm{TWR}}(x) \geq \mu \\ & \frac{1}{2}\widehat{x}^\top \Sigma \widehat{x} \leq \varrho. \end{aligned} \tag{4.4.30}$$

For simplicity, we here assume a positive definite covariance matrix $\Sigma = \mathrm{Cov}(\widehat{S}) \in \mathbb{R}^{M \times M}$ of a one-period financial market model as in Corollary 2.19. So the constraint $\mathfrak{r}_{\mathrm{Var}}(x) = \widehat{\mathfrak{r}}_{\mathrm{Var}}(\widehat{x}) := \frac{1}{2}\widehat{x}^\top \Sigma \widehat{x} \leq \varrho$ stands for a bounded portfolio variance. The portfolio variance could also be modeled in the d-period financial market of the log risk and utility functions $\mathfrak{r}_{\mathrm{drawdown}}$ and $\mathfrak{u}_{\ln\mathrm{TWR}}$, but the setup is more complex and therefore omitted (see Brenner [23, Proposition 5.1.18]).

Let us come back to (4.4.30). Since portfolios $x = (x_0, \widehat{x}^\top)^\top \in \mathcal{A}_1$ are allowed to have nontrivial bond component x_0, in particular the pure bond portfolio $x_{\text{pureBond}} := (1, \widehat{0}^\top)^\top$ is included in $\mathcal{A}_1$. So for $\varrho = \varrho_{\text{pureBond}} = 0$ and $\mu = \mu_{\text{pureBond}} := \mathfrak{u}_{\ln\text{TWR}}(x_{\text{pureBond}})$, obviously x_{pureBond} is a solution of (4.4.30) since $\mathfrak{r}_{\text{drawdown}}(x_{\text{pureBond}}) = 0$. Note that a similar situation happens in CAPM theory (see Maier-Paape and Zhu [78, Theorem 8]).

Furthermore, the existence of an efficient frontier for (4.4.30) with unique efficient portfolios for given points on $\mathcal{G}_{\text{eff}}^{(2d)} = \mathcal{G}_{\text{eff}}^{(2d)}(\mathfrak{r}_{\text{drawdown}}, \mathfrak{r}_{\text{Var}}, \mathfrak{u}_{\ln\text{TWR}}; \mathcal{A}_1)$ is no problem by Theorem 3.46, since under reasonable assumptions $\mathfrak{u}_{\ln\text{TWR}}$ is strictly proper concave (cf. Lemma 2.45) and $\mathfrak{r}_{\text{drawdown}}$ is proper convex (see Lemma 2.50), so Assumption 3.11 clearly holds (cf. Sect. 4.4.5 for a discussion of $\mathcal{G}_{\text{eff}}^{(2d)}$ from a qualitative point of view).

The goal here, however, is not to find exact solutions as in the previous subsections, but to reorganize (4.4.30) in such a way that a solution can be calculated with suitable numerical algorithms in reasonable time. The problem here is that, although (4.4.30) is a convex optimization problem, it contains the non-differentiable, convex risk function $\mathfrak{r}_{\text{drawdown}}$, and therefore typical convex programming algorithms like interior point methods are not applicable directly. To see this, we follow the discussion of Brenner [23, Theorem 5.2.11]. The non-differentiability can be see easily from

$$\mathfrak{r}_{\text{drawdown}}(x) := \mathrm{E}\left[\max\left\{0, \max_{1\le t\le d}\left[\sum_{s=t}^{d} -\ln\left(G_s^\top x\right)\right]\right\}\right], \quad x \in \mathcal{A}_{\text{TWR}}; \tag{4.4.31}$$

see Definition 2.48 and (2.2.15). Note that according to (2.2.35)

$$\text{dom}(\mathfrak{r}_{\text{drawdown}}) = \text{dom}(\mathfrak{u}_{\ln\text{TWR}}) = \mathcal{A}_{\text{TWR}},$$

so that the solution x of (4.4.30) has to be contained in $\mathcal{A}_{\text{TWR}} \subset \mathcal{A}_1$. Thus, $\mathfrak{r}_{\text{drawdown}}$ is represented by (4.4.31) and with (2.2.20), we have

$$\mathfrak{u}_{\ln\text{TWR}}(x) = \frac{1}{d}\sum_{s=1}^{d} \mathrm{E}\left[\ln\left(G_s^\top x\right)\right], \quad x \in \mathcal{A}_{\text{TWR}}.$$

Altogether, we can rearrange (4.4.30) equivalently as

$$\min_{x \in \mathbb{R}^{M+1}} \quad \mathrm{E}\left[\max\left\{\max_{1 \le t \le d}\left[\sum_{s=t}^{d} -\ln\left(G_s^\top x\right)\right]\right\}\right] \tag{4.4.32}$$

$$\text{subject to } \sum_{j=0}^{M} x_j = 1, \; \frac{1}{d}\sum_{s=1}^{d} \mathrm{E}\left[\ln\left(G_s^\top x\right)\right] \ge \mu, \tag{4.4.33}$$

$$\frac{1}{2}\widehat{x}^\top \Sigma \widehat{x} \le \varrho. \tag{4.4.34}$$

Since $\mathfrak{u}_{\ln\mathrm{TWR}}$ is differentiable in x, the only problem arises from $\mathfrak{r}_{\mathrm{drawdown}}$. To get more concrete, let us assume for simplicity that the underlying probability space is finite, say

$$\Omega = \{\omega_1, \ldots, \omega_m\}, \quad \text{i.e. } |\Omega| = m \quad \text{and} \quad \mathcal{P}(\omega_j) = \frac{1}{m}$$

for $j = 1, \ldots, m$. Then, $\mathfrak{r}_{\mathrm{drawdown}}$ on $x \in \mathcal{A}_{\mathrm{TWR}}$ can be represented alternatively as

$$\begin{aligned}\mathfrak{r}_{\mathrm{drawdown}}(x) &= \frac{1}{m}\sum_{j=1}^{m} \max\left\{0, \max_{1 \le t \le d}\left[\sum_{s=t}^{d} -\ln\left(G_s^\top(\omega_j)x\right)\right]\right\} \\ &= \min_{\varphi=(\varphi_1,\ldots,\varphi_m)\in\mathbb{R}^m_{\ge 0}} \left\{\frac{1}{m}\sum_{j=1}^{m}\varphi_j : \varphi_j \ge \sum_{s=t}^{d} -\ln\left(G_s^\top(\omega_j)\,x\right)\right. \\ &\qquad\qquad \left.\text{for all} \quad 1 \le t \le d \quad \text{and all} \quad 1 \le j \le m\right\}.\end{aligned} \tag{4.4.35}$$

Since the constraints for φ_j do not interfere each other, this simplifies to

$$\mathfrak{r}_{\mathrm{drawdown}}(x) = \frac{1}{m}\sum_{j=1}^{m} \psi(\omega_j, x), \quad x \in \mathcal{A}_{\mathrm{TWR}}, \tag{4.4.36}$$

where

$$\psi(\omega, x) := \min_{\varphi \in \mathbb{R}_{\ge 0}} \left\{\varphi : \varphi \ge \sum_{s=t}^{d} -\ln\left(G_s^\top(\omega)x\right) \text{ for all } 1 \le t \le d\right\}. \tag{4.4.37}$$

Hence, in order to get a smooth reformulation of (4.4.30), we only have to add the other constraints of (4.4.32)–(4.4.34). Therefore, for a given $\mu \in \mathbb{R}$ and $\varrho > 0$, we set

$$\begin{aligned}\widetilde{\psi}(\omega, x)[\mu, \varrho] &:= \min_{\varphi \in \mathbb{R}_{\geq 0}} \Bigg\{ \varphi : \varphi \geq \sum_{s=t}^{d} - \ln\left(G_s^\top(\omega) x\right) \text{ for all } 1 \leq t \leq d, \\ &\qquad \sum_{j=0}^{M} x_j = 1,\ \mathfrak{u}_{\ln \mathrm{TWR}}(x) \geq \mu \text{ and } \frac{1}{2} \widehat{x}^\top \Sigma \widehat{x} \leq \varrho \Bigg\} \\ &= \max_{\widetilde{\varphi} \in \mathbb{R}_{\geq 0}} \Bigg\{ \widetilde{\varphi} : \widetilde{\varphi} \leq \sum_{s=t}^{d} \ln\left(G_s^\top(\omega) x\right) \text{ for all } 1 \leq t \leq d, \\ &\qquad \sum_{j=1}^{M} x_j = 1,\ \mathfrak{u}_{\ln \mathrm{TWR}}(x) \geq \mu \text{ and } \frac{1}{2} \widehat{x}^\top \Sigma \widehat{x} \leq \varrho \Bigg\}.\end{aligned} \tag{4.4.38}$$

Note that this is a convex optimization problem with $d+3$ smooth constraints. Thus, bringing (4.4.36) and (4.4.38) together, we finally conclude that (4.4.30) for a given $\mu \in \mathbb{R}$ and $\varrho > 0$ is equivalent to

$$\min_{x \in \mathbb{R}^{M+1}} \frac{1}{m} \sum_{j=1}^{m} \widetilde{\psi}(\omega_j, x)[\mu, \varrho], \tag{4.4.39}$$

which can be solved numerically, e.g., with standard interior-point algorithms like the open-source optimization IPOPT (see, e.g., Boyd and Vandenberghe [22, Chapter 11] for more details). Finally, note that a similar approach also works for the other drawdown risk functions in Definition 2.48. Furthermore, instead of the expected drawdowns, other concepts are also possible, e.g., pathwise drawdown constraints, which would boil down to several drawdown constraints, one separately for each ω_j.

4.4.5 Efficient Frontier for Drawdown, Variance, and Log TWR

In this subsection, we reconsider the efficient frontier for Example 4.30 from a qualitative point of view. However, in contrast to (4.4.30), we here study

$$\begin{aligned} &\min_{x=(0,\widehat{x}^\top)^\top \in A} \quad \mathfrak{r}_{\mathrm{drawdown}}(x) \\ &\text{subject to } \mathfrak{u}_{\ln \mathrm{TWR}}(x) \geq \mu \\ &\qquad \frac{1}{2} \widehat{x}^\top \Sigma \widehat{x} \leq \varrho. \end{aligned} \tag{4.4.40}$$

for a given $\mu \in \mathbb{R}$ and $\varrho \geq 0$ fixed and with a restricted admissible portfolio set

$$A \subset \widehat{\mathcal{A}}_1 := \left\{ x = (x_0, \widehat{x}^\top)^\top \in \mathcal{A}_1 \, : \, x_0 = 0 \right\} \subset \mathcal{A}_1. \tag{4.4.41}$$

as defined in Assumption 2.41. In particular, A is admissible (cf. Definition 2.7), bounded, and $A \cap \mathbb{R}^{M+1}_{\geq 0} \neq \emptyset$. Note that the calculation procedure of Sect. 4.4.4 might still be applied to (4.4.40) when slightly adapted. We will show that the efficient frontier for (4.4.40) under generic assumptions is a bounded, closed, simply connected, and continuous two-dimensional surface in $\mathbb{R}^3$.

In order to get there, we need a lot of notation, but fortunately the theory developed in Chaps. 2 and 3 helps a lot. Firstly, we construct several curves in $\mathbb{R}^3$ which will later on turn out to be the boundary of the abovementioned surface. All these curves are related to $(1d)$-efficient frontiers. We will make similar assumptions as in Sect. 4.4.4, but additionally we assume for simplicity that $\mathfrak{r}_{\text{drawdown}}$ is strictly convex which eases the argumentation significantly.

Assumption 4.31 *Let $A \subset \widehat{\mathcal{A}}_1$ be as in* (4.4.41) *a restricted admissible portfolio set and let the underlying financial market be in such a way that*

(a) $\mathfrak{r}_{\text{drawdown}} \colon A \to \mathbb{R}_{\geq 0} \cup \{+\infty\}$ *is strictly convex and lower semi-continuous and has compact sublevel sets as in* (2.2.36).
(b) $\mathfrak{r}_{\text{Var}} \colon A \to \mathbb{R}_{\geq 0}$, $x = (0, \widehat{x}^\top)^\top \mapsto \frac{1}{2}\widehat{x}^\top \Sigma \widehat{x}$ *is strictly convex, where* Σ *is the positive definite covariance matrix from Corollary 2.19.*
(c) $\mathfrak{u}_{\ln\text{TWR}} \colon A \to \mathbb{R} \cup \{-\infty\}$ *is upper semi-continuous and strictly concave with compact superlevel sets as in* (2.2.26).

Remark 4.32

(a) Sufficient conditions for (a) and (c) in Assumption 4.31 have been given in Lemma 2.50, Remark 2.51, Lemmas 2.43, and 2.45.
(b) In case A is unbounded, but admissible and $A \cap \mathbb{R}^{M+1}_{\geq 0} \neq \emptyset$, by Remark 2.42, the generalization of the "no nontrivial riskless portfolio" condition to multi-period markets of Brenner [23, Lemma 2.2.3], may be utilized to replace A w.l.o.g. by $A \cap \mathcal{A}_{\text{TWR}}$ where the latter is a restricted admissible portfolio set.

Under Assumption 4.31, we can distinguish three portfolios in A. Firstly, as already known from the Markowitz setup (see Maier-Paape and Zhu [78, Section 4] or Remark 2.86), we have the minimum variance portfolio $x_{\text{minVar}} := (0, \widehat{x}^\top_{\text{minVar}})^\top \in \widehat{\mathcal{A}}_1$ as unique solution of

$$\min_{x \in \widehat{\mathcal{A}}_1} \mathfrak{r}_{\text{Var}}(x) = \frac{1}{2}\widehat{x}^\top \Sigma \widehat{x}. \tag{4.4.42}$$

We assume without loss of generality that A is large enough such that $x_{\text{minVar}} \in A$. Moreover, there exists a unique minimum drawdown portfolio $x_{\text{minDD}} = (0, \widehat{x}^\top_{\text{minDD}})^\top \in A$ which is the unique solution of

$$\min_{x \in A} \mathfrak{r}_{\text{drawdown}}(x), \tag{4.4.43}$$

since $\mathfrak{r}_{\text{drawdown}}$ is strictly convex and lower semi-continuous and has compact sublevels by Assumption 4.31 (a) (cf. Theorem A.8).

Similarly, the properties of Assumption 4.31 (c) imply that the problem

$$\max_{x \in A} \mathfrak{u}_{\ln\text{TWR}}(x) \tag{4.4.44}$$

also has a unique solution $x_{\text{maxGr}} = (0, \widehat{x}_{\text{maxGr}}^{\top})^{\top} \in A$, which is the growth optimal portfolio in A.

Remark 4.33 *The minimum variance portfolio x_{minVar} clearly is the unique global minimizer of (4.4.42) since $\mathfrak{r}_{\text{Var}}$ is strictly convex and coercive. Whether or not x_{minDD} and/or x_{maxGr} can also be chosen independent of A as global optimizers (for all A sufficiently large) is in general not so clear, but holds under generic assumptions as well (see Remark 4.32 (b)). With the above notation, we can formulate our next assumption.*

Assumption 4.34 *We assume besides Assumption 4.31 that the unique optimal solutions of (4.4.42), (4.4.43), and (4.4.44), i.e., x_{minVar}, x_{minDD} and x_{maxGr} are three different portfolios in A.*

Remark 4.35 *The reason why we study in (4.4.40) $A \subset \widehat{\mathcal{A}}_1$ instead of $A \subset \mathcal{A}_1$ is that x_{minVar} and x_{minDD} defined in $\mathcal{A}_1$ would both be equal to the pure bond portfolio $x_{\text{pureBond}} = (1, \widehat{0}^{\top})^{\top}$ and thus would not fit to Assumption 4.34.*

Definition 4.36 For further reference, we define the (vector risk, utility) values of x_{minVar}, x_{minDD}, and x_{maxGr}:

$$\begin{aligned}
&(\varrho_{\text{minVar}}, r_{\text{minVar}}, \mu_{\text{minVar}}) \\
&\quad := \Big(\mathfrak{r}_{\text{Var}}(x_{\text{minVar}}),\ \mathfrak{r}_{\text{drawdown}}(x_{\text{minVar}}),\ \mathfrak{u}_{\ln\text{TWR}}(x_{\text{minVar}})\Big) &(4.4.45)\\
&(\varrho_{\text{minDD}}, r_{\text{minDD}}, \mu_{\text{minDD}}) \\
&\quad := \Big(\mathfrak{r}_{\text{Var}}(x_{\text{minDD}}),\ \mathfrak{r}_{\text{drawdown}}(x_{\text{minDD}}),\ \mathfrak{u}_{\ln\text{TWR}}(x_{\text{minDD}})\Big) &(4.4.46)\\
&(\varrho_{\text{maxGr}}, r_{\text{maxGr}}, \mu_{\text{maxGr}}) \\
&\quad := \Big(\mathfrak{r}_{\text{Var}}(x_{\text{maxGr}}),\ \mathfrak{r}_{\text{drawdown}}(x_{\text{maxGr}}),\ \mathfrak{u}_{\ln\text{TWR}}(x_{\text{maxGr}})\Big) &(4.4.47)
\end{aligned}$$

With $A \subset \mathbb{R}^{M+1}$ admissible in the sense of Definition 2.7, we may take, for instance, $\mathfrak{r}_{\text{Var}}$ and $\mathfrak{u}_{\ln\text{TWR}}$ as risk and utility function for a scalar risk portfolio optimization problem with $\varrho \geq 0$:

$$\begin{aligned}
&\max_{x \in A}\ \mathfrak{u}_{\ln\text{TWR}}(x) \\
&\text{subject to } \mathfrak{r}_{\text{Var}}(x) \leq \varrho.
\end{aligned} \tag{4.4.48}$$

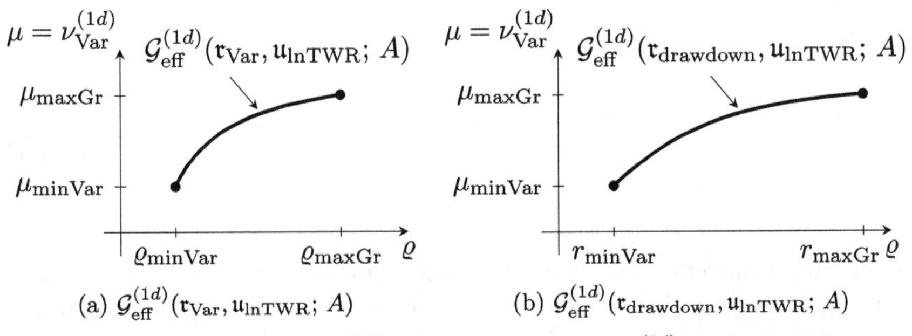

(a) $\mathcal{G}^{(1d)}_{\text{eff}}(\mathfrak{r}_{\text{Var}}, \mathfrak{u}_{\ln\text{TWR}};\ A)$ (b) $\mathcal{G}^{(1d)}_{\text{eff}}(\mathfrak{r}_{\text{drawdown}}, \mathfrak{u}_{\ln\text{TWR}};\ A)$

Fig. 4.6 The efficient frontiers $\mathcal{G}^{(1d)}_{\text{eff}}(\mathfrak{r}_{\text{Var}}, \mathfrak{u}_{\ln\text{TWR}};\ A)$ and $\mathcal{G}^{(1d)}_{\text{eff}}(\mathfrak{r}_{\text{drawdown}}, \mathfrak{u}_{\ln\text{TWR}};\ A)$

Clearly, Assumption 2.64 is satisfied for this setup and thus the theory of Chap. 2 for scalar risk applies. In particular, the efficient frontier for (4.4.48) denoted $\mathcal{G}^{(1d)}_{\text{eff}}(\mathfrak{r}_{\text{Var}}, \mathfrak{u}_{\ln\text{TWR}};\ A)$ can be represented as graph of a concave function $\nu = \nu^{(1d)}_{\text{Var}}$ on an interval $I = I_{\text{Var}} \subset \mathbb{R}$ (see Theorem 2.76). Apparently, x_{minVar} is an efficient portfolio for (4.4.48) with the least possible ϱ, i.e., $\varrho_{\text{minVar}} = \mathfrak{r}_{\text{Var}}(x_{\text{minVar}}) > 0$ is the left endpoint of I_{Var}. Similarly, x_{maxGr} has to be an efficient portfolio for (4.4.48) with the maximal possible utility value. Hence, $\varrho_{\text{maxGr}} = \mathfrak{r}_{\text{Var}}(x_{\text{maxGr}})$ is the right endpoint of I_{Var}. Both endpoints are included in I_{Var}, so I_{Var} is closed and bounded. Together, we get

$$\mathcal{G}^{(1d)}_{\text{eff}}(\mathfrak{r}_{\text{Var}}, \mathfrak{u}_{\ln\text{TWR}};\ A) = \operatorname{graph}\left(\nu^{(1d)}_{\text{Var}}\Big|_{[\varrho_{\text{minVar}}, \varrho_{\text{maxGr}}]}\right) \subset \mathbb{R}^2 \tag{4.4.49}$$

(see Fig. 4.6a). Note that

$$\nu^{(1d)}_{\text{Var}}(\varrho_{\text{minVar}}) = \mu_{\text{minVar}} \quad \text{and} \quad \nu^{(1d)}_{\text{Var}}(\varrho_{\text{maxGr}}) = \mu_{\text{maxGr}}. \tag{4.4.50}$$

Furthermore, with Theorem 2.79 we obtain a related continuous efficient portfolio map:

$$\begin{aligned} X^{(1d)}_{\text{Var}} : \mathcal{G}^{(1d)}_{\text{eff}}(\mathfrak{r}_{\text{Var}}, \mathfrak{u}_{\ln\text{TWR}};\ A) &\to A, \\ (\varrho, \mu) &\mapsto X^{(1d)}_{\text{Var}}(\varrho, \mu) \end{aligned} \tag{4.4.51}$$

with $X^{(1d)}_{\text{Var}}(\varrho, \mu)$ realizing the risk-utility value (ϱ, μ). Hence,

$$X^{(1d)}_{\text{Var}}\left(\varrho, \nu^{(1d)}_{\text{Var}}(\varrho)\right), \quad \varrho \in [\varrho_{\text{minVar}}, \varrho_{\text{maxGr}}] \tag{4.4.52}$$

represent all efficient portfolios for (4.4.48) with

$$\begin{aligned} X^{(1d)}_{\text{Var}}\left(\varrho_{\text{minVar}}, \nu^{(1d)}_{\text{Var}}(\varrho_{\text{minVar}})\right) &= x_{\text{minVar}} \quad \text{and} \\ X^{(1d)}_{\text{Var}}\left(\varrho_{\text{maxGr}}, \nu^{(1d)}_{\text{Var}}(\varrho_{\text{maxGr}})\right) &= x_{\text{maxGr}}. \end{aligned} \tag{4.4.53}$$

Setting

$$\Gamma_{\text{eff}}^{\text{Var,TWR}} := \left\{ \left(\varrho, \mathfrak{r}_{\text{drawdown}} \left(X_{\text{Var}}^{(1d)}(\varrho, \nu_{\text{Var}}^{(1d)}(\varrho)) \right), \nu_{\text{Var}}^{(1d)}(\varrho) \right), \right. \\ \left. \varrho \in [\varrho_{\text{minVar}}, \varrho_{\text{maxGr}}] \right\} \subset \mathbb{R}^3, \tag{4.4.54}$$

we obtain a continuous curve in $\mathbb{R}^3$ connecting the (vector risk, utility) values of x_{minVar} (see (4.4.45)) and of x_{maxGr} (see (4.4.47)). This curve will later on turn out to be a part of the efficient frontier for (4.4.40), i.e., of

$$\mathcal{G}_{\text{eff}}^{(2d)} = \mathcal{G}_{\text{eff}}^{(2d)}(\mathfrak{r}_{\text{Var}}, \mathfrak{r}_{\text{drawdown}}, \mathfrak{u}_{\ln\text{TWR}}; A) \subset \mathbb{R}^3. \tag{4.4.55}$$

Next, we can similarly consider (4.4.48) with $\mathfrak{r}_{\text{Var}}$ replaced by $\mathfrak{r}_{\text{drawdown}}$, i.e., for $r \geq 0$ let

$$\max_{x \in A} \mathfrak{u}_{\ln\text{TWR}}(x) \\ \text{subject to } \mathfrak{r}_{\text{drawdown}}(x) \leq r. \tag{4.4.56}$$

Assumption 2.64 is satisfied once more giving that $\mathcal{G}_{\text{eff}}^{(1d)}(\mathfrak{r}_{\text{drawdown}}, \mathfrak{u}_{\ln\text{TWR}}; A)$ can be represented as graph of a concave function $\nu = \nu_{\text{DD}}^{(1d)}$ on an interval $I = I_{\text{DD}} \subset \mathbb{R}$ (see Theorem 2.76). Again, x_{minDD} is an efficient portfolio for (4.4.56) with least possible r, i.e., $r_{\text{minDD}} = \mathfrak{r}_{\text{drawdown}}(x_{\text{minDD}})$, giving the left endpoint of I_{DD}. Furthermore, x_{maxGr} is efficient for (4.4.56) as well giving the right endpoint of the closed and bounded interval I_{DD} as $r_{\text{maxGr}} = \mathfrak{r}_{\text{drawdown}}(x_{\text{maxGr}})$. All together, we have

$$\mathcal{G}_{\text{eff}}^{(1d)}(\mathfrak{r}_{\text{drawdown}}, \mathfrak{u}_{\ln\text{TWR}}; A) = \text{graph}\left(\nu_{\text{DD}\,|\,[r_{\text{minDD}}, r_{\text{maxGr}}]}^{(1d)} \right) \subset \mathbb{R}^2, \tag{4.4.57}$$

(see Fig. 4.6b), where

$$\nu_{\text{DD}}^{(1d)}(r_{\text{minDD}}) = \mu_{\text{minDD}} \quad \text{and} \quad \nu_{\text{DD}}^{(1d)}(r_{\text{maxGr}}) = \mu_{\text{maxGr}}. \tag{4.4.58}$$

Using again Theorem 2.79, we get a continuous efficient portfolio map

$$X_{\text{DD}}^{(1d)} : \mathcal{G}_{\text{eff}}^{(1d)}(\mathfrak{r}_{\text{drawdown}}, \mathfrak{u}_{\ln\text{TWR}}; A) \to A, \\ (r, \mu) \mapsto X_{\text{DD}}^{(1d)}(r, \mu) \tag{4.4.59}$$

realizing the risk-utility value (r, μ) and

$$X_{\text{DD}}^{(1d)}\left(r, \nu_{\text{DD}}^{(1d)}(r) \right), \quad r \in [r_{\text{minDD}}, r_{\text{maxGr}}] \tag{4.4.60}$$

represent all efficient portfolios for (4.4.56), in particular

$$X_{\mathrm{DD}}^{(1d)}\left(r_{\mathrm{minDD}},\ \nu_{\mathrm{DD}}^{(1d)}\left(r_{\mathrm{minDD}}\right)\right)=x_{\mathrm{minDD}}\quad\text{and}$$
$$X_{\mathrm{DD}}^{(1d)}\left(r_{\mathrm{maxGr}},\ \nu_{\mathrm{DD}}^{(1d)}\left(r_{\mathrm{maxGr}}\right)\right)=x_{\mathrm{maxGr}}. \tag{4.4.61}$$

Setting

$$\Gamma_{\mathrm{eff}}^{\mathrm{DD,TWR}}:=\left\{\left(\mathfrak{r}_{\mathrm{Var}}\left(X_{\mathrm{DD}}^{(1d)}(r,\nu_{\mathrm{DD}}^{(1d)}(r))\right),r,\nu_{\mathrm{DD}}^{(1d)}(r)\right),\ r\in\left[r_{\mathrm{minDD}},r_{\mathrm{maxGr}}\right]\right\}\subset\mathbb{R}^3, \tag{4.4.62}$$

we obtain a curve connecting the (vector risk, utility) values of x_{minDD} (see (4.4.46)) and x_{maxGr} (see (4.4.47)). This curve is the second (proposed) boundary part of the efficient frontier for (4.4.40), cf. (4.4.55).

The third and last (proposed) boundary curve to construct connects the (vector risk, utility) values of x_{minVar} and x_{minDD}. For this, we view $-\mathfrak{r}_{\mathrm{drawdown}}$ as a utility function, i.e., we set

$$\mathfrak{u}_{\mathrm{drawdown}}(x):=-\mathfrak{r}_{\mathrm{drawdown}}(x),\quad x\in A, \tag{4.4.63}$$

which leads to another scalar risk problem for $s\geq 0$:

$$\max_{x\in A}\ \mathfrak{u}_{\mathrm{drawdown}}(x)$$
$$\text{subject to } \mathfrak{r}_{\mathrm{Var}}(x)\leq s. \tag{4.4.64}$$

Reasoning as before, we can represent $\mathcal{G}_{\mathrm{eff}}^{(1d)}(\mathfrak{r}_{\mathrm{Var}},-\mathfrak{r}_{\mathrm{drawdown}};A)$ as graph of a concave function $\nu_{\mathrm{RR}}=\nu_{\mathrm{RR}}^{(1d)}$ on an closed and bounded interval $I=I_{\mathrm{RR}}$ with left endpoint $s_{\mathrm{minVar}}=\mathfrak{r}_{\mathrm{Var}}\left(x_{\mathrm{minVar}}\right)=\varrho_{\mathrm{minVar}}$ and right endpoint $s_{\mathrm{minDD}}=\mathfrak{r}_{\mathrm{Var}}\left(x_{\mathrm{minDD}}\right)=\varrho_{\mathrm{minDD}}$, i.e.,

$$\mathcal{G}_{\mathrm{eff}}^{(1d)}\left(\mathfrak{r}_{\mathrm{Var}},-\mathfrak{r}_{\mathrm{drawdown}};A\right)=\operatorname{graph}\left(\nu_{\mathrm{RR}\,|\,[s_{\mathrm{minVar}},s_{\mathrm{minDD}}]}^{(1d)}\right)\subset\mathbb{R}^2, \tag{4.4.65}$$

(see Fig. 4.7a). Note that

$$\nu_{\mathrm{RR}}^{(1d)}\left(\varrho_{\mathrm{minVar}}\right)=-\mathfrak{r}_{\mathfrak{r}_{\mathrm{drawdown}}}\left(x_{\mathrm{minVar}}\right)=-r_{\mathrm{minVar}}\quad\text{and}$$
$$\nu_{\mathrm{RR}}^{(1d)}\left(\varrho_{\mathrm{minDD}}\right)=-\mathfrak{r}_{\mathfrak{r}_{\mathrm{drawdown}}}\left(x_{\mathrm{minDD}}\right)=-r_{\mathrm{minDD}}. \tag{4.4.66}$$

The related continuous efficient portfolio map now has the form

$$X_{\mathrm{RR}}^{(1d)}:\mathcal{G}_{\mathrm{eff}}^{(1d)}\left(\mathfrak{r}_{\mathrm{Var}},-\mathfrak{r}_{\mathrm{drawdown}};A\right)\to A,$$
$$(s,\mu)\mapsto X_{\mathrm{RR}}^{(1d)}(s,\mu), \tag{4.4.67}$$

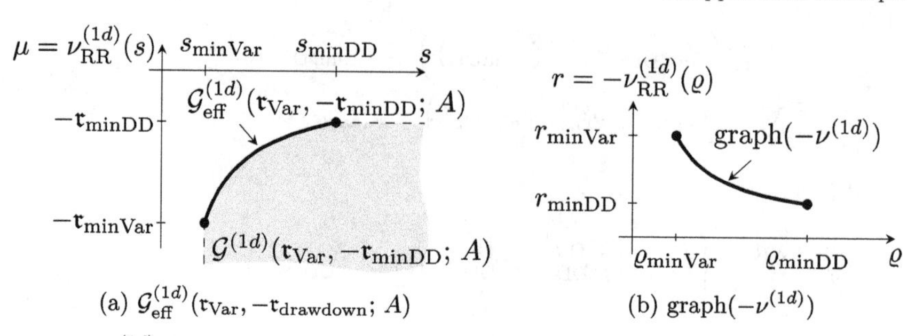

(a) $\mathcal{G}_{eff}^{(1d)}(\mathfrak{r}_{Var}, -\mathfrak{r}_{drawdown}; A)$ (b) graph$(-\nu^{(1d)})$

Fig. 4.7 $\mathcal{G}_{eff}^{(1d)}(\mathfrak{r}_{Var}, -\mathfrak{r}_{drawdown}; A)$ and its reflection at the s–axis

with $X_{RR}^{(1d)}(s, \mu)$ realizing the (risk = variance, utility = negative drawdown) value (s, μ). So

$$X_{RR}^{(1d)}\left(s, \nu_{RR}^{(1d)}(s)\right), \quad s \in [s_{minVar}, s_{minDD}] \tag{4.4.68}$$

represent all efficient portfolios for (4.4.64), where

$$\begin{aligned} X_{RR}^{(1d)}\left(s_{minVar}, \nu_{RR}^{(1d)}(s_{minVar})\right) &= x_{minVar} \quad \text{and} \\ X_{RR}^{(1d)}\left(s_{minDD}, \nu_{RR}^{(1d)}(s_{minDD})\right) &= x_{minDD}. \end{aligned} \tag{4.4.69}$$

Note that (4.4.65) can also be viewed in (variance, drawdown) space by reflecting $\nu_{RR}^{(1d)}$, i.e., with $-\nu_{RR}^{(1d)}$ convex (see Fig. 4.7b). We finally obtain the third (proposed) boundary curve:

$$\begin{aligned} \Gamma_{eff}^{Var,DD} := \Bigg\{ \Big(\varrho, -\nu_{RR}^{(1d)}(\varrho),\ \mathfrak{u}_{lnTWR}\big(X_{RR}^{(1d)}(\varrho, \nu_{RR}^{(1d)}(\varrho))\big) \Big), \\ \varrho \in [\varrho_{minVar}, \varrho_{minDD}] \Bigg\} \subset \mathbb{R}^3, \end{aligned} \tag{4.4.70}$$

which connects the (vector risk, utility) values of x_{minVar} and x_{minDD} (see (4.4.45) and (4.4.46)).

Corollary 4.37 *Using the setting of Definition 4.36, we obtain*

$$(\varrho_{minVar}, r_{minVar}, \mu_{minVar}) \in \Gamma_{eff}^{Var,TWR} \cap \Gamma_{eff}^{Var,DD}, \tag{4.4.71}$$

$$(\varrho_{minDD}, r_{minDD}, \mu_{minDD}) \in \Gamma_{eff}^{DD,TWR} \cap \Gamma_{eff}^{Var,DD}, \tag{4.4.72}$$

and

$$(\varrho_{maxGr}, r_{maxGr}, \mu_{maxGr}) \in \Gamma_{eff}^{Var,TWR} \cap \Gamma_{eff}^{DD,TWR}. \tag{4.4.73}$$

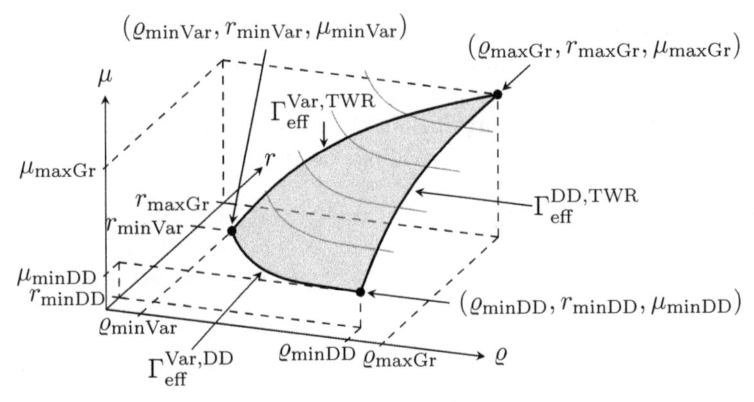

Fig. 4.8 The quasi-triangular-shaped surface $\mathcal{G}_{\text{eff}}^{(2d)}(\mathfrak{r}_{\text{Var}}, \mathfrak{r}_{\text{drawdown}}, \mathfrak{u}_{\ln\text{TWR}}; A)$

In order to keep our remaining discussion simple, we need a further assumption:

Assumption 4.38 *We assume besides Assumption 4.34 that the three points in* (4.4.71)–(4.4.73) *are the only points in the respective intersections, i.e., the three intersections in* (4.4.71)–(4.4.73) *excluding the endpoints of these curves are empty. Hence,*

$$\Gamma := \Gamma_{\text{eff}}^{\text{Var,TWR}} \cup \Gamma_{\text{eff}}^{\text{DD,TWR}} \cup \Gamma_{\text{eff}}^{\text{Var,DD}} \subset \mathbb{R}^3 \tag{4.4.74}$$

is a closed, continuous curve with no self-intersections.

Theorem 4.39 (Properties of $\mathcal{G}_{\text{eff}}^{(2d)}$ for Variance, Drawdown, and Log TWR) *Under Assumption 4.38, the efficient frontier for the vector risk problem* (4.4.40)*, i.e.,* $\mathcal{G}_{\text{eff}}^{(2d)} = \mathcal{G}_{\text{eff}}^{(2d)}(\mathfrak{r}_{\text{Var}}, \mathfrak{r}_{\text{drawdown}}, \mathfrak{u}_{\ln\text{TWR}}; A)$*, is graph of a continuous function. It is furthermore a closed set and a simply connected, two-dimensional bounded surface which (2d)-boundary is Γ, i.e., it consists of the three curves $\Gamma_{\text{eff}}^{\text{Var,DD}}$, $\Gamma_{\text{eff}}^{\text{Var,TWR}}$, and $\Gamma_{\text{eff}}^{\text{DD,TWR}}$ (cf. Fig. 4.8).*

The proof of Theorem 4.39 will be provided with several lemmata.

Lemma 4.40 (Properties of the Γ Curves) *Under Assumption 4.38, we have*

$$\Gamma = \Gamma_{\text{eff}}^{\text{Var,TWR}} \cup \Gamma_{\text{eff}}^{\text{DD,TWR}} \cup \Gamma_{\text{eff}}^{\text{Var,DD}} \subset \mathcal{G}_{\text{eff}}^{(2d)}(\mathfrak{r}_{\text{Var}}, \mathfrak{r}_{\text{drawdown}}, \mathfrak{u}_{\ln\text{TWR}}; A), \tag{4.4.75}$$

i.e., all these points lie on the (2d)*-efficient frontier of* (4.4.40) *(see Definition 3.8).*

Proof The endpoints of these three curves, i.e., the (vector risk, utility) values of $x_{\mathrm{minVar}}, x_{\mathrm{minDD}}$, and x_{maxGr} (see Definition 4.36), are clearly $(2d)$-efficient points because these portfolios are, as unique global optimizers of (4.4.42)–(4.4.44), efficient portfolios in the sense of Definition 3.7. Note that the uniqueness here is essential, yielding, for instance,

$$\mathfrak{u}_{\ln\mathrm{TWR}}(x) < \mathfrak{u}_{\ln\mathrm{TWR}}(x_{\mathrm{maxGr}}) \qquad \text{for all } x \in A \setminus \{x_{\mathrm{maxGr}}\}. \tag{4.4.76}$$

Exemplarily for all other points, let us consider some arbitrary point on $\Gamma^{\mathrm{Var,TWR}}$, say

$$(\varrho, r, \mu) \in \Gamma^{\mathrm{Var,TWR}}, \ \varrho \in (\varrho_{\mathrm{minVar}}, \varrho_{\mathrm{maxGr}}). \tag{4.4.77}$$

According to (4.4.54), we have $\mu = \nu^{(1d)}_{\mathrm{Var}}(\varrho)$. Therefore, we obtain $(\varrho, \mu) \in \mathcal{G}^{(1d)}_{\mathrm{eff}}(\mathfrak{r}_{\mathrm{Var}}, \mathfrak{u}_{\ln\mathrm{TWR}}; A)$ (see (4.4.49)) and $x^* := X^{(1d)}_{\mathrm{Var}}(\varrho, \mu) \in A$ is the unique efficient portfolio for (4.4.48) realizing the (variance, utility) value (ϱ, μ), i.e., $\mathfrak{r}_{\mathrm{Var}}(x^*) = \varrho$ and $\mathfrak{u}_{\ln\mathrm{TWR}}(x^*) = \mu$. Thus, by (4.4.54) $\mathfrak{r}_{\mathrm{drawdown}}(x^*) = r$. Note that this also implies that there cannot be any portfolio $\widetilde{x} \in A$ with

$$\left(\mathfrak{r}_{\mathrm{Var}}(\widetilde{x}), \mathfrak{r}_{\mathfrak{r}_{\mathrm{drawdown}}}(\widetilde{x}), \mathfrak{u}_{\ln\mathrm{TWR}}(\widetilde{x})\right) = (\varrho, \widetilde{r}, \mu) \quad \text{and} \quad \widetilde{r} \neq r. \tag{4.4.78}$$

Since x^* is $(1d)$-efficient for (4.4.48), we obtain from Definition 2.59 that there cannot be any $x' \in A$ such that either

$$\left[\mathfrak{r}_{\mathrm{Var}}(x') \leq \mathfrak{r}_{\mathrm{Var}}(x^*) = \varrho \quad \text{and} \quad \mathfrak{u}_{\ln\mathrm{TWR}}(x') > \mathfrak{u}_{\ln\mathrm{TWR}}(x^*) = \mu\right] \tag{4.4.79}$$

or

$$\left[\mathfrak{r}_{\mathrm{Var}}(x') < \mathfrak{r}_{\mathrm{Var}}(x^*) = \varrho \quad \text{and} \quad \mathfrak{u}_{\ln\mathrm{TWR}}(x') \geq \mathfrak{u}_{\ln\mathrm{TWR}}(x^*) = \mu\right] \tag{4.4.80}$$

Setting $\mathfrak{r} := (\mathfrak{r}_{\mathrm{Var}}, \mathfrak{r}_{\mathrm{drawdown}})$, we obtain from (4.4.79) that there cannot be any $x' \in A$ with

$$\mathfrak{r}(x') \leq \mathfrak{r}(x^*) \quad \text{and} \quad \mathfrak{u}_{\ln\mathrm{TWR}}(x') > \mathfrak{u}_{\ln\mathrm{TWR}}(x^*). \tag{4.4.81}$$

To show that x^* is indeed $(2d)$-efficient (and hence its (vector risk, utility) value $(\varrho, r, \mu) \in \mathcal{G}^{(2d)}_{\mathrm{eff}}(\mathfrak{r}_{\mathrm{Var}}, \mathfrak{r}_{\mathrm{drawdown}}, \mathfrak{u}_{\ln\mathrm{TWR}}; A)$ as claimed), it therefore remains to show that there cannot exist any $x' \in A$ with

$$\mathfrak{r}(x') \leq \mathfrak{r}(x^*), \ \mathfrak{r}(x') \neq \mathfrak{r}(x^*) \quad \text{and} \quad \mathfrak{u}_{\ln\mathrm{TWR}}(x') \geq \mathfrak{u}_{\ln\mathrm{TWR}}(x^*). \tag{4.4.82}$$

cf. Definition 3.7.

So assume for the contrary that $x' \in A$ satisfies (4.4.82). We cannot have $\mathfrak{r}_{\mathrm{Var}}(x') < \mathfrak{r}_{\mathrm{Var}}(x^*)$ according to (4.4.80). Hence, $\mathfrak{r}_{\mathrm{drawdown}}(x') < \mathfrak{r}_{\mathrm{drawdown}}(x^*)$. On the other hand,

$$\mathfrak{r}_{\mathrm{Var}}(x') \leq \mathfrak{r}_{\mathrm{Var}}(x^*) = \varrho \quad \text{and} \quad \mathfrak{u}_{\ln\mathrm{TWR}}(x') \geq \mathfrak{u}_{\ln\mathrm{TWR}}(x^*) = \mu \tag{4.4.83}$$

imply $x' = x^*$ since x^* is the unique portfolio with that property as $(\varrho, \mu) \in \mathcal{G}_{\mathrm{eff}}^{(1d)}(\mathfrak{r}_{\mathrm{Var}}, \mathfrak{u}_{\ln\mathrm{TWR}};\, A)$, a contradiction (see also (4.4.78)). The arguments for the other Γ curves are similar. □

Similar to (3.1.21), we define the projection

$$\begin{aligned} \mathrm{Proj}_{\mathbb{R}^2} : \mathbb{R}^3 &\to \mathbb{R}^2 \\ (\varrho, r, \mu) &\mapsto (\varrho, r) \end{aligned} \tag{4.4.84}$$

to the risk $\mathfrak{r} = (\mathfrak{r}_{\mathrm{Var}}, \mathfrak{r}_{\mathrm{drawdown}})$ space and the projection of the efficient frontier

$$N = N\left(\mathfrak{r}, \mathfrak{u}_{\ln\mathrm{TWR}};\, A\right) = \mathrm{Proj}_{\mathbb{R}^2}\left(\mathcal{G}_{\mathrm{eff}}^{(2d)}(\mathfrak{r}_{\mathrm{Var}}, \mathfrak{r}_{\mathrm{drawdown}}, \mathfrak{u}_{\ln\mathrm{TWR}};\, A)\right) \subset \mathbb{R}^2. \tag{4.4.85}$$

Let furthermore

$$\widetilde{\Gamma}_{\mathrm{eff}}^{\mathrm{Var,TWR}} := \mathrm{Proj}_{\mathbb{R}^2}\left(\Gamma_{\mathrm{eff}}^{\mathrm{Var,TWR}}\right) \subset N \subset \mathbb{R}^2, \tag{4.4.86}$$

$$\widetilde{\Gamma}_{\mathrm{eff}}^{\mathrm{DD,TWR}} := \mathrm{Proj}_{\mathbb{R}^2}\left(\Gamma_{\mathrm{eff}}^{\mathrm{DD,TWR}}\right) \subset N \subset \mathbb{R}^2 \tag{4.4.87}$$

and

$$\widetilde{\Gamma}_{\mathrm{eff}}^{\mathrm{Var,DD}} := \mathrm{Proj}_{\mathbb{R}^2}\left(\Gamma_{\mathrm{eff}}^{\mathrm{Var,DD}}\right) \subset N \subset \mathbb{R}^2 \tag{4.4.88}$$

be the respective projections of the three Γ curves. Moreover, using $\nu^{(2d)} : \mathbb{R}^2 \to \mathbb{R} \cup \{-\infty\}$ with

$$\nu^{(2d)}(\varrho, r) := \sup\left\{\mu \,:\, (\varrho, r, \mu) \in \mathcal{G}^{(2d)}\left(\mathfrak{r}_{\mathrm{Var}}, \mathfrak{r}_{\mathrm{drawdown}}, \mathfrak{u}_{\ln\mathrm{TWR}};\, A\right)\right\} \tag{4.4.89}$$

as in (3.1.8), we have the representation

$$\mathrm{graph}(\nu_{|N}^{(2d)}) = \mathcal{G}_{\mathrm{eff}}^{(2d)}\left(\mathfrak{r}_{\mathrm{Var}}, \mathfrak{r}_{\mathrm{drawdown}}, \mathfrak{u}_{\ln\mathrm{TWR}};\, A\right) \tag{4.4.90}$$

(see (3.1.23)). Note that the standing assumption of Sect. 3.1, i.e., Assumption 3.11, is satisfied for $\mathfrak{r} = (\mathfrak{r}_{\mathrm{Var}}, \mathfrak{r}_{\mathrm{drawdown}})$, $\mathfrak{u} = \mathfrak{u}_{\ln\mathrm{TWR}}$, and A due to Assumption 4.31.

Lemma 4.41 (Properties of the $\widetilde{\Gamma}$ Curves) *Under Assumption 4.38, we have*

$$\widetilde{\Gamma} := \widetilde{\Gamma}_{\mathrm{eff}}^{\mathrm{Var,TWR}} \cup \widetilde{\Gamma}_{\mathrm{eff}}^{\mathrm{DD,TWR}} \cup \widetilde{\Gamma}_{\mathrm{eff}}^{\mathrm{Var,DD}} \subset N \subset \mathbb{R}^2 \tag{4.4.91}$$

which is a continuous, closed curve in $\mathbb{R}^2$ with no self-intersections. Hence, according to Jordan's curve theorem, it has an interior domain $S \subset \mathbb{R}^2$ which is non-empty, simply connected, and open with boundary $\partial S = \widetilde{\Gamma}$ (cf. Fig. 4.9).

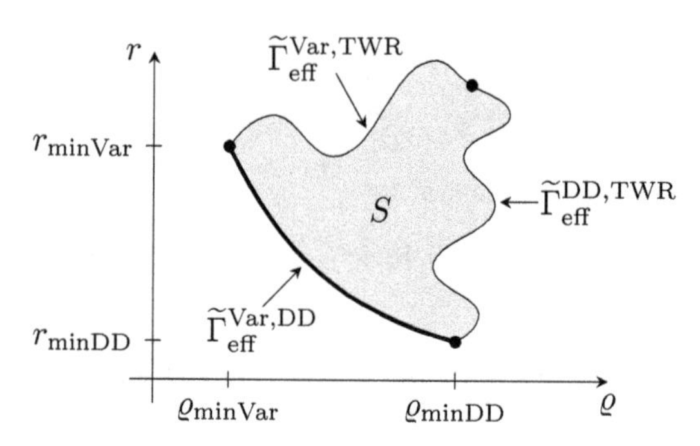

Fig. 4.9 Interior domain S of $\widetilde{\Gamma}$

Remark 4.42 *Note that only $\widetilde{\Gamma}_{\text{eff}}^{\text{Var,DD}}$ of* (4.4.88) *has monotonic r and ϱ components (cf.* (4.4.70) *and Fig. 4.7b). On the other hand, $\widetilde{\Gamma}_{\text{eff}}^{\text{Var,TWR}}$ has only monotonically increasing ϱ component, whereas $\widetilde{\Gamma}_{\text{eff}}^{\text{DD,TWR}}$ has only monotonically increasing r component (see Figs. 4.6a and 4.6b).*

Proof (of Lemma 4.41) Equation (4.4.91) is a direct consequence of (4.4.75) and (4.4.90). According to Assumption 4.38, Γ is a closed curve with no self-intersections. Thus, any self-intersections of $\widetilde{\Gamma} = \text{Proj}_{\mathbb{R}^2}(\Gamma) \subset N$ would contradict (4.4.75) and the fact that $\mathcal{G}_{\text{eff}}^{(2d)}(\mathfrak{r}_{\text{Var}}, \mathfrak{r}_{\text{drawdown}}, \mathfrak{u}_{\ln\text{TWR}}; A)$ equals the graph of $\nu_{|N}^{(2d)}$ (cf. (4.4.90)). So $\widetilde{\Gamma}$ is a continuous, closed curve in $\mathbb{R}^2$ with no self-intersections, as is Γ in $\mathbb{R}^3$. Therefore, Jordan's curve theorem is applicable. □

In order to prove our main result, Theorem 4.39, we next need to show that the closure of the interior domain S of Lemma 4.41 is in fact $N = N(\mathfrak{r}_{\text{Var}}, \mathfrak{r}_{\text{drawdown}}, \mathfrak{u}_{\ln\text{TWR}}; A)$ (cf. (4.4.85)). To start with, we show:

Lemma 4.43 *Under Assumption 4.38, we have $\overline{S}^c \cap N = \emptyset$.*

Proof Assume for the contrary that $(\varrho^*, r^*) \in \overline{S}^c \cap N$. Then, by (4.4.90) there exists some $\mu^* \in \mathbb{R}$ with $(\varrho^*, r^*, \mu^*) \in \mathcal{G}_{\text{eff}}^{(2d)}(\mathfrak{r}_{\text{Var}}, \mathfrak{r}_{\text{drawdown}}, \mathfrak{u}_{\ln\text{TWR}}; A)$. So let $x^* \in A$ be any efficient portfolio which realizes the (vector risk, utility) value (ϱ^*, r^*, μ^*), i.e.,

$$\mathfrak{r}_{\text{Var}}(x^*) = \varrho^*, \quad \mathfrak{r}_{\text{drawdown}}(x^*) = r^* \quad \text{and} \quad \mathfrak{u}_{\ln\text{TWR}}(x^*) = \mu^*. \tag{4.4.92}$$

We distinguish four regions where (ϱ^*, r^*) might be included (cf. Fig. 4.10).

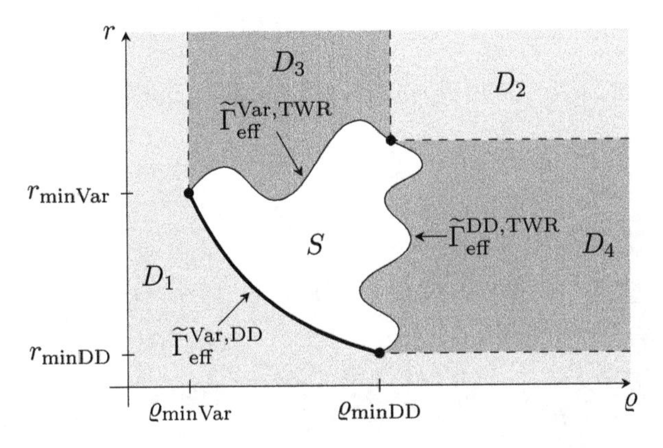

Fig. 4.10 Segmentation of $\overline{S}^c$

We may choose D_1 and D_2 open and $D_3 \dot\cup \widetilde{\Gamma}_{\text{eff}}^{\text{Var,TWR}}$ as well as $D_4 \dot\cup \widetilde{\Gamma}_{\text{eff}}^{\text{DD,TWR}}$ closed such that

$$\mathbb{R}^2 = \overline{S} \dot\cup D_1 \dot\cup D_2 \dot\cup D_3 \dot\cup D_4. \tag{4.4.93}$$

Case 1: $(\varrho^*, r^*) \in D_1$.
Going back to Fig. 4.7a, this means that

$$(\widetilde{s}, \widetilde{\mu}) := (\varrho^*, -r^*) \notin \mathcal{G}^{(1d)}\left(\mathfrak{r}_{\text{Var}}, -\mathfrak{r}_{\text{drawdown}};\, A\right) \tag{4.4.94}$$

which is impossible since $\mathfrak{r}_{\text{Var}}(x^*) = \varrho^*$ and $\mathfrak{r}_{\text{drawdown}}(x^*) = r^*$.

Case 2: $(\varrho^*, r^*) \in D_2$.
We here have $\varrho^* \geq \varrho_{\text{maxGr}}$ and $r^* \geq r_{\text{maxGr}}$ since the corner of D_2 stems from the risk values of x_{maxGr} (cf. (4.4.47)). Thus, using (4.4.75) we get $(\varrho_{\text{maxGr}}, r_{\text{maxGr}}, \mu_{\text{maxGr}}) \in \Gamma \subset \mathcal{G}_{\text{eff}}^{(2d)}\left(\mathfrak{r}_{\text{Var}}, \mathfrak{r}_{\text{drawdown}}, \mathfrak{u}_{\ln\text{TWR}};\, A\right)$ and it follows that

$$\begin{aligned} \mathfrak{u}_{\ln\text{TWR}}(x^*) &\overset{(4.4.92)}{=} \mu^* \overset{(4.4.90)}{=} \nu^{(2d)}(\varrho^*, r^*) \\ &\overset{\nu^{(2d)}\text{increasing}}{\geq} \nu^{(2d)}\left(\varrho_{\text{maxGr}}, r_{\text{maxGr}}\right) \overset{(4.4.90)}{=} \mu_{\text{maxGr}}. \end{aligned} \tag{4.4.95}$$

But since the solution x_{maxGr} of (4.4.44) is unique, we get $x^* = x_{\text{maxGr}}$, a contradiction.

Case 3: $(\varrho^*, r^*) \in D_3$.
Note that D_3 lies above $\widetilde{\Gamma}_{\text{eff}}^{\text{Var,TWR}} = \text{Proj}_{\mathbb{R}^2}\left(\Gamma_{\text{eff}}^{\text{Var,TWR}}\right)$. Thus, in this case, we have $\varrho^* \in \left[\varrho_{\text{minVar}}, \varrho_{\text{maxGr}}\right]$ and setting $\widetilde{x} := X_{\text{Var}}^{(1d)}\left(\varrho^*, \nu_{\text{Var}}^{(1d)}(\varrho^*)\right) \in A$, we get from (4.4.52)

$$r^* > \mathfrak{r}_{\text{drawdown}}(\widetilde{x}). \tag{4.4.96}$$

Also, $\widetilde{x}$ solves (4.4.48) with $\varrho = \varrho^*$, i.e., it is the unique solution of

$$\begin{aligned} &\max_{x\in A}\ \mathfrak{u}_{\ln\text{TWR}}(x) \\ &\text{subject to } \mathfrak{r}_{\text{Var}}(x) \leq \varrho^*. \end{aligned} \tag{4.4.97}$$

Now using that $\widetilde{x}$ has (vector risk, utility) values $\left(\varrho^*, \mathfrak{r}_{\text{drawdown}}(\widetilde{x}), \nu^{(1d)}_{\text{Var}}(\varrho^*)\right)$ (see (4.4.54)), we obtain

$$\begin{aligned} \mathfrak{u}_{\ln\text{TWR}}(x^*) &\overset{(4.4.92)}{=} \mu^* \overset{(4.4.90)}{=} \nu^{(2d)}(\varrho^*, r^*) \overset{\nu^{(2d)}\text{increasing}}{\geq} \nu^{(2d)}(\varrho^*, \mathfrak{r}_{\text{drawdown}}(\widetilde{x})) \\ &= \nu^{(2d)}(\mathfrak{r}_{\text{Var}}(\widetilde{x}), \mathfrak{r}_{\text{drawdown}}(\widetilde{x})) \geq \mathfrak{u}_{\ln\text{TWR}}(\widetilde{x}) = \nu^{(1d)}_{\text{Var}}(\varrho^*). \end{aligned} \tag{4.4.98}$$

But since x^* is also eligible for (4.4.97) which has the unique solution $\widetilde{x}$, we must have $x^* = \widetilde{x}$, a contradiction.

Case 4: $(\varrho^*, r^*) \in D_4$.
This case is completely symmetric with $\mathfrak{r}_{\text{Var}}$ and $\mathfrak{r}_{\text{drawdown}}$ exchanged. We leave the details to the reader. □

The next lemma finally yields $\overline{S} = N$ where S is defined in Lemma 4.41.

Lemma 4.44 *Under Assumption 4.38, we get that* $\nu^{(2d)}\colon \overline{S} \to \mathbb{R},\ (\varrho, r) \mapsto \nu^{(2d)}(\varrho, r)$ *from* (4.4.89) *is*

(a) continuous and strictly increasing in ϱ *(for fixed* r*).*
(b) continuous and strictly increasing in r *(for fixed* ϱ*).*

Moreover, we have

(c) $\overline{S} = N = N(\mathfrak{r}_{\text{Var}}, \mathfrak{r}_{\text{drawdown}}, \mathfrak{u}_{\ln\text{TWR}};\ A)$ *(see* (4.4.85)*).*

Proof Since (a) and (b) work similarly, we only prove (b).
ad (b). Let $\varrho_0 \in \mathbb{R}$ be fixed such that $S_{\varrho_0} := \{r : (\varrho_0, r) \in S\} \neq \emptyset$. We set

$$r_* := \inf S_{\varrho_0} > -\infty \quad \text{and} \quad r^* := \sup S_{\varrho_0} < +\infty, \tag{4.4.99}$$

cf. Fig. 4.11.

Note that both (r_*, ϱ_0) and (r^*, ϱ_0) lie on $\widetilde{\Gamma} = \text{Proj}_{\mathbb{R}^2}(\Gamma) \subset N$ defined in (4.4.91) and thus are vector risk values of points on the efficient frontier $\mathcal{G}^{(2d)}_{\text{eff}} = \mathcal{G}^{(2d)}_{\text{eff}}(\mathfrak{r}_{\text{Var}}, \mathfrak{r}_{\text{drawdown}}, \mathfrak{u}_{\ln\text{TWR}};\ A)$; see (4.4.75). We have to show that $\nu^{(2d)}(\varrho_0, \cdot)$ defined in (4.4.89) is continuous and strictly increasing on S_{ϱ_0}, where increasing generally holds (see Proposition 3.12 (b)), only strictness is the issue. We may assume $r_* < r^*$, because otherwise there is nothing to show. According to (3.1.9), we can rewrite

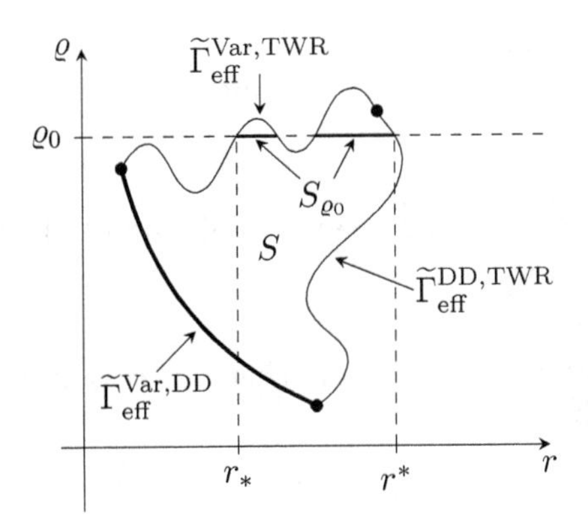

Fig. 4.11 The sets S and S_{ϱ_0}

$$\nu^{(2d)}(\varrho_0, r) = \sup\{\mathfrak{u}_{\ln\mathrm{TWR}}(x) : \mathfrak{r}_{\mathrm{Var}}(x) \le \varrho_0,\ \mathfrak{r}_{\mathrm{drawdown}}(x) \le r,\ x \in A\} \tag{4.4.100}$$

$$= \sup\{\mathfrak{u}_{\ln\mathrm{TWR}}(x) : \mathfrak{r}_{\mathrm{drawdown}}(x) \le r,\ x \in A_{\varrho_0}\}, \tag{4.4.101}$$

where we set $A_{\varrho_0} := \{x \in A : \mathfrak{r}_{\mathrm{Var}}(x) \le \varrho_0\}$, which is convex, closed, and non-empty, hence admissible in the sense of Definition 2.7. Moreover, with Assumption 4.31, we obtain the standing assumption of the scalar risk case, Assumption 2.64, for $A_{\varrho_0}, \mathfrak{r} = \mathfrak{r}_{\mathrm{drawdown}}$ and $\mathfrak{u} = \mathfrak{u}_{\ln\mathrm{TWR}}$. Therefore, problem (4.4.101) has an $(1d)$-efficient frontier $\mathcal{G}_{\mathrm{eff}}^{(1d)}(\mathfrak{r}_{\mathrm{drawdown}}, \mathfrak{u}_{\ln\mathrm{TWR}}; A_{\varrho_0})$ which by Theorem 2.76 (c) can be parameterized as a graph of a concave function $\nu = \nu_{\mathrm{DD},\varrho_0}^{(1d)}$ on an intervall $I = I_{\mathrm{DD},\varrho_0} \subset \mathbb{R}$, i.e.,

$$\mathcal{G}_{\mathrm{eff}}^{(1d)}(\mathfrak{r}_{\mathrm{drawdown}}, \mathfrak{u}_{\ln\mathrm{TWR}}; A_{\varrho_0}) = \operatorname{graph}\left(\nu_{\mathrm{DD},\varrho_0}^{(1d)}\Big|_{I_{\mathrm{DD},\varrho_0}}\right). \tag{4.4.102}$$

Since (r_*, ϱ_0) and (r^*, ϱ_0) are vector risk values of $(2d)$-efficient portfolios of (4.4.100), clearly r_* and r^* are risk = drawdown values of the same portfolios, now $(1d)$-efficient for (4.4.101) (see also Lemma 3.22). Therefore, both r_* and r^* have to lie in $I_{\mathrm{DD},\varrho_0}$, i.e., $[r_*, r^*] \subset I_{\mathrm{DD},\varrho_0}$. However, according to Theorem 2.76, $\nu_{\mathrm{DD},\varrho_0}^{(1d)}$ is continuous and strictly increasing on $I_{\mathrm{DD},\varrho_0}$ and with (2.3.14), we have

$$\nu_{\mathrm{DD},\varrho_0}^{(1d)}(r) = \sup\{\mathfrak{u}_{\ln\mathrm{TWR}}(x) : \mathfrak{r}_{\mathrm{drawdown}}(x) \le r,\ x \in A_{\varrho_0}\}, \tag{4.4.103}$$

so that $\nu^{(2d)}(\varrho_0, r)$, $r \in [r_*, r^*]$ is continuous and strictly increasing in r as well (see (4.4.100) and (4.4.101)). This concludes the proof since $S_{\varrho_0} \subset [r_*, r^*]$.

ad (c). We already know that $\overline{S}^c \cap N = \emptyset$ (Lemma 4.43) and $\partial S \subset N$ (Lemma 4.41). It only remains to show that $S \overset{!}{\subset} N$. But from (a) and (b), we get that $\nu^{(2d)}$ is strictly increasing in both variables for all $(\varrho, r) \in S$.

Hence, given $(\varrho, r) \in S$ arbitrary, $\left(\varrho, r, \nu^{(2d)}(\varrho, r)\right)$ are $(2d)$-efficient points as a direct consequence of the definition of efficient portfolios (cf. Definition 3.7). That is,

$$\left(\varrho, r, \nu^{(2d)}(\varrho, r)\right) \in \mathcal{G}_{\text{eff}}^{(2d)}\left(\mathfrak{r}_{\text{Var}}, \mathfrak{r}_{\text{drawdown}}, \mathfrak{u}_{\ln\text{TWR}}; A\right) \tag{4.4.104}$$

and therefore $(\varrho, r) \in N$ follows. □

We have by now collected all ingredients necessary for the proof of the main theorem in this subsection.

Proof (of Theorem 4.39) Using $\overline{S} = N$ from the above lemma and (4.4.90), we obtain

$$\mathcal{G}_{\text{eff}}^{(2d)} = \mathcal{G}_{\text{eff}}^{(2d)}\left(\mathfrak{r}_{\text{Var}}, \mathfrak{r}_{\text{drawdown}}, \mathfrak{u}_{\ln\text{TWR}}; A\right) = \text{graph}\left(\nu^{(2d)}_{|\overline{S}}\right). \tag{4.4.105}$$

Furthermore, $\nu^{(2d)}_{|\overline{S}}$ is continuous according to Lemma 4.44 (a) and (b). Since $\overline{S} = S \,\dot{\cup}\, \widetilde{\Gamma}$ is a closed set, simply connected, and with non-empty interior (cf. Lemma 4.41), by (4.4.105), $\mathcal{G}_{\text{eff}}^{(2d)}$ is a two-dimensional surface, which is also a closed set and simply connected and has $\Gamma = \Gamma_{\text{eff}}^{\text{Var,DD}} \cup \Gamma_{\text{eff}}^{\text{Var,TWR}} \cup \Gamma_{\text{eff}}^{\text{DD,TWR}}$ as boundary, since $\text{Proj}_{\mathbb{R}^2}(\Gamma) = \widetilde{\Gamma}$ (see (4.4.86)–(4.4.88)). □

Remark 4.45 *The properties of* $\mathcal{G}_{\text{eff}}^{(2d)}$ *in Theorem 4.39 are much better than in the general theory of Sect. 3.1 (simply connected and closed set vs. just path-connected (cf. Theorem 3.43) and not necessarily closed (see Remark 3.18); also,* $\nu^{(2d)}_{|N}$ *here is even continuous on* $\overline{N}$ *not only on the relative interior of* N *and some parts of the boundary (see Corollary 3.24)).*

Remark 4.46 *As already noted before, we get that both* (4.4.43) *and* (4.4.44) *with* A *replaced by* $\widehat{\mathcal{A}}_1$ *(cf.* (4.4.41)*) also have unique global optimizers under generic conditions (see Remark 4.33). In this case, then Theorem 4.39 holds for* $A = \widehat{\mathcal{A}}_1$ *as well. Furthermore,* $\mathcal{G}_{\text{eff}}^{(2d)} = \mathcal{G}_{\text{eff}}^{(2d)}(\mathfrak{r}_{\text{Var}}, \mathfrak{r}_{\text{drawdown}}, \mathfrak{u}_{\ln\text{TWR}}; A)$ *is even independent of* A*, provided* A *is sufficiently large.*

Remark 4.47 *An alternative of* $\mathfrak{u}_{\ln\text{TWR}}$ *as utility in* (4.4.40) *which guarantees a unique global maximizer is the expected* log *utility*

$$\mathfrak{u}_{\ln\text{EXP}}(x) := \text{E}\left[\ln(S_1^\top x)\right], \quad x \in \widehat{\mathcal{A}}_1,$$

of a one-period financial market with no nontrivial riskless portfolio (see Lemma 2.34 and [78, Theorem 12]). Similarly, the tracking error

$$\mathfrak{r}_{\text{Track}}(x) := \frac{1}{2}(\widehat{x} - \widehat{x}^*)^\top \Sigma (\widehat{x} - \widehat{x}^*), \quad x \in \widehat{\mathcal{A}}_1$$

with a prescribed benchmark $\widehat{x}^* \in \mathbb{R}^M$ *and the positive definite covariant matrix* Σ *has a unique global minimizer. Therefore, if we replace* (4.4.40) *by*

$$\begin{aligned} \min_{x=(0,\widehat{x}^\top)^\top \in A} \quad & \mathfrak{r}_{\text{Track}}(x) \\ \text{subject to } & \mathfrak{u}_{\text{lnEXP}}(x) \geq \mu \\ & \mathfrak{r}_{\text{Var}}(x) := \frac{1}{2}\widehat{x}^\top \Sigma \widehat{x} \leq \varrho, \end{aligned} \tag{4.4.106}$$

both risk and utility functions have global optimizers and a similar version of Theorem 4.39 holds accordingly for $\mathcal{G}_{\text{eff}}^{(2d)} = \mathcal{G}_{\text{eff}}^{(2d)}(\mathfrak{r}_{\text{Var}}, \mathfrak{r}_{\text{Track}}, \mathfrak{u}_{\text{lnEXP}}; A)$*. Clearly,* $\mathcal{G}_{\text{eff}}^{(2d)}$ *is independent for all sufficiently large* A*. In particular, it holds for* $A = \widehat{\mathcal{A}}_1$ *and gives a bounded efficient frontier* $\mathcal{G}_{\text{eff}}^{(2d)}$ *in this case as well. Furthermore, note that* (4.4.106) *for* $A = \widehat{\mathcal{A}}_1$ *has essentially the same efficient frontier (with just exchanged axes) as*

$$\begin{aligned} \frac{1}{2}\sigma^2 = \min_{x=(0,\widehat{x}^\top)^\top \in \widehat{\mathcal{A}}_1} \quad & \mathfrak{r}_{\text{Var}}(x) \\ \text{subject to } & \mathfrak{u}_{\text{lnEXP}}(x) \geq \mu \\ & \mathfrak{r}_{\text{Track}}(x) \leq \varrho, \end{aligned} \tag{4.4.107}$$

which is very similar to Example 3.50 with only $\mathfrak{u}_{\text{lnEXP}}$ *replaced by the linear expected utility* $\mathfrak{u}(x) = \mathrm{E}\left[S_1^\top(x)\right]$*. According to Theorem 4.39, and in particular according to Lemma 4.44, the projection of the efficient frontier for* (4.4.107) *to the risk space, i.e.,* $N = N(\mathfrak{r}_{\text{Var}}, \mathfrak{r}_{\text{Track}}, \mathfrak{u}_{\text{lnEXP}}; \widehat{\mathcal{A}}_1)$*, has a form similar to Figure 4.9, cf. Fig. 4.12. The efficient frontier in Example 3.50, however, completely differs from that of* (4.4.107)*. In particular, the efficient frontier in Example 3.50 is unbounded (see Fig. 3.13b for a comparison of the* N *set).*

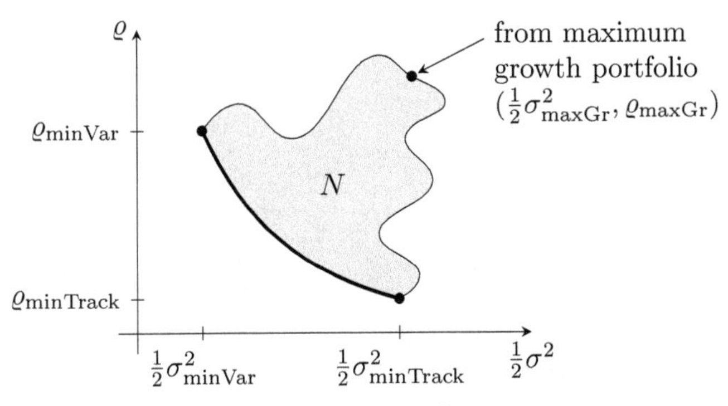

Fig. 4.12 Sketch of $N = N(\mathfrak{r}_{\text{Var}}, \mathfrak{r}_{\text{Track}}, \mathfrak{u}_{\text{lnEXP}}; \widehat{\mathcal{A}}_1)$

Chapter 5
Conclusion

As we are approaching the end of the description of this stage of our interdisciplinary collaboration, a look back may help the reader to obtain a more global view of the material provided in this monograph and furthermore have a glimpse of the road ahead.

A bank balance sheet management problem is what motivated us to start this line of investigation. Such a problem is different from the well-studied portfolio optimization problem in that it involves the control of multiple types of risk and, as a result, leads to a mathematical model of multi-objective optimization problems with convex data. Although numerical solutions to such problems do exist, we quickly realized that often such numerical solutions on their own are not what bank managers can effectively use in their practice. As explained in Sect. 1.1, the main reasons are twofold: firstly, a brute force numerical optimization of a bank balance sheet may very well lead to overfitting and secondly, balance sheet managers need to explain their decision and method to the administration and board. It is never persuasive to tell people that the balance sheet construction is produced by a black-box optimization algorithm.

So, what is helpful? For practicing bank balance sheet managers, useful mathematical models should provide guidance that points to the general principle and directions toward decision making. Furthermore, such guidance should have clear financial meaning that can be effectively communicated. This realization guides our endeavor in studying the Pareto efficient frontier which turns out to be the multidimensional generalization of the Markowitz bullet in the high-dimensional space where trade-off between reward and risks happens. The results reported in this monograph are thus naturally organized in two blocks: the theoretical discussion on the topological structure of the Pareto efficient frontier in Chaps. 2 and 3 and its applications including financial interpretations via convex duality in Chap. 4. The former tells us what is possible mathematically and the latter tells us how to implement this in a way that is intuitive to practitioners. The contents of these two blocks of

S. Maier-Paape et al., *Scalar and Vector Risk in the General Framework of Portfolio Theory*, CMS/CAIMS Books in Mathematics 9,
https://doi.org/10.1007/978-3-031-33321-7_5

material have already been highlighted in Sect. 1.4. We further want to point out that, for example, the corresponding discussions of Sects. 3.3.3 and 4.3.3 nicely illustrate how the theoretical discussion and practical application come together. Similarly, the example discussed in Sect. 4.4.5 renders beautifully how powerful the interplay of theory and practice can be. Moreover, Chap. 4 contains a bunch of explicitly solvable multi-risk problems, all of which heavily benefiting from the convex duality machinery.

As is often the case, this research project also leads to more questions as we are answering some of the initial motivating ones. We elaborate a few here that we find particularly interesting.

Firstly, our focus is mainly from the perspective of a bank balance sheet manager as already alluded to in Sect. 1.2. From this point of view, the main focus is how to trade off between risks and reward. There is, however, a different yet complementary perspective. That is the perspective of a regulator. Here the main concern is to control the risks of bank balance sheets so that they do not cause harmful interruptions to the financial markets. For this purpose, one needs to be able to describe safe or acceptable portfolios that set boundaries for bank managers. Typical results are those elaborated in Jouini, Meddeb, and Touzi [63] and Hamel and Heyde [56] as well as Hamel, Heyde, and Rudloff [57]. They extend convex analysis and duality tools to vector-valued and set-valued mappings to achieve the goal. How these two different perspectives do interact is an interesting direction worthy of further exploration.

Secondly, roughly speaking, the center theoretical result in Chap. 3 asserts that for a multi-objective optimization model related to bank balance sheet problems with multiple types of risks, the Pareto efficient set where the trade-off between reward and risks happens is path-connected. In the bank balance sheet management problem, this tells us that one can trade off reward and risks continuously. This is similar to the model that involves only one risk. But unlike the one risk model, somewhat surprisingly, the multi-risk Pareto efficient set may not have a representation in terms of convex functions. Thus, however, such a multi-risk trade-off may not enjoy stronger conditions such as a local Lipschitz condition that are typically associated with a convex representation. Moreover, the lack of convexity in representation also means that there may not be an easy way of finding a direct parameterization of transitioning from one Pareto efficient point to another, although granted by theory. However, for the linear model discussed in Sect. 4.2, the corresponding Pareto efficient set, in fact, has a piecewise linear representation. Likewise, the efficient frontier for the tracking error approach also bears nice structure (see Sect. 4.3.3). Thus, finding better structure for concrete balance sheet management problems is important for applying the general theory. This requires case studies related to the concrete structure of the risk and reward functions.

Furthermore, although our investigation is mainly motivated by the bank balance sheet management problem, the methods and results described here should also be applicable to other problems such as investment funds with or without illiquid assets, multiple risks arising in corporations, household finance risks, and liability management for governments (see Examples 4.1–4.4). For instance, fund managers or even private investors with large portfolios may benefit from multi-risk approaches, e.g., the here demonstrated ansatz with the tracking error.

Finally, we hope that this monograph will stimulate further interest in such interdisciplinary research.

Appendix A
Convex Programming Problems

In financial problems, one often has to deal with convex risk measures and concave utility functions. Thus, convex analysis becomes a crucial tool. In this appendix for convenience of the reader, we collect some well-known theory from convex analysis (cf. e.g., Rockafellar [90], Borwein and Lewis [18], Borwein and Zhu [19] and Carr and Zhu [25]). In particular, we will discuss the subjects semi-continuity, convexity, and duality in more detail.

A.1 Semi-continuity

Let for simplicity $\mathcal{X}$ be a finite-dimensional (real) Banach space, i.e., essentially $\mathcal{X} = \mathbb{R}^m$ and $B_\delta(x_0) := \{y \in \mathcal{X} : \|y - x_0\| < \delta\}$ is a ball around x_0 with radius $\delta > 0$. Recall that an *extended-valued function* $f\colon \mathcal{X} \to \mathbb{R} \cup \{\pm\infty\}$ is called *lower semi-continuous* (l.s.c.) at $x_0 \in \mathcal{X}$, if

$$f(x_0) \leq \liminf_{y \searrow x_0} f(y) \overset{\text{def}}{=} \lim_{\delta \searrow 0} \left[\inf \{f(y) : y \in B_\delta(x_0) \setminus \{x_0\}\}\right], \quad \text{(A.1.1)}$$

and f is called lower semi-continuous if it is l.s.c. for all $x \in \mathcal{X}$.

Alternative, but equivalent, characterizations for lower semi-continuity as stated in Lemma A.1 and Lemma A.2 are straightforward (see, e.g., [90, Section 7]).

Lemma A.1 *Let $f\colon \mathcal{X} \to \mathbb{R} \cup \{\pm\infty\}$ and $x_0 \in \mathcal{X}$ be given. Then the following properties are equivalent:*

(a) *f is lower semi-continuous at x_0.*

(b) $$f(x_0) = \lim_{\delta \searrow 0} \left[\inf \{f(y) : y \in B_\delta(x_0)\}\right] \quad \text{(A.1.2)}$$

S. Maier-Paape et al., *Scalar and Vector Risk in the General Framework of Portfolio Theory*, CMS/CAIMS Books in Mathematics 9,
https://doi.org/10.1007/978-3-031-33321-7

Lemma A.2 *An extended-valued function* $f : \mathcal{X} \to \mathbb{R} \cup \{\pm\infty\}$ *is lower semi-continuous at* $x \in \mathcal{X}$, *if and only if*

$$f(x) \leq \lim_{i\to\infty} f(x_i) \tag{A.1.3}$$

for every sequence $(x_i)_{i\in\mathrm{N}} \subset \mathcal{X}$ *with* $\lim_{i\to\infty} x_i = x$ *and for which the limit of* $(f(x_i))_{i\in\mathrm{N}}$ *exists in* $\mathbb{R} \cup \{\pm\infty\}$.

The following theorem characterizes lower semi-continuity in terms of sublevel sets or epigraphs.

Theorem A.3 (Rockafellar [90, Theorem 7.1]) *Let* f *be any extended-valued function* $f : \mathcal{X} \to \mathbb{R} \cup \{\pm\infty\}$. *Then the following conditions are equivalent:*

(a) f *is lower semi-continuous on the whole of* $\mathcal{X}$.
(b) Sublevel sets $\{x \in \mathcal{X} : f(x) \leq \alpha\} \subset \mathcal{X}$ *are closed for all* $\alpha \in \mathbb{R}$.
(c) The epigraph *of* f,

$$\operatorname{epi}(f) := \{(x, r) \in \mathcal{X} \times \mathbb{R} : f(x) \leq r\}, \tag{A.1.4}$$

is a closed set in $\mathcal{X} \times \mathbb{R}$.

Remark A.4 *A similar theorem holds true for a lower semi-continuous function* $g : \mathcal{Y} \to \mathbb{R} \cup \{\pm\infty\}$ *when* $\mathcal{Y} \subset \mathcal{X}$ *is a closed subset of* $\mathcal{X}$. *To see this, just consider* $f : \mathcal{X} \to \mathbb{R} \cup \{\pm\infty\}$ *defined by*

$$f(x) := \begin{cases} g(x), & x \in \mathcal{Y}, \\ \infty, & x \in \mathcal{Y}^c, \end{cases}$$

which is lower semi-continuous on the whole of $\mathcal{X}$ *(e.g., by Lemma A.2), and thus Theorem A.3 applies. In particular, for instance, the sublevel sets* $\{y \in \mathcal{Y} : g(y) \leq \alpha\}$ *are always closed subsets of* $\mathcal{X}$.

On the other hand, closed epigraphs provide lower semi-continuous functions.

Proposition A.5 ([78, Proposition 1]) *Let* F *be a closed subset of* $\mathcal{X} \times \mathbb{R}$ *such that*

$$\inf\{r : (x, r) \in F\} > -\infty \quad \textit{for all} \quad x \in \mathcal{X}. \tag{A.1.5}$$

Then the following are equivalent:

(i) F *is the epigraph of a lower semi-continuous function* $f : \mathcal{X} \to \mathbb{R} \cup \{+\infty\}$, *i.e.,* $F = \operatorname{epi}(f)$.
(ii)
$$(x, r) \in F \Rightarrow (x, r + k) \in F, \qquad \textit{for all} \;\; k > 0. \tag{A.1.6}$$

Moreover, the function f can be defined by

$$f(x) := \inf \{r : (x,r) \in F\}. \tag{A.1.7}$$

Definition A.6 *An extended-valued function $f\colon \mathcal{X} \to \mathbb{R} \cup \{\pm\infty\}$ is called* upper semi-continuous *if $g := -f$ is lower semi-continuous.*

Hence, for the *hypograph*

$$\mathrm{hypo}(g) := \{(x,r) \in \mathcal{X} \times \mathbb{R} : g(x) \geq r\} \tag{A.1.8}$$

of upper semi-continuous functions, a symmetric characterization holds true.

Proposition A.7 ([78, Proposition 2]) *Let G be a closed subset of $\mathcal{X} \times \mathbb{R}$ such that*

$$\sup \{r : (x,r) \in G\} < +\infty \quad \textit{for all} \quad x \in \mathcal{X}. \tag{A.1.9}$$

Then the following are equivalent:

(i) *G is the hypograph of an upper semi-continuous function $g\colon \mathcal{X} \to \mathbb{R} \cup \{-\infty\}$, i.e., $G = \mathrm{hypo}(g)$.*

(ii)
$$(x,r) \in G \Rightarrow (x, r-k) \in G, \qquad \textit{for all}\ \ k > 0. \tag{A.1.10}$$

Moreover, the function g can be defined by

$$g(x) := \sup \{r : (x,r) \in G\}. \tag{A.1.11}$$

The following theorem is helpful to obtain global minimizers.

Theorem A.8 (cf. Barbu and Precupanu [8, Theorem 2.8]) *Let $M \subset \mathbb{R}^n$ be compact and $f\colon M \to \mathbb{R}$ be lower semi-continuous. Then f attains its global minimum on M.*

A.2 Convexity

We next focus our interest on convexity: Let in the following $C \subset \mathcal{X}$ be a *convex set*, i.e.,

$$x, y \in C,\ s \in [0,1] \Rightarrow sx + (1-s)y \in C. \tag{A.2.1}$$

Definition A.9 (Domain of an Extended-Valued Function) *For an extended-valued function $f\colon C \to \mathbb{R} \cup \{\pm\infty\}$, we define its* domain *by*

$$\mathrm{dom}(f) := \{x \in C : f(x) \in \mathbb{R}\}. \tag{A.2.2}$$

Definition A.10 (Convex Functions)

(a) An extended-valued function $f: C \to \mathbb{R} \cup \{+\infty\}$ *is called* convex, *if for all* $x, y \in C$ *and* $\lambda \in (0, 1)$

$$f(\lambda x + (1-\lambda)y) \leq \lambda f(x) + (1-\lambda) f(y) \tag{A.2.3}$$

holds.

(b) $f: C \to \mathbb{R} \cup \{+\infty\}$ *is called* proper convex *if it is convex and* $\operatorname{dom}(f) = \{x \in C : f(x) < \infty\} \neq \emptyset$.

(c) It is called strictly convex *if it is proper convex and*

$$f(\lambda x + (1-\lambda)y) < \lambda f(x) + (1-\lambda) f(y)$$

whenever $x \neq y$, $\lambda \in (0, 1)$ *and* $x, y \in \operatorname{dom}(f)$.

Remark A.11

(a) Note that from Definition A.10 (a) immediately follows that for every convex function f, *its domain* $\operatorname{dom}(f)$ *is always a convex set.*

(b) Also note that convexity only makes sense for functions $f: C \to \mathbb{R} \cup \{+\infty\}$, *because any function with* (A.2.3) *and a value* $-\infty$ *at a single point cannot take any finite values elsewhere.*

(c) Clearly, any convex function $f: C \to \mathbb{R} \cup \{+\infty\}$ *can be extended trivially to the whole Banach space* $\mathcal{X}$ *by setting* $f_{\mathcal{X} \setminus C} := +\infty$, *and the extended function* $f: \mathcal{X} \to \mathbb{R} \cup \{+\infty\}$ *is still convex with the same domain as the original function.*

Example A.12 *The function* f, *defined by*

$$f(x) = \begin{cases} +\infty, & \text{for } x < 0, \\ 1, & \text{for } x = 0, \\ x^2, & \text{for } x > 0, \end{cases}$$

is (strictly) convex with $\operatorname{dom}(f) = [0, \infty)$, *but* f *is not lower (or upper) semi-continuous at* $x = 0$.

Lemma A.13 *A function* $f: \mathcal{X} \to \mathbb{R} \cup \{+\infty\}$ *is convex if and only if its epigraph* $\operatorname{epi}(f) \subset \mathcal{X} \times \mathbb{R}$ *is convex.*

Proof This is left to the reader as an exercise. □

Fortunately, problems with continuity (or lower semi-continuity) of convex functions only occur near points where f assumes $+\infty$, i.e., at the boundary of its domain.

In the next theorem, $\dim \mathcal{Z} = \infty$ is allowed.

Theorem A.14 (Borwein and Zhu [19, Theorem 4.1.3]; Rockafellar [90, Theorem 10.4])

(a) Let $\mathcal{Z}$ be a Banach space and let $f\colon \mathcal{Z} \to \mathbb{R} \cup \{+\infty\}$ be convex and lower semi-continuous. Then f is locally Lipschitz continuous on the interior of its domain, $\operatorname{int}(\operatorname{dom}(f))$.

(b) Let $\mathcal{X} \cong \mathbb{R}^m$ and $f\colon \mathcal{X} \to \mathbb{R} \cup \{+\infty\}$ be convex. Then f is locally Lipschitz continuous on $\operatorname{int}(\operatorname{dom}(f))$.

Remark A.15

(a) Hence, the discontinuities as in Example A.12 can only occur at the boundary of $\operatorname{dom}(f)$.

(b) In particular, $f\colon \mathbb{R}^m \to \mathbb{R}$ convex is always continuous.

Remark A.16 (Concave Functions) *We call an extended-valued function $f\colon \mathcal{X} \to \mathbb{R} \cup \{-\infty\}$* (proper, strictly) concave *if $-f$ is (proper, strictly) convex. The domain (cf.* (A.2.2)*) of a concave function thus is $\operatorname{dom}(f) = \{x \in \mathcal{X} : f(x) > -\infty\}$. Furthermore, due to Lemma A.13, f is concave if and only if $\operatorname{hypo}(f) \subset \mathcal{X} \times \mathbb{R}$ is convex.*

Example A.17 (Natural Logarithm) *The extended-valued function $f\colon \mathbb{R} \to \mathbb{R} \cup \{-\infty\}$, defined by*

$$f(x) := \begin{cases} \ln(x), & x > 0, \\ -\infty, & x \leq 0, \end{cases}$$

is concave (and upper/lower semi-continuous) with $\operatorname{dom}(f) = \mathbb{R}_{>0}$.

A.3 Convex Programming Problems

Since utility functions in portfolio problems are concave and risk functions are convex, the analysis of a general trade-off between utility and risk naturally leads to a *convex programming problem*:

$$v(y, z) := \inf_{x \in \mathcal{X}} \left[f(x) \,:\, g(x) \leq y \text{ and } h(x) = z \right] \text{ for fixed } y \in \mathbb{R}^k,\ z \in \mathbb{R}^n, \tag{A.3.1}$$

where f, g, and h satisfy:

Assumption A.18 (For Convex Programming Problem) *Assume that as always $\dim \mathcal{X} < \infty$ and furthermore:*

(i) $f\colon \mathcal{X} \to \mathbb{R} \cup \{+\infty\}$ is a lower semi-continuous extended-valued convex function.

(ii) $g\colon \mathcal{X} \to \mathbb{R}^k$ *is a vector-valued function with convex components and* $h\colon \mathcal{X} \to \mathbb{R}^n$, $x \mapsto Ax + b$, *is an affine mapping, for* $k, n \in \mathbb{N}$ *natural numbers,* $A \in \mathcal{L}(\mathcal{X}, \mathbb{R}^n)$ *and* $b \in \mathbb{R}^n$.

(iii) Moreover, at least one of the components of g *has compact sublevel sets, i.e.,* $\exists\, i_0$ *with* $1 \leq i_0 \leq k$ *such that* $g_{i_0}\colon \mathcal{X} \to \mathbb{R}$ *satisfies:*

$$\{x : g_{i_0}(x) \leq \alpha\} \subset \mathcal{X} \qquad \textit{is compact for all} \quad \alpha \in \mathbb{R}.$$

Under these assumptions, we get:

Proposition A.19 (Properties of Optimal Value Function, cf. Borwein and Zhu [20, Lemma 1]) *Let* $f, g,$ *and* h *satisfy Assumption A.18.*

Then the optimal value function $v\colon \mathbb{R}^k \times \mathbb{R}^n \to \mathbb{R} \cup \{+\infty\}$ *in the convex programming problem* (A.3.1) *is convex and lower semi-continuous. Furthermore, for any* $(y, z) \in \operatorname{dom}(v)$, *there exists a solution* $\overline{x} \in \mathcal{X}$ *to* (A.3.1), *i.e.,* $f(\overline{x}) = v(y, z)$ *with* $g(\overline{x}) \leq y$ *and* $h(\overline{x}) = z$.

Proof We leave the well-definedness of $v\colon \mathbb{R}^k \times \mathbb{R}^n \to \mathbb{R} \cup \{+\infty\}$, the lower semi-continuity of v, and the existence part to the reader. Here we show that v is convex: Consider

$$\left(y^{(i)}, z^{(i)}\right) \in \operatorname{dom}(v) = \left\{(y, z) \in \mathbb{R}^k \times \mathbb{R}^n \text{ with } v(y, z) < \infty\right\}, \; i = 1, 2.$$

For $\lambda \in (0, 1)$ we need to show

$$v\left(\lambda\left(y^{(1)}, z^{(1)}\right) + (1 - \lambda)\left(y^{(2)}, z^{(2)}\right)\right) \overset{!}{\leq} \lambda v\left(y^{(1)}, z^{(1)}\right) + (1 - \lambda)\, v\left(y^{(2)}, z^{(2)}\right). \tag{A.3.2}$$

Note that for $\left(y^{(i)}, z^{(i)}\right) \notin \operatorname{dom}(v)$, (A.3.2) is clear. Let $\varepsilon > 0$ be given and find $x_\varepsilon^{(i)} \in \mathcal{X}$, $i = 1, 2$, feasible for the constrained problem $v\left(y^{(i)}, z^{(i)}\right)$, i.e.,

$$g\left(x_\varepsilon^{(i)}\right) \leq y^{(i)}, \quad h\left(x_\varepsilon^{(i)}\right) = z^{(i)}, \tag{A.3.3}$$

such that

$$f\left(x_\varepsilon^{(i)}\right) < v\left(y^{(i)}, z^{(i)}\right) + \varepsilon. \tag{A.3.4}$$

Since f is convex, we get for $\lambda \in (0, 1)$

$$\begin{aligned} f\left(\lambda x_\varepsilon^{(1)} + (1 - \lambda) x_\varepsilon^{(2)}\right) &\leq \lambda f\left(x_\varepsilon^{(1)}\right) + (1 - \lambda)\, f\left(x_\varepsilon^{(2)}\right) \\ &\overset{(A.3.4)}{<} \lambda v\left(y^{(1)}, z^{(1)}\right) + (1 - \lambda)\, v\left(y^{(2)}, z^{(2)}\right) + \varepsilon. \end{aligned} \tag{A.3.5}$$

Convexity of g and affine linearity of $h(x) = Ax + b,\ x \in \mathcal{X}$, imply that $\lambda x_\varepsilon^{(1)} + (1-\lambda)\, x_\varepsilon^{(2)}$ is feasible for $v\left(\lambda\left(y^{(1)}, z^{(1)}\right) + (1-\lambda)\left(y^{(2)}, z^{(2)}\right)\right) = v\left(\lambda y^{(1)} + (1-\lambda) y^{(2)},\ \lambda z^{(1)} + (1-\lambda) z^{(1)}\right)$, i.e.,

$$\begin{aligned} g\left(\lambda x_\varepsilon^{(1)} + (1-\lambda)\, x_\varepsilon^{(2)}\right) &\overset{(A.3.3)}{\leq} \lambda y^{(1)} + (1-\lambda)\, y^{(2)} \qquad \text{and} \\ h\left(\lambda x_\varepsilon^{(1)} + (1-\lambda)\, x_\varepsilon^{(2)}\right) &= A\left[\lambda x_\varepsilon^{(1)} + (1-\lambda)\, x_\varepsilon^{(2)}\right] + b \\ &\overset{(A.3.3)}{=} \lambda z^{(1)} + (1-\lambda)\, z^{(2)}. \end{aligned}$$

Therefore,

$$v\left(\lambda\left(y^{(1)}, z^{(1)}\right) + (1-\lambda)\left(y^{(2)}, z^{(2)}\right)\right) \overset{(A.3.1)}{\leq} f\left(\lambda x_\varepsilon^{(1)} + (1-\lambda)\, x_\varepsilon^{(2)}\right)$$

and combined with (A.3.5)

$$\begin{aligned} &v\left(\lambda\left(y^{(1)}, z^{(1)}\right) + (1-\lambda)\left(y^{(2)}, z^{(2)}\right)\right) \\ &\qquad\qquad \leq \lambda v\left(y^{(1)}, z^{(1)}\right) + (1-\lambda)\, v\left(y^{(2)}, z^{(2)}\right) + \varepsilon. \end{aligned}$$

Now let $\varepsilon \searrow 0$ and (A.3.2) follows. □

In order to solve the convex programming problem (A.3.1), we will use Lagrange multipliers. This method tells us that (under mild assumptions), we can expect a *Lagrange multiplier* $\lambda = (\lambda_y, \lambda_z) \in \mathbb{R}^k \times \mathbb{R}^n$ with $\lambda_y \geq 0$ such that with (y, z) fixed, $\overline{x}$ is a solution of (A.3.1) if and only if it is a solution of the *unconstrained optimization problem* of minimizing the *Lagrangian* $L(\cdot, \lambda)\colon \mathcal{X} \to \mathbb{R} \cup \{+\infty\}$:

$$\begin{aligned} L(x, \lambda) &:= f(x) + \langle \lambda, [g(x) - y,\ h(x) - z] \rangle_{\mathbb{R}^k \times \mathbb{R}^n} \\ &= f(x) + \langle \lambda_y, g(x) - y \rangle_{\mathbb{R}^k} + \langle \lambda_z, h(x) - z \rangle_{\mathbb{R}^n}. \end{aligned} \tag{A.3.6}$$

To see that, we need the concept of a subdifferential.

Definition A.20 (Subdifferential) *Let $\mathcal{X}$ be a finite-dimensional Banach space and $\mathcal{X}^*$ be its dual space. The subdifferential of a lower semi-continuous convex function $\Phi\colon \mathcal{X} \to \mathbb{R} \cup \{+\infty\}$ at $x \in \operatorname{dom}(\Phi)$ is defined by*

$$\partial\Phi(x) := \{x^* \in \mathcal{X}^* : \Phi(y) - \Phi(x) \geq \langle x^*, y - x \rangle \ \textit{for all}\ y \in \mathcal{X}\}. \tag{A.3.7}$$

Remark A.21 *Note that if $\Phi\colon \mathbb{R}^m \to \mathbb{R}$ is convex and differentiable at $x \in \mathbb{R}^m$, then $\partial\Phi(x) = \{\nabla\Phi(x)\}$ is the gradient at x.*

Theorem A.22 (Lagrange Multiplier Theorem [25, Theorem 1.2.15]) *Let $v\colon \mathbb{R}^k \times \mathbb{R}^n \to \mathbb{R} \cup \{+\infty\}$ be the optimal value function of the constrained optimization problem* (A.3.1) *for $y \in \mathbb{R}^k,\ z \in \mathbb{R}^n$, with f, g, and h satisfying Assumption A.18. Suppose that for some fixed $(y, z) \in \mathbb{R}^k \times \mathbb{R}^n$,*

$$-\lambda = (-\lambda_y, -\lambda_z) \in \partial v(y, z) \subset \mathbb{R}^k \times \mathbb{R}^n \tag{A.3.8}$$

and $\overline{x} \in \mathcal{X}$ *is a solution of* (A.3.1), *i.e.,*

$$f(\overline{x}) = v(y,z) \quad \textit{and} \quad g(\overline{x}) \leq y,\ h(\overline{x}) = z. \tag{A.3.9}$$

Then λ is called *Lagrange multiplier* and the following holds:

(i) $\lambda_y \geq 0$, (componentwise)
(ii) the Lagrangian $L = L(\cdot, \lambda)$ (see (A.3.6)) attains a global minimum at $\overline{x}$, i.e., $L(\overline{x}, \lambda) \leq L(x, \lambda)$ for all $x \in \mathcal{X}$ and
(iii) λ satisfies the *complementary slackness condition*

$$\langle \lambda,\ [g(\overline{x}) - y, h(\overline{x}) - z]\rangle_{\mathbb{R}^k \times \mathbb{R}^n} = \langle \lambda_y,\ g(\overline{x}) - y\rangle_{\mathbb{R}^k} = 0, \tag{A.3.10}$$

i.e., for all components $i \in \{1, \ldots, k\}$ either $\lambda_y^i = 0$ or $g_i(\overline{x}) - y_i = 0$.

Remark A.23 *When there is an additional set constraint* $x \in C$ *in Problem* (A.3.1), *where* C *is a closed convex subset of* $\mathcal{X}$, *we can always replace the objective function* f *by* $f + \iota_C$ *to remove it, where* ι_C *is the indicator function of the set* C *(cf. Example A.26 in Appendix A.4). This is the same as using the original objective function* f *but replacing* $\mathcal{X}$ *by its subset* C.

Remark A.24 *Theorem A.22 shows the existence of a Lagrange multiplier when* (A.3.1) *has a solution* $\overline{x} \in \mathcal{X}$ *and if* $\partial v(y,z) \neq \emptyset$. *However, calculating* $\partial v(y,z)$ *requires knowing the value of* v *in a neighborhood of* (y,z), *which is not realistic.*

But fortunately the well-known Fenchel-Rockafellar theorem yields (cf. [19, Theorem 4.2.8])

$$(y,z) \in \operatorname{int}(\operatorname{dom}(v)) \Rightarrow \partial v(y,z) \neq \emptyset.$$

A.4 Duality

Duality is a powerful tool in convex analysis based on the concept of conjugate functions introduced by Fenchel. The *Fenchel conjugate* of a function (not necessarily convex) $f\colon \mathcal{X} \to [-\infty, +\infty]$ is the function $f^*\colon \mathcal{X}^* \to [-\infty, +\infty]$ defined on the dual space $\mathcal{X}^*$ by

$$f^*(x^*) := \sup_{x \in \mathcal{X}} \{\langle x^*, x\rangle - f(x)\}. \tag{A.4.1}$$

The operation $f \to f^*$ is also called a Fenchel-Legendre transform. As the supremum of a class of linear functions, the function f^* is always convex and lower semi-continuous [98, Proposition 2.2.3]. Moreover, if the domain of f is non-empty in $\mathcal{X}$, then f^* never takes the value $-\infty$. Clearly, the conjugate operation is *order-reversing*: for functions $f, g\colon \mathcal{X} \to [-\infty, +\infty]$, the inequality $f \geq g$ implies $f^* \leq g^*$.

An elementary but important result that relates the conjugate operation with the subgradient of convex functions is the Fenchel-Young inequality, whose proof follows directly from the definition of the Fenchel conjugate and convex subdifferential (see, e.g., [25, Proposition 1.3.1]).

Proposition A.25 (Fenchel-Young Inequality) *Let $f\colon \mathcal{X} \to \mathbb{R} \cup \{+\infty\}$ be a convex function. Suppose that $x^* \in \mathcal{X}^*$ and $x \in \operatorname{dom}(f)$. Then*

$$f(x) + f^*(x^*) \geq \langle x^*, x\rangle. \tag{A.4.2}$$

Equality holds if and only if $x^* \in \partial f(x)$.

We can consider the conjugate of f^* called the *biconjugate* of f and denote it f^{**}. This is a function on $\mathcal{X}^{**}$. When $\mathcal{X}$ is a reflexive Banach space, i.e., $\mathcal{X} = \mathcal{X}^{**}$, it follows from the Fenchel-Young inequality (A.4.2) that $f^{**} \leq f$. The function f^{**} is the largest among all the convex functions dominated by f and is called the *convex hull* of f. Many important convex functions f on $\mathcal{X} = \mathbb{R}^N$ equal to their biconjugate f^{**}. Such functions thus occur as natural pairs, $f = g^*$ and $f^* = g$, where both f and g are lower semi-continuous convex functions. Table A.1 gives a few examples on $\mathcal{X} = \mathbb{R}$.

Note that the first four functions in Table A.1 are special cases of indicator functions on $\mathbb{R}$. A more general result is Example A.26.

Example A.26 (Conjugate of Indicator Function) *Let C be a closed convex set in the reflexive Banach space $\mathcal{X}$. Then $\iota_C^* = \sigma_C$ and $\sigma_C^* = \iota_C$. Here ι_C is the* indicator function *of set C defined by $\iota_C(x) = 0$ if $x \in C$ and $+\infty$ otherwise and σ_C is the* support function *of set C defined by $\sigma_C(x^*) = \sup_{x\in C}\langle x^*, x\rangle$. The first four lines of Table A.1 (c.f. [18, Section 3.3]) describe four different indicator functions and their conjugate functions.*

$f(x) = g^*(x)$	$\operatorname{dom}(f)$	$g(y) = f^*(y)$	$\operatorname{dom}(g)$
0	$\mathbb{R}$	0	$\{0\}$
0	$[0, \infty)$	0	$(-\infty, 0]$
0	$[-1, 1]$	$\lvert y\rvert$	$\mathbb{R}$
0	$[0, 1]$	$y^+ = \max(y, 0)$	$\mathbb{R}$
$\lvert x\rvert^p/p,\ \ p > 1$	$\mathbb{R}$	$\lvert y\rvert^q/q\ \ (\frac{1}{p} + \frac{1}{q} = 1)$	$\mathbb{R}$
$\lvert x\rvert^p/p,\ \ p > 1$	$[0, +\infty)$	$\lvert y^+\rvert^q/q\ \ (\frac{1}{p} + \frac{1}{q} = 1)$	$\mathbb{R}$
$-x^p/p,\ \ 0 < p < 1$	$[0, +\infty)$	$-(-y)^q/q\ \ (\frac{1}{p} + \frac{1}{q} = 1)$	$(-\infty, 0)$
$-\log x$	$(0, +\infty)$	$-1 - \log(-y)$	$(-\infty, 0)$
e^x	$\mathbb{R}$	$\begin{cases} y\log y - y & (y > 0) \\ 0 & (y = 0) \end{cases}$	$[0, +\infty)$

Table A.1 Conjugate pairs of convex functions f and g on $\mathcal{X} = \mathbb{R}$

Example A.27 (Conjugate of Transform) *Next let us assume that h is a lower semi-continuous function. Then the effect of some simple transforms on the conjugate is summarized in Table A.2 (see [18, Section 3.3] for details).*

$f(x)$	$f^*(y)$
$h(ax)\ \ (a \neq 0)$	$h^*(y/a)$
$h(x+b)$	$h^*(y) - \langle b, y\rangle$
$ah(x)\ \ (a > 0)$	$ah^*(y/a)$

Table A.2 Transformed conjugates

The Fenchel-Young inequality tells us that $f^{**}(x) = f(x)$ exactly holds when $\partial f(x) \neq \emptyset$. A sufficient condition is (see Remark A.24)

$$x \in \operatorname{int}(\operatorname{dom}(f)). \tag{A.4.3}$$

Condition (A.4.3) is often referred to as a *constraint qualification* condition.

We summarize:

Theorem A.28 (Biconjugate Duality [25, Theorem 1.3.5]) *Let $\mathcal{X}$ be a finite-dimensional Banach space and let $f\colon \mathcal{X} \to \mathbb{R} \cup \{+\infty\}$ be a lower semi-continuous convex function. Then we always have the* week duality $f^{**} \leq f$ *in* $\operatorname{dom}(f)$. *Moreover,* strong duality, *that is, equality holds at points where the constraint qualification condition* (A.4.3) *is given.*

Remark A.29 *The constraint qualification condition* (A.4.3) *is a sufficient condition that can be relaxed. For example, when the* $\operatorname{span}(\operatorname{dom}(f))$ *is not the entire space, we can replace the interior of* $\operatorname{dom}(f)$ *by the relative interior ri*$[\operatorname{dom}(f)]$ *which is the interior with respect to* $\operatorname{span}(\operatorname{dom}(f))$. *This is because outside* $\operatorname{span}(\operatorname{dom}(f))$, $f = +\infty$. *Thus, only the behavior of f on* $\operatorname{span}(\operatorname{dom}(f))$ *is relevant. Another useful situation is related to polyhedral functions. We say a function f is polyhedral if the boundary of* $\operatorname{epi}(f)$ *consists of finitely many hyperplanes. For a polyhedral function f, we can replace the constraint qualification condition* (A.4.3) *by* $0 \in \operatorname{dom}(f)$ *(see [18, Section 5.1] for details).*

Using the Fenchel-Young inequality for each constrained optimization problem, we can write its companion *dual problem*. We discuss two different but equivalent perspectives following the exposition in [18] and [19].

Fenchel Duality

Let $\mathcal{X}$ and $\mathcal{Y}$ be two finite-dimensional Banach spaces, $f\colon \mathcal{X} \to \mathbb{R} \cup \{+\infty\}$, $g\colon \mathcal{Y} \to \mathbb{R} \cup \{+\infty\}$ lower semi-continuous convex functions, and $A\colon \mathcal{X} \to \mathcal{Y}$

a linear mapping. Consider the primal problem

$$p = v(0) = \inf_x [f(x) + g(Ax)], \tag{A.4.4}$$

as a special case of a class of convex optimization problems

$$v(y) := \inf_x [f(x) + g(Ax + y)], \tag{A.4.5}$$

with $v : \mathcal{Y} \to \mathbb{R} \cap \{+\infty\}$ convex and lower semi-continuous. To derive the dual problem, we calculate

$$v^*(-y^*) = \sup_{x,y} [\langle -y^*, y\rangle - f(x) - g(Ax + y)].$$

Letting $u = Ax + y$. Then $u \in \mathcal{Y}$ is arbitrary when y is and we have

$$\begin{aligned} v^*(-y^*) &= \sup_{x,u} [\langle -y^*, u - Ax\rangle - f(x) - g(u)] \\ &= \sup_x [\langle y^*, Ax\rangle - f(x)] + \sup_u [\langle -y^*, u\rangle - g(u)] \\ &= f^*(A^* y^*) + g^*(-y^*). \end{aligned}$$

Thus, the Fenchel dual problem is defined by

$$d = v^{**}(0) = \sup_{y^*} [-f^*(A^* y^*) - g^*(-y^*)], \tag{A.4.6}$$

where $A^* : \mathcal{Y}^* \to \mathcal{X}^*$ is the dual operator of A defined by $\langle y^*, Ax\rangle = \langle A^* y^*, x\rangle$ for all $x \in \mathcal{X}$ and $y^* \in \mathcal{Y}^*$. The inequality $v^{**}(0) \le v(0)$ tells us that the value of the primal problem dominates that of the dual. The difference $p - d$ is referred to as the *duality gap*.

By assumption, both f and g are convex functions, and so is v defined by (A.4.5). Moreover, it is not hard to check that $\operatorname{dom}(v) = \operatorname{dom}(g) - A \operatorname{dom}(f)$. Thus, by Remark A.29 and (A.4.3), a sufficient condition for the existence of Lagrange multipliers for the primal problem, i.e., $\partial v(0) \neq \emptyset$ [90, Section 31.1], is

$$0 \in \operatorname{ri}[\operatorname{dom}(v)] = \operatorname{ri}[\operatorname{dom}(g) - A \operatorname{dom}(f)]. \tag{A.4.7}$$

Condition (A.4.7) is a *constraint qualification* for the Fenchel duality. Enforcing such constraint qualification conditions we have

Theorem A.30 (Strong Duality [19, Theorem 4.4.3]) *If the lower semi-continuous convex functions f, g, and the linear operator A satisfy the constraint qualification condition* (A.4.7)*, then there is a zero duality gap between the primal and dual problems,* (A.4.4) *and* (A.4.6)*, and the dual problem has a solution.*

Lagrange Duality

For the convex programming problem (A.3.1), define the *Lagrangian* as in (A.3.6), i.e., with $\lambda = (\lambda_y, \lambda_z) \in \mathbb{R}^k_{\geq 0} \times \mathbb{R}^n$, $x \in \mathcal{X} = \mathbb{R}^m$ and $(y, z) \in \mathbb{R}^{k+n}$ and functions f, g, and h satisfying Assumption A.18, we have

$$L(\lambda, x; (y, z)) = f(x) + \langle \lambda, (g(x) - y, h(x) - z) \rangle.$$

Then

$$\sup_{\lambda \in \mathbb{R}^k_{\geq 0} \times \mathbb{R}^n} L(\lambda, x; (y, z)) = \begin{cases} f(x), & \text{if } g(x) \leq y, h(x) = z, \\ +\infty, & \text{otherwise.} \end{cases}$$

Thus, for $(y, z) = 0 \in \mathbb{R}^{k+n}$, problem (A.3.1) can be written as

$$p = v(0) = \inf_{x \in \mathcal{X}} \sup_{\lambda \in \mathbb{R}^k_{\geq 0} \times \mathbb{R}^n} L(\lambda, x; 0). \tag{A.4.8}$$

We can calculate

$$\begin{aligned} v^*(-\lambda) &= \sup_{y,z} [\langle -\lambda, (y, z) \rangle - v(y, z)] \\ &= \sup_{y,z} [\langle -\lambda, (y, z) \rangle - \inf_{x \in \mathcal{X}} \{f(x) : g(x) \leq y, h(x) = z\}] \\ &= \sup_{x \in \mathcal{X}, y, z} \{\langle -\lambda, (y, z) \rangle - f(x) : g(x) \leq y, h(x) = z\}. \end{aligned}$$

Letting $\xi = y - g(x) \in \mathbb{R}^k_{\geq 0}$, we can rewrite the expression above as

$$\begin{aligned} v^*(-\lambda) &= \sup_{x \in \mathcal{X}, \xi \in \mathbb{R}^k_{\geq 0}} [\langle -\lambda, (g(x), h(x)) \rangle - f(x) + \langle -\lambda, (\xi, 0) \rangle] \\ &= - \inf_{x \in \mathcal{X}, \xi \in \mathbb{R}^k_{\geq 0}} [L(x, \lambda; 0) + \langle \lambda, (\xi, 0) \rangle] \\ &= \begin{cases} -\inf_{x \in \mathcal{X}} L(x, \lambda; 0), & \text{if } \lambda \in \mathbb{R}^k_{\geq 0} \times \mathbb{R}^n, \\ +\infty, & \text{otherwise.} \end{cases} \end{aligned}$$

Thus, the dual problem is

$$d = v^{**}(0) = \sup_{\lambda} -v^*(-\lambda) = \sup_{\lambda \in \mathbb{R}^k_{\geq 0} \times \mathbb{R}^n} \inf_{x \in \mathcal{X}} L(\lambda, x; 0). \tag{A.4.9}$$

As a consequence of Theorem A.28, we have

Theorem A.31 (Lagrange Duality (See [18, Corollary 4.3.6], [25, Section 1.4.3])) *Suppose that functions f, g, and h satisfy Assumption A.18. Then we always have the Lagrange weak duality:*

$$\inf_{x\in\mathcal{X}} \sup_{\lambda\in\mathbb{R}^k_{\geq 0}\times\mathbb{R}^n} L(\lambda, x; 0) \geq \sup_{\lambda\in\mathbb{R}^k_{\geq 0}\times\mathbb{R}^n} \inf_{x\in\mathcal{X}} L(\lambda, x; 0). \tag{A.4.10}$$

Moreover, Lagrange strong duality, i.e., equality in (A.4.10) *holds if and only if* $\partial v(0) \neq \emptyset$.

Remark A.32 *A commonly used sufficient condition for* $\partial v(u) \neq \emptyset$ *is (see [90, Section 31.1])*

$$u \in \text{ri dom}(v). \tag{A.4.11}$$

For problem (A.3.1), *one often uses the* Slater condition, *i.e., there exists a feasible point* $x \in \text{dom}(f)$ *such that* $g(x) < y$. *Note that since* x *is feasible, we have* $h(x) = z$. *This implies that the affine set* $\{t : h(t) = z\}$ *has a non-empty intersection with the open set* $\{t : g(t) < y\}$. *Thus, this will hold also for a neighborhood of* (y, z) *and, therefore,* $\partial v(y, z) \neq \emptyset$.

Example A.33 (Classical Linear Programming Duality) *Consider a linear programming problem*

$$\begin{aligned} &\max\ \langle c, x\rangle \\ &\text{subject to } Ax \leq b, x \geq 0 \end{aligned} \tag{A.4.12}$$

where $x \in \mathbb{R}^N$, $b \in \mathbb{R}^M$, A *is a* $M \times N$ *matrix. Note that here the sign constraint* $x \geq 0$ *can be viewed as a set constraint* $x \in C = \mathbb{R}^N_{\geq 0}$ *(cf. Remark A.23). Then by the Lagrangian duality, the dual problem is*

$$\begin{aligned} &\min\ \langle b, \lambda\rangle \\ &\text{subject to } A^\top \lambda \geq c, \lambda \geq 0. \end{aligned} \tag{A.4.13}$$

To derive this in terms of convex duality, we deal with the minimizing problem

$$p = \min[\langle -c, x\rangle : Ax \leq b, x \geq 0] = -\max[\langle c, x\rangle : Ax \leq b, x \geq 0],$$

a primal problem. We write the Lagrangian

$$L(\lambda, x) = \langle -c, x\rangle + \langle \lambda, Ax - b\rangle. \tag{A.4.14}$$

Then the primal problem is

$$p = \inf_{x\geq 0} \sup_{\lambda\geq 0} L(\lambda, x),$$

and the dual problem is

$$d = \sup_{\lambda\geq 0} \inf_{x\geq 0} L(\lambda, x).$$

We can see that

$$\inf_{x\geq 0} L(\lambda, x) = \inf_{x\geq 0} \langle -c + A^\top \lambda, x\rangle - \langle \lambda, b\rangle = \begin{cases} -\langle \lambda, b\rangle & \text{if } A^\top\lambda \geq c \\ -\infty & \text{otherwise.} \end{cases}$$

So we have

$$\begin{aligned} \max[\langle c, x\rangle : Ax \leq b, x \geq 0] &= -p = -d \\ &= -\max_{\lambda\geq 0}[-\langle \lambda, b\rangle : A^\top\lambda \geq c] \\ &= \min[\langle \lambda, b\rangle : A^\top\lambda \geq c, \lambda \geq 0]. \end{aligned}$$

Clearly, all the functions involved here are polyhedral. Applying the constraint qualification condition for polyhedral functions, we can conclude that if either the primal problem or the dual problem is feasible, then there is no duality gap. Moreover, when the common optimal value is finite, then both problems have optimal solutions.

Linear programming duality is well-known before the development of convex analysis. Here we follow the exposition in [25, Section 1.4.3]. The hard work in Example A.33 was hidden in establishing that the constraint qualification (A.4.7) is sufficient, but unlike many applied developments, we have *rigorously* recaptured linear programming duality within the framework of convex duality.

Note that the primal Lagrange multiplier λ is the dual solution and vice versa. Table A.3 ([25, Table 1.3]) can help us in formulating the dual problem.

Primal constraint	Dual variable	Primal variable	Dual constraint
$Ax \leq b$	$\lambda \geq 0$	$x \geq 0$	$A^\top\lambda \geq c$
$Ax = b$	λ free	x free	$A^\top\lambda = c$
$Ax \geq b$	$\lambda \leq 0$	$x \leq 0$	$A^\top\lambda \leq c$

Table A.3 Transformed conjugates

References

1. Acerbi, C., Tasche, D.: Expected shortfall: a natural coherent alternative to value at risk. Economic Notes **31**(2), 379–388 (2002). https://doi.org/10.1111/1468-0300.00091
2. Adam, A.: Handbook of Asset and Liability Management: From Models to Optimal Return Strategies. John Wiley & Sons, Hoboken, NJ (2008)
3. Aït-Sahalia, Y., Cacho-Diaz, J., Hurd, T.R.: Portfolio choice with jumps: A closed-form solution. The Annals of Applied Probability **19**(2), 556–584 (2009)
4. Artzner, P., Delbaen, F., Eber, J.M., Heath, D.: Coherent measures of risk. Mathematical Finance **9**(3), 203–228 (1999). https://doi.org/10.1111/1467-9965.00068
5. Bailey, D., Borwein, J., López de Prado, M.: Portfolio design and backtest overfitting. Journal of Investment Management **15**, 1–13 (2017)
6. Bailey, D., Borwein, J., López de Prado, M., Zhu, Q.J.: Pseudo-mathematics and financial charlatanism: the effect of backtest overfitting on out-of-sample performance. Notice of Amer. Math. Soc. **61**(5), 458–471 (2014)
7. Bailey, D., Borwein, J., López de Prado, M., Zhu, Q.J.: The probability of backtest overfitting. Journal of Computational Finance **20**, 1–31 (2016)
8. Barbu, V., Precupanu, T.: Convexity and Optimization in Banach Spaces. Springer Monographs in Mathematics. Springer, Heidelberg (2012)
9. Basel Committee on Banking Supervision: Fundamental review of the trading book: a revised market risk framework - consultative document (2013). https://www.bis.org/publ/bcbs265.pdf
10. Basel Committee on Banking Supervision: Interest rate risk in the banking book (2016). https://www.bis.org/bcbs/publ/d368.pdf

S. Maier-Paape et al., *Scalar and Vector Risk in the General Framework of Portfolio Theory*, CMS/CAIMS Books in Mathematics 9,
https://doi.org/10.1007/978-3-031-33321-7

11. Basel Committee on Banking Supervision: Climate related financial risks - measurement methodologies (2021). https://www.bis.org/bcbs/publ/d518.pdf
12. Battiston, S., Mandel, A., Monasterolo, I., Schütze, F., Visentin, G.: A climate stress-test of the financial system. Nature Climate Change **7**(4), 283–288 (2017)
13. Bertsimas, D., Lauprete, G., Samarov, A.: Shortfall as a risk measure: properties, optimization and applications. Working paper, Sloan School of Management, MIT, Cambridge (2000)
14. Bessis, J.: Risk Management in Banking, 4th edn. John Wiley & Sons, Hoboken, NJ (2015)
15. Bielecki, T.R., Rutkowski, M.: Credit Risk: Modeling, Valuation and Hedging. Springer Finance. Springer, Heidelberg (2013). https://doi.org/10.1007/978-3-662-04821-4
16. Black, F., Litterman, R.: Asset allocation: combining investor views with market equilibrium. The Journal of Fixed Income **1**(2), 7–18 (1991). https://doi.org/10.3905/jfi.1991.408013
17. Bohn, J., Crosbie, P.: Modeling Default Risk. Moody's KMV Company (2003)
18. Borwein, J.M., Lewis, A.S.: Convex Analysis and Nonlinear Optimization. Springer, Heidelberg (2000). https://doi.org/10.1007/978-0-387-31256-9
19. Borwein, J.M., Zhu, Q.J.: Techniques of Variational Analysis. Springer, Heidelberg (2005). https://doi.org/10.1007/0-387-28271-8
20. Borwein, J.M., Zhu, Q.J.: A variational approach to lagrange multipliers. Journal of Optimization Theory and Applications **171**, 727–756 (2016). https://doi.org/10.1007/s10957-015-0756-2
21. Boyd, S., Busseti, E., Diamond, S., Kahn, R.N., Koh, K., Nystrup, P., Speth, J.: Multi-period trading via convex optimization. Foundations and Trends in Optimization **3**(1), 1–76 (2017). https://doi.org/10.1561/2400000023
22. Boyd, S., Vandenberghe, L.: Convex optimization. Cambridge University Press (2004)
23. Brenner, R.: Consistency in portfolio optimization: A new approach using the general framework of portfolio theory. Ph.D. thesis, RWTH Aachen (2021)
24. Cariño, D.R., Ziemba, W.T.: Formulation of the Russell-Yasuda Kasai financial planning model. Operations Research **46**(4), 433–604 (1998). https://doi.org/10.1287/opre.46.4.433
25. Carr, P., Zhu, Q.J.: Convex Duality and Financial Mathematics. SpringerBriefs in Mathematics. Springer, Heidelberg (2018). https://doi.org/10.1007/978-3-319-92492-2
26. Chekhlov, A., Uryasev, S., Zabarankin, M.: Drawdown measure in portfolio optimization. International Journal of Theoretical and Applied Finance **8**(1), 13–58 (2005). https://doi.org/10.1142/S0219024905002767

27. Choudhry, M.: Bank Asset and Liability Management: Strategy, Trading, Analysis. John Wiley & Sons, Hoboken, NJ (2011)
28. Committee of Sponsoring Organizations of the Treadway Commission: Enterprise risk management - integrated framework (2004). https://erm.ncsu.edu/library/article/coso-erm-framework
29. Committee of Sponsoring Organizations of the Treadway Commission: Enterprise risk management - integrating with strategy and performance (2017). https://www.coso.org/sitepages/guidance-on-enterprise-risk-management.aspx
30. Cont, R., Kotlicki, A., Valderrama, L.: Liquidity at risk: Joint stress testing of solvency and liquidity. Journal of Banking & Finance **118**, 105871 (2020). https://doi.org/10.1016/j.jbankfin.2020.105871
31. Cornuéjols, G., Peña, J.F., Tütüncü, R.: Optimization methods in finance. Cambridge University Press (2018)
32. Cox, J., Ross, S.: The valuation of options for alternative stochastic processes. Journal of Financial Economics **3**, 145–166 (1976). https://doi.org/10.1016/0304-405X(76)90023-4
33. Cox, J., Ross, S., Rubinstein, M.: Options pricing: a simplified approach. Journal of Financial Economics **7**, 229–263 (1979). https://doi.org/10.1016/0304-405X(79)90015-1
34. Crouhy, M., Galai, D., Mark, R.: A comparative analysis of current credit risk models. Journal of Banking & Finance **24**(1–2), 59–117 (2000). https://doi.org/10.1016/S0378-4266(99)00053-9
35. CVX Research, I.: CVX: Matlab Software for Disciplined Convex Programming. http://cvxr.com/cvx/ (2020)
36. Delbaen, F., Schachermayer, W.: A general version of the fundamental theorem of asset pricing. Math. Ann. **300**, 463–520 (1994)
37. Dermine, J., Bissada, Y.F.: Asset and Liability Management: The Banker's Guide to Value Creation and Risk Control. Financial Times Prentice Hall, Hoboken, NJ (2007)
38. Dewasurendra, S., Júdice, P., Zhu, Q.J.: The optimum leverage level of the banking sector. Risks **7**(2), 51 (2019). https://doi.org/10.3390/risks7020051
39. Dietz, S., Bowen, A., Dixon, C., Gradwell, P.: Climate value at risk of global financial assets. Nature Climate Change **6**(7), 676–679 (2016). https://doi.org/10.1038/nclimate2972
40. Dunbar, N.: Inventing money: The story of long–term capital management and the legends behind it. John Wiley & Sons, Hoboken, NJ (2000)
41. El-Hassan, N., Kofman, P.: Tracking error and active portfolio management. Australian Journal of Management **28**(2), 183–207 (2003). https://doi.org/10.1177/031289620302800204
42. El Karoui, N., Kapoudjian, C., Pardoux, E., Peng, S., Quenez, M.C.: Reflected solutions of backward sde's, and related obstacle problems for pde's. Ann. Probability **25**, 702–737 (1997)

43. El Karoui, N., Peng, S., Quenez, M.C.: Backward stochastic differential equations in finance. Math. Finance **7**, 1–71 (1997)
44. Elton, E.J., Gruber, M.J.: Portfolio theory when investment relatives are lognormally distributed. The Journal of Finance **29**(4), 1265–1273 (1974). https://doi.org/10.1111/j.1540-6261.1974.tb03103.x
45. Elton, E.J., Gruber, M.J., Brown, S.J., Goetzmann, W.N.: Modern Portfolio Theory and Investment Analysis, 9th edn. John Wiley & Sons, Hoboken, NJ (2014)
46. Embrechts, P.: Extreme value theory: potential and limitations as an integrated risk management tool. Working paper, ETH Zürich (2000)
47. Entrop, O., Memmel, C., Ruprecht, B., Wilkens, M.: Determinants of bank interest margins: Impact of maturity transformation. Journal of Banking & Finance **54**, 1–19 (2015). https://doi.org/10.1016/j.jbankfin.2014.12.001
48. Fenchel, W.: Convex cones, sets and functions, Lecture Notes. Princeton University, Princeton, NJ (1953)
49. Fernholz, E.R.: Stochastic Portfolio Theory. Springer, Heidelberg (2002)
50. Föllmer, H., Schied, A.: Stochastic Finance: An Introduction in Discrete Time, 4th edn. De Gruyter Studies in Mathematics. De Gruyter (2016). https://doi.org/10.1515/9783110212075
51. Fouque, J.P., Sircar, R., Zariphopoulou, T.: Portfolio optimization and stochastic volatility asymptotics. Mathematical Finance **27**(3), 704–745 (2017)
52. Goldberg, L.R., Mahmoud, O.: Drawdown: from practice to theory and back again. Mathematics and Financial Economics **11**(3), 275–297 (2017). https://doi.org/10.1007/s11579-016-0181-9
53. Gordy, M.B.: A comparative anatomy of credit risk models. Journal of Banking & Finance **24**(1–2), 119–149 (2000). https://doi.org/10.1016/S0378-4266(99)00054-0
54. Gupton, G., Finger, C., Bhatia, M.: CreditMetrics—technical document. J.P. Morgan (1997)
55. Haase, D., Platen, A.: Momentum vs. Volatilität. VTAD (2019). https://www.vtad.de/fa/momentum-vs-volatilitaet
56. Hamel, A., Heyde, F.: Duality for set-valued measures of risk. SIAM Journal of Financial Math. **1**(1), 66–95 (2010)
57. Hamel, A.H., Heyde, F., Rudloff, B.: Set-valued risk measures for conical market models. Mathematics and Financial Economics **5**, 1–28 (2011)
58. Harrison, J.M., Kreps, D.M.: Martingales and arbitrage in multiperiod securities markets. J. Econom. Theory **20**, 381–408 (1979). https://doi.org/10.1016/0022-0531(79)90043-7
59. Harrison, J.M., Pliska, S.: Martingales and stochastic integrals in the theory of continuous trading. Stochastic Processes Appl. **11**, 215–260 (1981). https://doi.org/10.1016/0304-4149(81)90026-0

60. Henkes, J.: Log Risiko- und Nutzenfunktionen für die fraktionale Trading Strategie im "General Framework of Portfolio Theory". Bachelor Thesis. RWTH Aachen University (2021)
61. Hull, J.C.: Risk Management and Financial Institutions, 3rd edn. John Wiley & Sons, Hoboken, NJ (2012)
62. Jorion, P.: Value at Risk: The New Benchmark for Managing Financial Risk, 3rd edn. McGraw Hill, New York, NY (2006)
63. Jouini, E., Meddeb, M., Touzi, N.: Vector-valued coherent risk measures. Finance Stoch. **8**, 531–552 (2004)
64. Júdice, P., Vazifedan, M., Zhu, Q.J.: A general duality approach to bank asset-liability management (2022). Working paper
65. Júdice, P., Zhu, Q.J.: Bank balance sheet risk allocation. Journal of Banking & Finance **133**, 106257 (2021). https://doi.org/10.1016/j.jbankfin.2021.106257
66. Kac, M.: On distributions of certain Wiener functionals. Trans. AMS **65**, 1–13 (1949). https://doi.org/10.2307/1990512
67. Kassouf, S.T., Thorp, E.O.: Beat the Market. Random House, New York, NY (1967)
68. Keller, A.A.: Multi-Objective Optimization in Theory and Practice I: Classical Mathods. Bentham Science Publishers Ltd. (2017). https://doi.org/10.2174/97816810856851170101
69. Keller, A.A.: Multi-Objective Optimization in Theory and Practice II: Metaheuristic Algorithms. Bentham Science Publishers Ltd. (2019). https://doi.org/10.2174/97816810870541190101
70. Kelly, Jr., J.L.: A new interpretation of information rate. Bell System Technical Journal **35**(4), 917–926 (1956)
71. Lintner, J.: The valuation of risk assets and the selection of risky investments in stock portfolios and capital budgets. The Review of Economics and Statistics **47**(1), 13–37 (1965). https://doi.org/10.2307/1924119
72. Lipton, A.: Modern monetary circuit theory, stability of interconnected banking network, and balance sheet optimization for individual banks. International Journal of Theoretical and Applied Finance **19**(06), 1650034 (2016)
73. López de Prado, M., Vince, R., Zhu, Q.J.: Optimal risk budgeting under finite investment horizon. Risks **7**(3), 86 (2019). https://doi.org/10.3390/risks7030086
74. Lubinska, B.: Asset Liability Management Optimisation: A Practitioner's Guide to Balance Sheet Management and Remodelling. John Wiley & Sons, Hoboken, NJ (2020)
75. Ma, J., Yong, J.: Forward-Backward Stochastic Differential Equations and their Applications, vol. 1702. Springer (2007). Lecture Notes in Mathematics
76. MacLean, L.C., Thorp, E.O., Ziemba, W.T.: The Kelly Capital Growth Investment Criterion: Theory and Practice. World Scientific, Singapore (2011). https://doi.org/10.1142/7598

77. Maier-Paape, S., Platen, A., Zhu, Q.J.: A general framework for portfolio theory. part III: Multi-period markets and modular approach. Risks **7**(2), 60 (2019). https://doi.org/10.3390/risks7020060
78. Maier-Paape, S., Zhu, Q.J.: A general framework for portfolio theory. part I: Theory and various models. Risks **6**(2), 53 (2018). https://doi.org/10.3390/risks6020053
79. Maier-Paape, S., Zhu, Q.J.: A general framework for portfolio theory. part II: Drawdown risk measures. Risks **6**(3), 76 (2018). https://doi.org/10.3390/risks6030076
80. Markowitz, H.M.: Portfolio selection. The Journal of Finance **7**(1), 77–91 (1952). https://doi.org/10.2307/2975974
81. Markowitz, H.M.: Portfolio Selection, *Cowles Monograph*, vol. 16. John Wiley & Sons, Hoboken, NJ (1959)
82. Matz, L., Neu, P.: Liquidity Risk Measurement and Management: A practitioner's guide to global best practices, vol. 408. John Wiley & Sons (2006)
83. McNeil, A.J., Frey, R., Embrechts, P.: Quantitative Risk Management: Concepts, Techniques and Tools - Revised Edition. Princeton Series in Finance. Princeton University Press (2015)
84. Merton, R.C.: Lifetime portfolio selection under uncertainty: the continuous-time case. The Review of Economics and Statistics **51**(3), 247–257 (1969). https://doi.org/10.2307/1926560
85. Merton, R.C.: Optimum consumption and portfolio rules in a continuous-time model. Journal of Economic Theory **3**(4), 373–413 (1971). https://doi.org/10.1016/0022-0531(71)90038-X
86. Merton, R.C.: Continuous-Time Finance. Blackwell Boston (1992)
87. Moreau, J.J.: Fonctionelles convexes, Lecture Notes. College de France (1967)
88. Phelps, E.S.: The accumulation of risky capital: A sequential utility analysis. Econometrica **30**(4), 729–743 (1962). https://doi.org/10.2307/1909322
89. Platen, A.: Modular portfolio theory: A general framework with risk and utility measures as well as trading strategies on multi-period markets. Ph.D. thesis, RWTH Aachen (2018). https://doi.org/10.18154/RWTH-2018-230691
90. Rockafellar, R.T.: Convex Analysis, 2. edn. Princeton landmarks in mathematics and physics. Princeton University Press (1972)
91. Rockafellar, R.T., Uryasev, S.: Optimization of conditional value-at-risk. Journal of Risk **2**(3), 21–41 (2000). https://doi.org/10.21314/JOR.2000.038
92. Rockafellar, R.T., Uryasev, S.: Conditional value-at-risk for general loss distributions. Journal of Banking & Finance **26**(7), 1443–1471 (2002). https://doi.org/10.1016/S0378-4266(02)00271-6
93. Rockafellar, R.T., Uryasev, S., Zabarankin, M.: Master funds in portfolio analysis with general deviation measures. Journal of Banking & Fi-

nance **30**(2), 743–778 (2006). https://doi.org/10.1016/j.jbankfin.2005.04.004

94. Roman, S.: Introduction to the Mathematics of Finance. Undergraduate Texts in Mathematics. Springer (2012). https://doi.org/10.1007/978-1-4614-3582-2
95. Rosenberg, J.V., Schuermann, T.: A general approach to integrated risk management with skewed, fat-tailed risks. Journal of Financial Economics **79**(3), 569–614 (2006). https://doi.org/10.1016/j.jfineco.2005.03.001
96. Samuelson, P.A.: Lifetime portfolio selection by dynamic stochastic programming. The Review of Economics and Statistics **51**(3), 239–246 (1969). https://doi.org/10.2307/1926559
97. Saunders, A., Cornett, M.: Financial Institutions Management: a Risk Management Approach, 9th edn. McGraw Hill, New York, NY (2017)
98. Schirotzek, W.: Nonsmooth Analysis. Universitext. Springer, Heidelberg (2007)
99. Scholes, M., Black, F.: The pricing of options and corporate liabilities. Journal of Political Economy **81**(3), 637–654 (1973)
100. Schönbucher, P.J.: Credit Derivatives Pricing Models: Models, Pricing and Implementation. John Wiley & Sons, Hoboken, NJ (2003)
101. Shannon, C., Weaver, W.: The Mathematical Theory of Communication. University of Illinois Press, Urbana (1949)
102. Sharpe, W.F.: A simplified model for portfolio analysis. Management science **9**(2), 277–293 (1963)
103. Sharpe, W.F.: Capital asset prices: A theory of market equilibrium under conditions of risk. The Journal of Finance **19**(3), 425–442 (1964). https://doi.org/10.1111/j.1540-6261.1964.tb02865.x
104. Sharpe, W.F.: Mutual fund performance. Journal of Business **39**(1), 119–138 (1966)
105. Shreve, S.E.: Stochastic Calculus for Finance I. Springer Finance. Springer, Heidelberg (2004). https://doi.org/10.1007/978-0-387-22527-2
106. Steinbach, M.C.: Markowitz revisited: Mean-variance models in financial portfolio analysis. SIAM Review **43**(1), 31–85 (2001). https://doi.org/10.1137/S0036144500376650
107. Thorp, E.O.: Beat the dealer: a winning strategy for the game of twenty-one. Vintage Books, New York, NY (1962)
108. Tobin, J.: Liquidity preference as behavior towards risk. Review of Economic Studies **25**(1), 65–86 (1958)
109. Trenor, J.L.: Towards a theory of market value of risky assets. In: R.A. Korajczyk (ed.) Asset pricing and portfolio performance: models, strategy and performance metrics, pp. 15–22. Risk Books, London (1999)
110. Vasicek, O.: The distribution of loan portfolio value. Risk **15**(12), 160–162 (2002)

111. Vince, R.: The Mathematics of Money Management: Risk Analysis Techniques for Traders. John Wiley & Sons, Hoboken, NJ (1992)
112. Vince, R.: The Leverage Space Trading Model: Reconciling Portfolio Management Strategies and Economic Theory. John Wiley & Sons, Hoboken, NJ (2009)
113. Vince, R., Zhu, Q.J.: Optimal betting sizes for the game of blackjack. Journal of Investment Strategies **4**, 53–75 (2012). https://doi.org/10.21314/JOIS.2015.059
114. van Vliet, P., de Koning, J.: High Returns from Low Risk: A Remarkable Stock Market Paradox. John Wiley & Sons, Hoboken, NJ (2016)
115. Whalley, M., Guzelian, C.: The Legal Risk Management Handbook: An International Guide to Protect Your Business from Legal Loss. Kogan Page Publishers (2016)
116. Wilde, T.: CreditRisk+: A Credit Risk Management Framework. Credit Suisse Financial Products (1997)
117. Wilson, T.: Portfolio credit risk I. Risk **10**(9) (1997)

Index

S. Maier-Paape et al., *Scalar and Vector Risk in the General Framework of Portfolio Theory*, CMS/CAIMS Books in Mathematics 9,
https://doi.org/10.1007/978-3-031-33321-7